Polyurethane Shape Memory Polymers

W. M. Huang
Bin Yang
Yong Qing Fu

CRC Press
Taylor & Francis Group
Boca Raton   London   New York

CRC Press is an imprint of the
Taylor & Francis Group, an **informa** business

CRC Press
Taylor & Francis Group
6000 Broken Sound Parkway NW, Suite 300
Boca Raton, FL 33487-2742

First issued in paperback 2017

© 2012 by Taylor & Francis Group, LLC
CRC Press is an imprint of Taylor & Francis Group, an Informa business

Version Date: 20111003

ISBN 13: 978-1-4398-3800-6 (hbk)
ISBN 13: 978-1-138-07500-9 (pbk)

Library of Congress Cataloging-in-Publication Data

Huang, Wei Min.
    Polyurethane shape memory polymers / authors, W.M. Huang, Bin Yang, Yong Qing Fu.
      p. cm.
    "A CRC title."
    Includes bibliographical references.
    ISBN 978-1-4398-3800-6 (hardcover : alk. paper)
    1. Shape memory polymers. 2. Polyurethanes. I. Yang, Bin. II. Fu, Yong Qing. III. Title.

TA455.P585H83 2012
620.1'9204232--dc23   2011031714

Visit the Taylor & Francis Web site at
http://www.taylorandfrancis.com

and the CRC Press Web site at
http://www.crcpress.com

Contents

Forewords

This book, *Polyurethane Shape Memory Polymers*, authored by Dr. Wei Min Huang, Dr. Bin Yang, and Dr. Yong Qing Fu, is the first book dedicated to polyurethane shape memory polymers—some of the most important types of shape memory polymers in existence. As a senior researcher with more than two decades of experience in this field, I am very happy to see the publication of this book by CRC Press/Taylor & Francis Group.

This book is suitable not only for macromolecule engineers and researchers working on developing new shape memory polymers and improving the performances of existing ones, but also for students studying polymer chemistry to acquire a good understanding of the fundamentals of shape memory polymers. On one hand, this book is a useful reference for research and development; on the other, it serves as a perfect introduction for anyone who is interested in learning more about this fantastic material.

The content of this book is carefully structured with plenty of application examples. I am sure that readers will be fascinated by the wonderful world of shape memory polymers.

Dr. Shunichi Hayashi
SMP Technologies, Inc.
Japan

In recent years, intelligent or smart materials have attracted worldwide attention. One of the main types that stimulated research into intelligent materials is the shape memory alloy. The polyurethane shape memory polymer, in particular, also has been applied practically in a wide variety of fields. Dependence of the elastic modulus and yield stress on temperature in a shape memory polymer is quite the opposite of the principle of a shape memory alloy. The very large volume change of a shape memory polymer foam makes this material valuable for inflatable structures in the aerospace field. The gas permeability of the shape memory polymer thin film varies significantly above and below the glass transition temperature—a property with many applications in the medical and textile fields. If both of these outstanding qualities of the shape memory alloy and the shape memory polymer are combined, it becomes possible to develop a shape memory composite that exhibits completely new high-performance functions.

In order to develop shape memory polymer elements, it is also important to understand the thermomechanical properties of these materials. The thermomechanics of shape memory polymers are very complex because they depend on hysteresis. The most basic and important properties of the polyurethane shape memory polymer are introduced in this fine book that is certain to be of interest for researchers, students, and engineers in numerous fields of materials science and engineering.

Professor Hisaaki Tobushi
Aichi Institute of Technology
Toyota-city, Japan

Shape memory polymers (SMPs) are some of the most important and valuable new materials developed in the last 25 years. The ability to respond and recover a large deformation with the application of a particular external stimulus such as heat, light, or moisture is of great scientific and technological significance. Furthermore, these materials exhibit enormous innovation potential. Because of their novel properties and behavior, they can be utilized in a broad range of applications and address challenges in advanced aerospace, commercial, and biomedical technologies. In addition, SMPs have advantages over their older shape memory alloy cousins—their light weight, high strain and shape recovery ability, ease of processing, low cost, and other properties may be tailored for a variety of applications.

Although the focus of this book is on one commercially available thermoplastic polyurethane SMP, the main features, mechanisms, and applications discussed here are largely applicable to other SMPs. Based on the remarkable properties of SMPs and their great potential, technologists have developed many applications ranging from outer space exploration to the interior of the human body. As a result of their considerable shape memory effects (SMEs), SMPs are in the process of reshaping our thinking, approaches, and design methods in many ways that conventional materials and traditional approaches do not allow.

In the aerospace arena, SMPs formed as foams, composites, and hybrid structures used with cold hibernated elastic memory (CHEM) technology have the potential to provide innovative self-deployable space structures with significantly better reliability, lower cost, and simplicity than other expandable and deployable structures. Some advanced SMP space concepts represent a new generation of deployable structures. If developed, these innovative technologies will introduce a new paradigm for defining configurations for space-based structures and future mission architectures.

The unique attributes, biocompatibility, and other properties make polyurethane SMP technology viable for self-deployable stents and other medical products. Recently developed SMP foams combined with CHEM processing expand their potential medical applications even further. SMP materials will significantly and positively impact the medical device industry. They have unique characteristics that will revolutionize the manufacture of medical devices and usher in an era of simple, low-cost, self-deployable medical devices.

This book provides a comprehensive discussion of SMPs from a brief introduction to SMPs and their position in the world of materials, through the details of shape memory behavior, fabrication, and characterization of composites, fabrication of porous materials and their applications, investigation of SMEs at micro and nano scales, biomedical applications, fundamentals of multi-SMEs, and the future of polyurethane SMPs.

Dr. Witold M. Sokolowski
California Institute of Technology
Pasadena, California

Shape memory polymers (SMPs) belong to a novel class of intelligent ("smart") polymers introduced in the mid-1980s in Japan that have gained much attention in Japan and in the United States. SMPs are stimuli-responsive polymers that function by changing their moduli and shapes on exposure to external stimuli such as heat, light, and chemicals. This class of polymers developed rapidly in the past two decades, and many articles about various aspects of SMPs have been published in recent years including review articles focusing on material systems. Since then, continuous development efforts by various organizations and university groups have expanded the applications of SMPs into diverse fields such as morphing aircraft structures, textiles, and biomedical devices.

Among various classes of SMPs reported to date, polyurethane-based SMPs are the most extensively investigated; they were also the first SMPs to be commercialized. The scientific knowledge revealed by investigation of their molecular architectures and structure–property relationships provided a sound foundation for the understanding of the unique functionalities of these materials. The translation of such knowledge into the development of new capabilities will lead to innovations that will benefit society.

This book provides a foundation for better understanding of various aspects of SMPs, including shape memory mechanisms and mechanics. In addition, various application concepts are introduced to establish a good framework for assessing the potential utility of this class of materials in modern society. The authors provide numerous examples to illustrate the unique functionalities that SMPs can bring about in engineering designs for future applications, such as the discussions on fabrication of micro- and nano-sized elements in Chapter 9, large-scale surface pattern generation in Chapter 10, and multi-shape memory effect in Chapter 12.

The material in this book is valuable for university students, research scientists, and engineers. As an active participant in the SMP technology field in the past decade, I am convinced that this book will play a valuable role in generating a greater awareness of the numerous possibilities presented by this class of unique materials. SMPs are organic polymeric materials that offer actuation, shape memory, and stimulus responsiveness enabling new capabilities for engineering design. The "smart" nature of these materials represents a new design paradigm and invites us to consider them for use in applications that are not feasible with conventional polymers.

Dr. Tat Hung Tong
Cornerstone Research Group, Inc.
Dayton, Ohio

Preface

The discovery of materials exhibiting shape memory effects (SMEs)—so-called shape memory materials (SMMs)—opened an exciting field and made a significant breakthrough in the development of materials that complement or supplant the traditional materials and approaches for a variety of engineering applications and also influenced the way many products are designed.

At present, shape memory polymers (SMPs) are undergoing rapid growth and becoming the leading members of the fantastic world of SMMs. Based on their advantages over other SMMs, in particular low cost and high versatility, SMPs are more accessible and more flexible, thus showing greater potential for numerous applications, from deployable structures in outer space to medical devices such as stents and sutures.

Although a few types of SMPs, e.g., the polystyrene SMP invented by Dr. Tat Hung Tong and now marketed by Cornerstone Research Group, are gaining in popularity, polyurethane SMPs are to date the most extensively studied. The polyurethane SMP invented by Dr. Shunichi Hayashi, and subsequently developed into a range of products by him and his co-workers, is seemingly the most successful SMP in the market at present.

This book, although focused on polyurethane, reveals the fascinating aspects of SMPs in a systematic way, from fundamentals to applications, from macro scale to submicron scale, from the history to the future. By focusing on a particular SMP that is available commercially, we are able to reveal its technical details and features. The many illustrations and vivid pictures included will help readers to instantly visualize and grasp the basic concepts and mechanisms. We hope that, after reading this book, readers will be ready to explore their own designs.

Chapter 1 introduces SMMs and SMPs and describes their mechanisms and general applications. Thereafter, the focus is on polyurethane SMPs. Chapters 2 to 5 present the thermal- and moisture-responsive features, electrically conductive composites, and thermomechanical properties of these remarkable materials. Chapter 6 elucidates the fabrication and characterization of magnetic SMP composites. A more extensive and systematic review of SMP composites can be found in Chapter 7. Chapter 8 focuses on porous and foam SMPs. Chapter 9 discusses SMEs at micro and nano scales. Chapter 10 is about wrinkling atop SMPs. Novel biomedical applications are revealed in Chapter 11. Chapter 12 explains the fundamentals of the multi-SMEs and how these effects add a new dimension to SMP applications. The concluding chapter covers the future of polyurethane SMPs.

We greatly appreciate the kind help of Dr. Shunichi Hayashi, Prof. Hisaaki Tobushi, Dr. Witold M. Sokolowski, and Dr. Tat Hung Tong in providing constructive comments and writing the Forewords. The origin of this book may be traced back more than 12 years, when Dr. Sokolowski passed a few pieces of SMP foams to W.M. Huang. We thank him for introducing us to this magic material.

W.M. Huang would like to thank all current and previous members of his research group for their support. The content of this book is largely based on their hard work over several years. It is indeed enjoyable to work with them and share both sad and exciting moments.

Last but not least, Hendra Purnawali helped edit, compile, and finalize this book. Without his help and patience, the book would not have been ready for publishing. Many big thanks to him.

W.M. Huang, B. Yang, and Y.Q. Fu

Authors

Wei Min Huang is an associate professor at the School of Mechanical and Aerospace Engineering at Nanyang Technological University (NTU), Singapore. He was awarded his PhD from Cambridge University (United Kingdom) in 1998 and has published approximately 100 journal papers, mainly in the field of shape memory materials. His current research interests include the fundamentals of the shape–temperature memory effect, shape memory materials and technologies, and their applications.

Bin Yang obtained his PhD from the School of Mechanical and Aerospace Engineering, Nanyang Technological University, Singapore, in 2007. His PhD work focused on the polyurethane shape memory polymer and its composites, under the supervision of Professor W.M. Huang and Professor C. Li. Dr. Yang is now a chief engineer at Helvoet Rubber & Plastic Technologies Pte. Ltd. (Singapore).

Richard Yong Qing Fu is a reader at the Thin Film Centre at the University of the West of Scotland in the United Kingdom. He was a lecturer at Heriot-Watt University, Edinburgh, from 2007 to 2010. He earned a PhD from Nanyang Technological University, Singapore, and then worked as a research fellow at the Singapore–Massachusetts Institute of Technology Alliance and as a research associate at the University of Cambridge. Dr. Fu has extensive experience with microelectromechanical systems (MEMS), thin films, surface coatings, shape memory alloys, smart materials, and nanotechnology. He has published a book on thin film shape memory alloys, about 10 book chapters, and more than 150 refereed journal papers.

1 Introduction

1.1 SHAPE MEMORY MATERIALS AND SHAPE MEMORY POLYMERS

There are plenty of fascinating materials in the world, and we still see many new materials invented almost every day—largely driven by the needs for new materials from a variety of engineering applications. One group of materials can respond accordingly to a particular stimulus by means of altering certain physical or chemical properties. The types of stimuli include heat (thermo-responsive); stress and/or pressure (mechano-responsive); electrical current and/or voltage (electro-responsive); magnetic field (magneto-responsive); change of pH, solvent, or moisture level (chemo-responsive); light (photo-responsive); and other factors.

Technically speaking, members of this group are known as stimulus-responsive materials (SRMs; Figure 1.1). SRMs, in particular stimulus-responsive polymers (including polymeric gels) and their composites, have become very hot topics in recent years due to a wide range of potential applications, from functional nano-composites to controlled and targeted drug and gene delivery therapies (Zhao et al. 2009, Kostanski et al. 2009, Onaca et al. 2009, Bajpai et al. 2008, Schmaljohann 2006, Stratakis et al. 2010, Kundys et al. 2010, Huck 2008, Caruso et al. 2009, Stuart et al. 2010, Chaterji et al. 2007, Osada and Gong 1998, Thornton et al. 2004).

One group of SRMs exhibits the ability to change shape in the presence of the right stimulus (Lendlein 2010). If the shape change is spontaneous and instant when the stimulus is presented, the result is a shape change material (SCM); electro-active polymers (EAPs) and piezoelectrical materials (such as PZTs) are two typical examples of SCMs (Aschwanden and Stemmer 2006, Haertling 1999, Yu et al. 2003). A second group designated shape memory materials (SMMs) maintain their temporary shapes virtually forever until the right stimulus is applied. All SMMs are characterized by the shape memory effect (SME), defined as the ability to recover original shape in the presence of the right stimulus after severe and quasi-plastic distortion (Huang et al. 2010a).

A few major types of SMMs have been developed. Probably the most important ones at present are shape memory alloys (SMAs) and shape memory polymers (SMPs; Wei et al. 1998, Feninat et al. 2002, Hornbogen 2006, Gunes and Jana 2008). While the mechanism behind the SMEs in SMAs is the reversible martensitic transition, the dual-segment/domain system is the mechanism in SMPs (Funakubo 1987, Sun and Huang 2010a, 2010b). Probably the newest type of SMM, the shape memory hybrid (SMH) composed of at least two components and lacking SMEs as an individual compound, shares the same mechanism as SMPs (Fan et al. 2011, Huang et al. 2010a). Shape memory ceramics (SMCs) exhibit the same working principle as SMAs, i.e.,

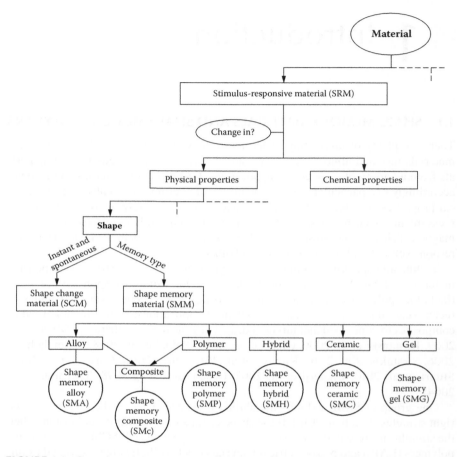

FIGURE 1.1 Location of shape memory polymers within the world of materials.

reversible phase transformation, or have multi-phase systems and are thus similar to SMPs (Wills 1977, Swain 1986, Pandit et al. 2004). Due to swelling effects and/ or electrical charges, gels are normally considered typical SCMs. Some gels exhibit SMEs due to, for instance, a reversible order–disorder transition (Osada and Gong 1998, Mitsumata et al. 2001, Liu et al. 2007a, Gong 2010). It is interesting that the swelling effects have been utilized recently to actuate SMPs (Lv et al. 2008). A shape memory composite (SMc) is defined as a composite with at least one SMM (most likely SMA or SMP) as its component (Liang et al. 1997, Wei et al. 1998, Tobushi et al. 2009). From this view, SMcs do not constitute an independent subgroup of SMMs, even for shape memory bulk metallic glass composites (Hofmann 2010).

Although SMEs were found in a gold cadmium alloy as early as 1932, the attraction of this phenomenon was not apparent until 1971 when significant recoverable strain was observed in a nickel titanium alloy at the US Naval Ordnance Laboratories (Funakubo 1987). At present, a few SMA systems have been developed and are commercially available (Huang 2002). Thin film SMAs have become promising actuation materials for microelectromechanical systems (MEMS) (Miyazaki et al. 2009).

If heat shrinkable polymers (HSPs) are classified as SMPs, we can trace the history of SMPs back to 1906 (Gunes and Jana 2008) before the invention of SMAs. Water shrinkable polymers (WSPs) are other examples (Willett 2008). However, both HSPs and WSPs are limited to shrinkage only; all current SMPs can recover from any type of quasi-plastic distortion. According to Liang et al. (1997), we regard the polynorbornene-based SMP invented in 1984 by Nippon Zeon Co. in Japan as the first generation of SMPs. This SMP and two more invented soon after it (trans-isopolypreme based, developed by Kuraray Company, and styrene–butadiene based, developed by Asahi Company, both in Japan) all suffer from the same problem of limited processability. The thermoplastic polyurethane SMP invented by Dr. S. Hayashi does not have this problem (Hayashi 1990) and has been developed into a full range of products that are now commercially available. Another relatively recent and successfully marketed SMP from Cornerstone Research Group in the US is thermoset polystyrene based (Dietsch and Tong 2007).

Note that despite possible differences in origins, after proper programming and training processes, most polymers exhibit certain levels of SMEs. For example, the SME in polytetrafluoroethylene (PTFE), although limited, has been reported in Hornbogen (1978); poly(methylmethacrylate) (PMMA) shows much better performance. Figure 1.2 reveals SMEs in acrylonitrile butadiene styrene (ABS) and ethylene-vinyl acetate (EVA) materials, which in terms of the shape recovery ability is not ideal when compared with that ability in traditional SMPs. Conventionally, only those with substantial recoverable strain and shape recovery ability are of interest and considered SMPs.

A number of SMP systems have been developed and we still see considerable effort aimed at discovering new SMPs (Liu et al. 2006, 2007a, Mather et al. 2009, Behl et al. 2010, Lendlein 2010, Meng and Hu 2010). As compared with their alloy counterparts, the major advantages of SMPs are

1. Low density (e.g., the typical bulk density of a polyurethane SMP is 1.25 g/cm^3; that of a NiTi SMA is 6.4 g/cm^3) and low cost (for raw material and also for fabrication and processing; Hayashi 1990, Yang 2007, Huang 2002).
2. Ease in producing high quality and specifically shaped materials including thin and ultrathin films and wires, foams with different porosities, and others at different scales using various traditional and advanced polymer processing technologies (injection molding, extrusion, dip coating, spin coating, water float casting; Hayashi 1990, Gunes and Jana 2008, Sun and Huang 2010b, Huang et al. 2010b).
3. Extremely high recoverable strain (on the order of 100% in tension for solids (Figure 1.3) and over 95% in compression for foams (Tey et al. 2001, Huang et al. 2006), as compared with a maximum recoverable strain below 10% for conventional SMAs.
4. Ease in tailoring of thermomechanical properties (e.g., by blending with different types of fillers or varying their compositions; Gall et al. 2002, Liu and Mather 2002, Yang et al. 2005b, Cao and Jana 2007, Yang 2007, Rezanejad and Kokabi 2007, Meng and Hu 2008, Ratna and Karger-Kocsis

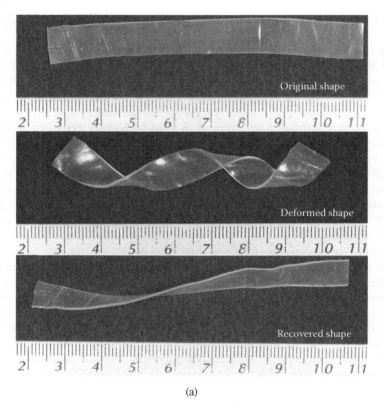

(a)

FIGURE 1.2   Shape memory effects in EVA (a) and ABS (b).

    2008, Pan et al. 2008, Gunes and Jana 2008, Yakacki et al. 2008, Xie and Rousseau 2009, Meng and Hu 2009, Xu et al. 2010).

5. The abilities to be always transparent (Yakacki et al. 2008, Jung et al. 2010b), electrically conductive (Leng et al. 2008a, 2008b, Lu et al. 2010), magnetic (Mohr et al. 2006), and even contain threshold temperature sensors (Kunzelman et al. 2008).

6. Wide shape recovery temperature range (from –20°C to +150°C).

7. Convenient gradient function (Huang et al. 2005, DiOrio et al. 2010).

8. High damping ratio within transition range (Yang 2007).

9. High potential for recycle and reuse at low cost (Inoue et al. 2009).

10. Excellent chemical stability, biocompatibility and even biodegradability (Han et al. 2007, Choi and Lendlein 2007, Yakacki et al. 2008, Lendlein and Langer 2002, Nardo et al. 2009, Luo et al. 2008, Zhang et al. 2010). The degradation rate can be adjusted if required (Kelch et al. 2007, Knight et al. 2009).

11. A variety of different stimulating methods, including heat (direct, Joule, induction, infrared and radiation, laser, and other methods), moisture, solvent addition, change in pH value, and light, are applicable to SMPs (Jung et al. 2006, Lendlein et al. 2005, Havens et al. 2005, Razzaq et al. 2007, Liu

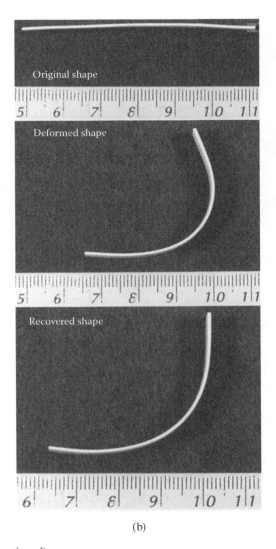

(b)

FIGURE 1.2 (Continued)

et al. 2007a, Du and Zhang 2010). Stimuli for SMAs are limited to heat and magnetic fields.

12. Great convenience for truly wireless and/or contactless operation inside a human body (Mohr et al. 2006, Buckley et al. 2006) based on remote contactless actuation, for example, by applying an alternating magnetic field for induction heating.

13. Response to multiple stimuli (Huang et al. 2005, Yang et al. 2005b, Du and Zhang 2010), even with temperature sensing function by means of color change (Kunzelman et al. 2008).

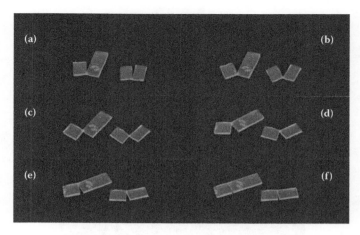

FIGURE 1.3 Shape recovery in severely pre-distorted polyurethane SMP plates upon heating. (From Huang WM, in *Shape Memory Polymers and Multifunctional Composites*, Taylor & Francis, 2010. With permission.)

14. Multiple-shape recovery ability (the multi-SME allows recovery from the temporary to the original shape through one or more intermediate shapes even within a single transition in a polymer or polymer composite (Bellin et al. 2006, 2007, Liu et al. 2005, Xie 2010, Sun and Huang 2010a, Luo and Mather 2010, Pretsch 2010a). Thus the recovery sequence is programmable (Huang et al. 2005). Multi-stimuli or functionally gradient SMPs have more advantages and flexibility for multi-SMEs following a required recovery sequence.

1.2 MECHANISMS OF SHAPE MEMORY EFFECTS IN SHAPE MEMORY POLYMERS

Despite possible differences in synthesis and the resulting differences in morphologies (Gunes and Jana 2008) and the required stimuli, the fundamental mechanism of SMEs in SMPs is the dual-segment (or domain) system in which one segment or domain is always elastic within the whole application range of concern and the other is transitionable in the presence of the right stimulus (Huang et al. 2010a).

Figure 1.4 illustrates the mechanisms of thermo-responsive SMPs. At low temperatures (Figure 1.4a), both the elastic and transition segments are hard. Upon heating above the transition temperature (phase–glass transition or melting of the segment), the transition segment becomes soft and can be easily deformed (Figure 1.4b). Accordingly, the elastic segment is deformed and elastic energy is built up and stored in it. Cooling below the transition temperature triggers the hardening of the transition segment. If the deformed (temporary) shape is held (Figure 1.4c) during cooling, it will be largely maintained even after the removal of the constraint (Figure 1.4d). This is because the transition segment is hard below the transition temperature and thus prevents the elastic recovery of the elastic segment. This is

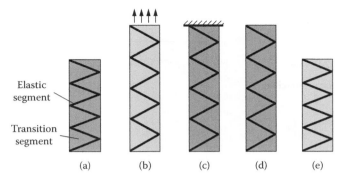

FIGURE 1.4 Mechanisms of shape memory effects in SMPs. (From Huang WM et al., *Materials Today,* 13: 54–61, 2010. With permission.)

the standard procedure to set temporary shapes of SMPs in the training and programming processes. After reheating above the transition temperature, the transition segment softens and thus loses its ability to hold the elastic segment in place. The freestanding piece of SMP returns back to the original shape due to the release of elastic energy in the elastic segment (Figure 1.4e). This discussion is applicable to the dual-domain system if *domain* replaces *segment*. Additional segments or domains may be added to enhance performance (strength, electrical conductivity, multiple transitions, etc.). The SMEs for other types of SMPs basically follow similar mechanisms. The driving force behind the SMEs in SMPs is the elastic energy stored in the elastic segment, whereas for SMAs, the force is the reversible martensitic transformation (Huang et al. 2010a).

Thermomechanical characterizations of various SMPs are well documented (Liu and Mather 2002, Liu et al. 2003, Prima et al. 2010a, Prima et al. 2010c). For a list of standard experiments and their technical details, see Hu (2007) and Lendlein (2010). In recent years, unconventional techniques including nanoindentation, bulging and point membrane deflection tests, and other methods have been applied to small SMP samples (Poilane et al. 2000, Fulcher et al. 2010, Huang et al. 2010a).

As with polymers, SMPs are of two types: thermoset and thermoplastic. While a thermoset requires proper molds to fix the original shape and the original shape after fixing should be permanent (although that is not always true, depending on the polymer and synthesis procedure), a thermoplastic has the flexibility to easily alter its original shape (upon heating above its melting temperature) and is thus more convenient for shape setting in some engineering applications (Huang 2010).

Another important issue relates to the training and programming processes for setting temporary shapes. As noted previously, all thermo-responsive SMPs are deformed easily above their transition temperatures (Figure 1.4). However, certain SMPs that are not so brittle may be deformed at low temperatures to set their temporary shapes (e.g., the polyurethane SMP from SMP Technologies, Japan). Technically speaking, for both types of materials, the need is to ensure that microfractures that deteriorate the SMEs, in particular during cyclic actuation, are prevented or at least minimized during training and programming.

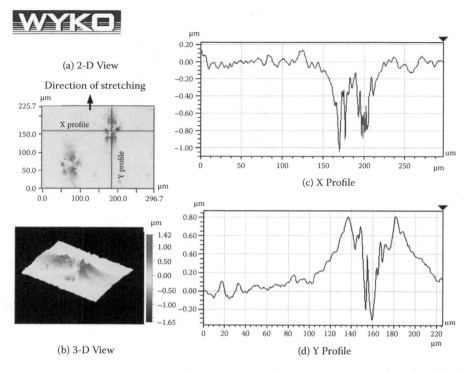

FIGURE 1.5 Butterfly-like micro features after stretching a piece of polystyrene SMP 50% at a temperature slightly lower than the transition finish temperature. (a) Two-dimensional view of two butterfly-like micro features. (b) Three-dimensional view of one of the butterfly-like micro features. (c) and (d) Profiles along X and Y directions as marked in (a) of typical butterfly-like micro feature.

Figure 1.5 reveals many micro-sized butterfly-like features on the surface of a piece of pre-polished polystyrene SMP (from Cornerstone Research Group) after it was stretched 50% above its T_g. After polishing followed by heating, the butterfly-like features appeared again but all switched direction by 90 degrees (Figure 1.6). A careful investigation revealed that the butterfly-like features formed when the hard polymer network broke under stretching. The practical way to avoid such micro-fractures, in particular for brittle SMPs, is to ensure that the SMP is deformed above its transition temperature range (the polymer is fully softened) and within its deformation limit.

A few constitutive models have been proposed to simulate the thermomechanical behavior of SMPs, from early work by Tobushi et al. (1997; mainly phenomenological in nature) to recent works focusing on the underlying mechanisms, for instance, Srivastava et al. (2010) and Qi et al. (2008) for SMP solids, and Prima et al. (2010a, 2010b) and Xu and Li (2010) for SMP foams. From an engineering application view, we need to know the long-term stability of SMPs in terms of shape recovery ratio and shape recovery stress following a particular training, programming, and actuation procedure. Preliminary investigations and numerical modeling were conducted on

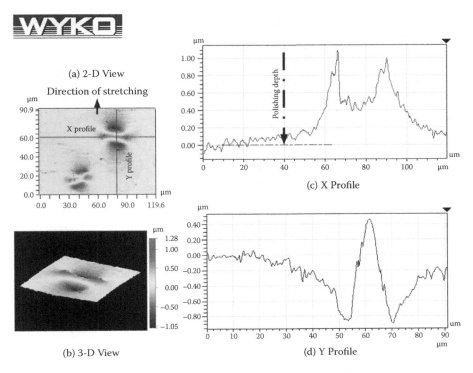

FIGURE 1.6 Switching directions of butterflies. (a) Two-dimensional view of two butterfly-like micro features. (b) Three-dimensional view of one of the butterfly-like micro features. (c) and (d) Profiles along X and Y directions as marked in (a) of typical butterfly-like micro feature.

some SMPs (Tey et al. 2001, Tobushi et al. 2004, Pretsch et al. 2009, Pretsch 2010b, Pretsch and Muller 2010, Muller and Pretsch 2010).

1.3 TYPICAL APPLICATIONS OF SHAPE MEMORY POLYMERS

We have seen a wide range of applications of various types of SMPs and their composites, from smart cloth and reusable mandrels to sensors for tracking cumulative environmental exposures (Everhart and Stahl 2005, Everhart et al. 2006, Dietsch and Tong 2007, Snyder et al. 2010, Hu 2007, Leng and Du 2010). We now present some novel examples of thermo-responsive SMPs as showcases for revealing the great potential of SMPs.

As we know, Braille is a form of printing that enables blind or partially sighted people to read and write by means of touching. Braille writing consists of patterns of raised dots in a configuration of a 3 × 2 matrix. The dots are made manually by a stylus and slate (Figure 1.7a) or automatically by a thermoset typing machine to the surface of a conventional paper or plastic sheet (called Braille paper or braillon) and are difficult to remove. A need existed to develop a new type of Braille paper that would allow easy

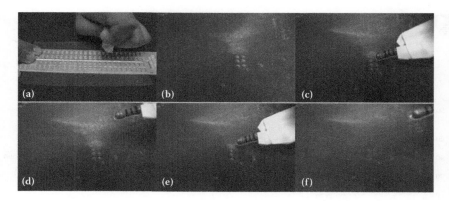

FIGURE 1.7 SMP Braille paper. (a) Writing. (b) through (f) Removing dots using a point heater in a one-by-one manner. (From Huang WM, in *Shape-Memory Polymers and Multifunctional Composites*, Taylor & Francis, 2010. With permission.)

removal of typographical errors and retyping just like normal writing with a pencil and eraser. Thin paper made of polyurethane SMP (SMP Technologies, Japan) was developed for such a purpose to help the blind. Any errors (small protrusive dots) may be removed precisely by a point heater as demonstrated in Figure 1.7. Subsequent corrections may be made at the same location. In addition, heating the entire sheet allows recycling again and again. The low cost of SMPs makes this innovation a very cost-effective and convenient approach for the blind. Additionally, *refreshable* SMP paper can be integrated into a machine for automatic and continuous display.

Screws are normally required for the assembly of electrical devices and also for implants in medical applications to firmly hold components together. Conventionally, different sized screws are required for different sized holes. However, SMP screws can serve as one-for-all solutions for a range of different sized holes, with or without thread, and no screwdriver is needed for tightening. This concept, known as active assembly, is demonstrated in Figure 1.8. For an SMP screw fitted into a threaded hole, a screwdriver can be used to loosen the screw for disassembly.

We have seen many electrical devices such as hand-held phones, video and audio players, and personal computers used in our daily lives become obsolete and they now constitute a serious environmental threat. We must use fewer materials and recycle and

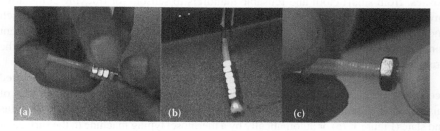

FIGURE 1.8 SMP screw for active assembly. (a) Sliding nuts onto a pre-stretched SMP rod. (b) Heating the SMP rod for shape recovery; nuts are firmly fitted. (c) Loosening nuts manually.

reuse them to minimize the threat. From a practical view, we need to develop a cost-effective approach to massively disassemble such obsolete devices. Manual disassembly and sorting in many developed countries have become impossible due to high labor costs, although those countries make the most significant contributions to the global proliferation of obsolete electrical devices. Active disassembly using shape memory materials (both SMAs and SMPs) has been proposed in the UK (Chiodo et al. 2001, 2002, Chiodo and Boks 2002). This concept is illustrated in Figure 1.9. An SMP rod can be easily transformed into a screw at a high temperature (Figure 1.9a) and may

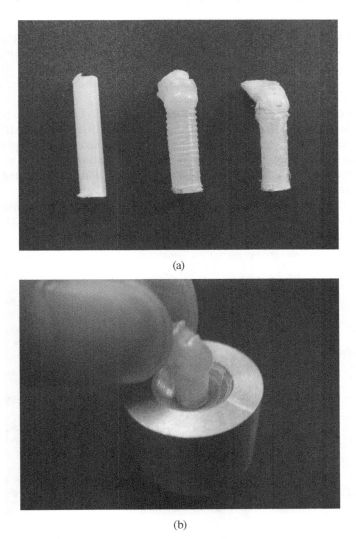

(a)

(b)

FIGURE 1.9 SMP for active disassembly. (a) Left to right: initial shape, deformed shape (screw), and recovered shape of SMP rod. (b) SMP screw in use. (From Huang WM, in *Shape-Memory Polymers and Multifunctional Composites,* Taylor & Francis, 2010. With permission.)

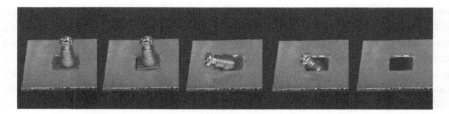

FIGURE 1.10   SMP for automatic hole opening upon heating.

be used as an ordinary plastic screw in an electrical device as shown in Figure 1.9b. Upon heating, the threads of the SMP screw disappear and it recovers its original rod shape. Consequently, active disassembly can then be achieved by means of slight shaking without additional physical touch. This is a convenient and cost-effective approach for massive disassembly of electrical devices.

In addition, an SMP plate (for instance, as a part of a cover case) can be used to open a hole automatically as shown in Figure 1.10. By utilizing multi-SMEs, different SMP pieces may be removed following a predetermined sequence upon heating to different temperatures. The increase in temperature allows different parts to be actively separated from the device, so that they can be automatically sorted and collected for reuse or recycling.

SMPs, their foams, and composites have been proposed for aerospace applications in deployable structures (Sokolowski and Tan 2007; Figure 1.11) and for

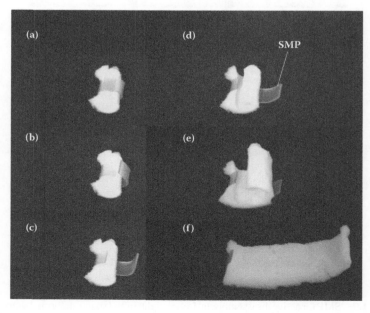

FIGURE 1.11   Deployment sequence of piece of folded sponge when SMP wrapper is heated for shape recovery.

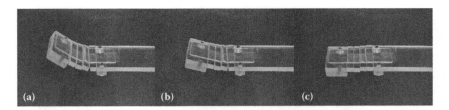

FIGURE 1.12 Thin-film SMP actuated hinge. Upon heating, the hinge straightens due to contraction (shape recovery) of the prestretched film.

actuation, e.g., a thin-film SMP actuated hinge for morphing wings as demonstrated in Figure 1.12. Deployable structures including wheels, masts, solar sails, solar arrays, and antennas must be compact because of limited space inside launching rockets, then must take on expanded configurations when deployed. Also, modern unmanned aviation vehicles (UAVs) have different configurations for achieving high performance for different missions. The ability to reconfigure during flight is called morphing. Morphing wings are required because changing the shapes and/or sizes of wings based on flight speed can reduce drag remarkably.

SMP is an ideal material for surface patterning to achieve different surface morphologies for micro scale applications (Liu et al. 2007b, Sun et al. 2009, Huang et al. 2010a). As evidenced by many natural products such as lotus leaves and shark skins, well-patterned surfaces can significantly enhance many surface-related properties. Applying a laser through a micro-lens array for local heating at required points produces required protrusions over a large area instantly (Figure 1.13).

Probably the hottest application for SMPs at present is in biomedical engineering. SMPs have been considered promising as materials for degradable and functional cardiovascular implants and other therapeutic applications (Jung et al. 2010a, Karp and Langer 2007). A stent (Figure 1.14 shows its working principle) is one of the most attractive micro devices subjected to intensive research in recent years (Yakacki et al. 2007). The advantage of degradable SMP stents is that no follow-up operation is required for removal (Chen et al. 2007a). Sutures made of such SMPs do not have to be removed (Lendlein and Langer 2002). Both biodegradable and nondegradable SMPs have been used for controlled drug release (Wache et al. 2003, Chen et al. 2009b, Wischke et al. 2009, Zhang et al. 2010). SMPs are now used in endovascular thrombectomy devices (Wilson et al. 2007a), active microfluidic reservoirs (Gall et al. 2004), ocular implants (Song et al. 2010), and self-deploying neuronal electrodes (Sharp et al. 2006).

1.4 POLYURETHANE SHAPE MEMORY POLYMERS

It is well known that polyurethanes are made of tangled long linear chains consisting of two types of alternatively connected segments. Flexible, soft segments (typically diisocyanate-coupled low-melting polyester or polyether chains) alternate with elastic and relatively hard segments (typically diurethane bridges resulting from the reaction of a diisocyanate with a small-molecule glycol chain extender). By selecting a proper diisocyanate and polyol combination and/or soft segment-to-hard segment ratio, one can easily tailor required properties such as elasticity,

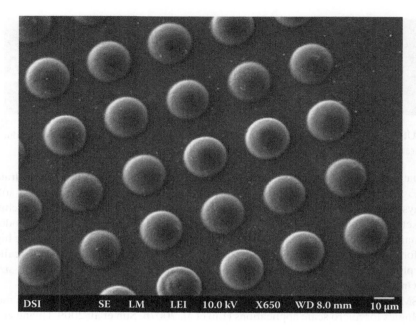

FIGURE 1.13 Micro scaled protrusion array atop a piece of pre-compressed polystyrene SMP produced by laser heating through a micro lens array. (From Huang WM, in *Shape-Memory Polymers and Multifunctional Composites*, Taylor & Francis 2010. With permission.)

crystallization temperature range, and melting point. Refer to the SME mechanism for SMPs discussed in Section 1.2. It is obvious that we can utilize the soft segments of polyurethanes as transition segments for SMEs.

A number of polyurethane SMPs have been developed to date (Lee et al. 2001, Ping et al. 2005, Xu et al. 2006, Buckley et al. 2007, Chen et al. 2007b, Chen et al.

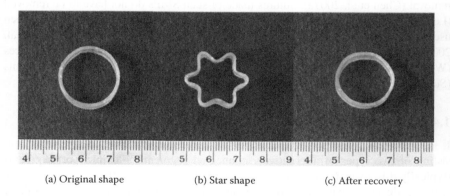

(a) Original shape   (b) Star shape   (c) After recovery

FIGURE 1.14 SMP stent. The original circular-shaped stent (a) is deformed into a star-like shape (b), then compacted (by means of folding) into a catheter. After release at the required location, the stent recovers its original circular shape upon heating (c).

2010). To further reduce densities and enhance compression ratios (and consequently abilities to recover from compression), polyurethane SMP foams have been fabricated using different techniques and characteristics (Lee et al. 2007, Chung and Park 2010). Physical aging is an important issue, particularly in medical applications, and has been studied preliminarily by Lorenzo et al. (2009). Thermo-responsive braided stents have been investigated numerically (Kim et al. 2010).

Electrically conductive polyurethane SMPs for Joule heating have been achieved by blending with various types of conductive fillers (Li et al. 2000, Sahoo et al. 2007, Gunes et al. 2009, Cho et al. 2005, Jung et al. 2010c). After mixing with magnetite particles, induction heating by means of applying a magnetizing field of 4.4 kA/m at a frequency of 50 Hz has been demonstrated to successfully trigger shape recovery in a polyurethane SMP–magnetite composite (Razzaq et al. 2007). Furthermore, electrically conductive, optically transparent, and mechanically strong polyurethane SMP films have been developed by incorporating photochemically surface-modified multiwalled carbon nanotubes for optically actuated actuation (Jung et al. 2010b).

Different types of fillers such as silica, carbon nano fiber, silicon carbide, carbon black, clay, and others have been added to polyurethane SMP composites for reinforcement (Gunes et al. 2008, Cao and Jana 2007, Jang et al. 2009, Kuriyagawa et al. 2010). According to Deka et al. (2010), hyperbranched polyurethane–multi-walled carbon nanotube composites exhibit enhanced biodegradability as compared to the pristine polymer. Cytocompatibility testing based on hemolysis of red blood cells revealed a lack of cytotoxicity.

Although most of these polyurethane SMPs are thermo-responsive, water-responsive materials have been achieved by means of modification using polyhedral oligomeric silsesquioxane or pyridine moieties (Jung et al. 2006, Chen et al. 2009a).

While we have seen a few types of polyurethane SMPs documented in the literature, an extensive review and market search reveal that the polyurethane SMP invented by Dr. S. Hayashi (Hayashi 1990, Hayashi et al. 1995) using the basic synthesis process as follows:

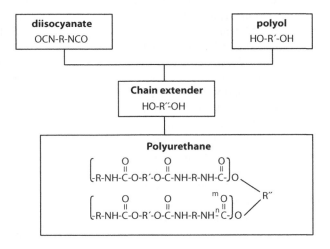

Dr. Hayashi's polyurethane SMP is currently the most popular; successfully marketed and thoroughly investigated.

The excellent SME of this SMP is demonstrated in Figure 1.3. Two pieces of SMP plates are partially cut and bent 180 degrees in the in-plane direction at room temperature (about 23°C). Subsequently, they are placed atop a hot plate for thermally induced shape recovery. At the end of the heating process, as we can see clearly, both plates almost fully recover their original straight shapes. At present, a full range of products (mainly thermoplastics) including pellet (MM), resin and hardener (MP), solution (MS), and foam (MF) forms have been developed and are available through SMP Technologies, Japan. Interested readers and potential users may refer to the website (http://www.smptechno.com/index_en.html) for technical details.

Very recently, micro beads, micro springs, thin wires, and thin and ultrathin films have been produced from this SMP (Huang et al. 2010b, 2010c). Because they all display excellent SMEs, we expect them to have great potential for micro and nano devices (Huang et al. 2010a).

The thermomechanical behavior of this SMP was systematically investigated by Tobushi et al. (1996, 1998, 2000, 2001b, 2001c, 2004, 2006) in addition to their great efforts at phenomenological modeling (Tobushi et al. 1997, 2001a).

As compared with other SMPs, this polyurethane SMP is generally ductile even at low temperatures. Therefore, programming can be done at low or high temperatures, as revealed in Figure 1.15a. During stretching at high temperatures ($T_g + 15$°C), the deformation is virtually uniform along the central part of the sample, but we can clearly see the necking and propagation phenomena during stretching at low (room) temperatures, at which the deformation is not uniform. This is a point that we should bear in mind. After heating to $T_g + 15$°C, both samples recover their original dogbone shapes (Figure 1.15b). A closer look at the process of shape recovery upon heating reveals that whatever the deformation is, uniform or non-uniform, the shape recovery upon heating is spontaneous and uniform everywhere (Figure 1.16).

Based on the characteristics of polyurethane SMP foam, Dr. W.M. Sokolowski proposed the concept of cold hibernated elastic memory (CHEM) and has been actively working on its space and medical applications (Sokolowski and Chmielewski 1999, Sokolowski et al. 2007, Metcalfe et al. 2003). Hybrid polyurethane SMP composites have been investigated by Liang et al. (1997). In recent years, carbon black and nickel powder have been blended into this SMP to achieve Joule heating as done with SMAs because such heating is more convenient in engineering practices (Leng et al. 2008a, 2008c, Huang and Yang 2010). Figure 1.17 demonstrates the concept of wing morphing by passing an electrical current through a piece of conductive SMP (carbon black is used as the conductive filler).

To enhance the strength of this polyurethane SMP among others, attapulgite (palygorskite) clays (heat treated and untreated)—found in nature in nano fiber form and very poor in electrical conductivity—have been used as cost-effective alternatives (Pan et al. 2008, Xu et al. 2009). As seen in Figure 1.18, the composite maintains excellent shape memory ability.

Although this SMP was originally designed to be thermo-responsive only, Yang et al. (2004) found that moisture can significantly reduce its T_g. This means that instead of heating above the T_g point to achieve shape recovery, the T_g of the SMP can be

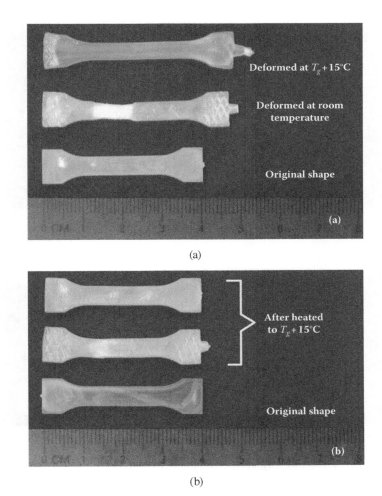

FIGURE 1.15 Stretching at high and low temperatures (a) and after heating for shape recovery (b).

reduced upon immersion in water as an alternative to trigger shape recovery, i.e., this polyurethane SMP is actually thermo- and moisture responsive (Yang et al. 2006). This unique feature, which is also applicable to the material's electrically conductive and nonconductive composites (Yang et al. 2005a, 2005b, Huang et al. 2010b), has been used for water-driven programmable actuation (Huang et al. 2005, Sun and Huang 2010b). Figure 1.19 shows the recovery process of a piece of polyurethane SMP–attapulgite composite upon immersion into room-temperature water.

At macroscopic scale, autochokes for engines, intravenous cannulae, spoon and fork handles for those unable to grasp objects, and sportswear are some traditional applications of this SMP (Tobushi et al. 1996). At micro scale, among other applications, wrinkles of submicron wavelength (Sun et al. 2009, Zhao et al. 2010) have

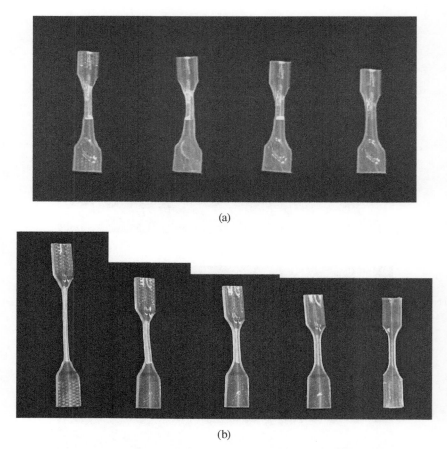

(a)

(b)

FIGURE 1.16 Shape recovery upon heating. (a) Partially pre-stretched sample. (From Huang WM, in *Shape-Memory Polymers and Multifunctional Composites*, Taylor & Francis 2010. With permission.) (b) Fully pre-stretched sample.

been produced atop this SMP by heating (Figure 1.20) and by immersion into room-temperature water.

Since excellent biocompatibility was proven, a range of medical applications have been developed (Sokolowski et al. 2007, Metcalfe et al. 2003, Nardo et al. 2009). Some typical biomedical uses of this polyurethane SMP in solid or foam form are listed as follows (refer to the website of SMP Technologies, Japan, and the cited references for product numbers and technical details):

- A vascular stent (MM5520) fabricated and deployed *in vitro* by laser heating (Baer et al. 2007a).
- A 0.35 mm diameter wire (MM5520) produced by extrusion and used for thrombus removal activated by 800 nm diode laser light (Metzger et al. 2005).

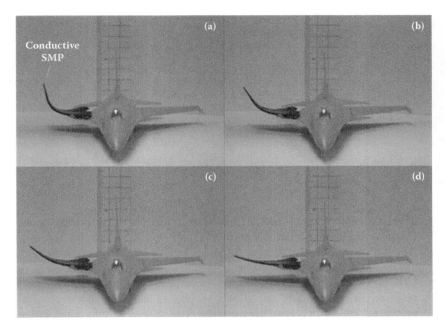

FIGURE 1.17 Electrically conductive SMP for morphing wing. (From Huang WM, in *Shape-Memory Polymers and Multifunctional Composites*, Taylor & Francis, 2010. With permission.)

- A dialysis needle adapter (MM5520) for reducing hemodynamic stress within arteriovenous grafts (Ortega et al. 2007).
- Proposed use of MM6520 for infrared diode laser-activated intravascular thrombectomy to remove blood clots (Wilson et al. 2005).
- Possible use of MP 3510, MP 4510, and MP 5510 for deployable biomedical devices for minimally invasive surgery (Baer et al. 2007b).

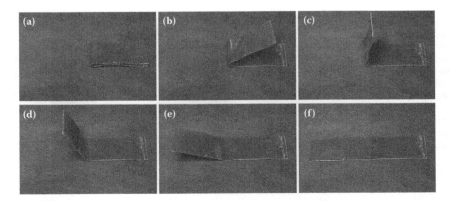

FIGURE 1.18 Shape recovery of SMP and clay composite upon heating.

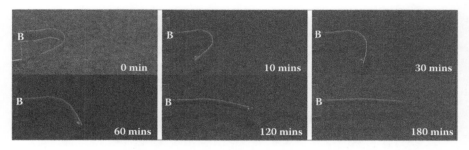

FIGURE 1.19 Shape recovery of a piece of SMP–attapulgite composite upon immersion into room-temperature water.

- Thermoset MP 4510- and MP 5510-based micro actuators for treating stroke (Metzger et al. 2002, Maitland et al. 2002).
- An SMP foam (MM5520) stent for endovascular embolization of fusifom aneurysms (Buckley et al. 2007).
- CHEM 3520 and CHEM 5520 foams have been tested for cerebral aneurysm repair, particularly the effects of plasma sterilization on physical properties and cytocompatibility (Nardo et al. 2009). Results suggest that CHEM foam may be advantageous for manufacturing devices for minimally invasive embolizations of aneurysms.
- Utilizing the thermo- and moisture-responsive features of MM5520 and MM3520 for retractable stents, self-tightening sutures, and other biomedical devices (Sun and Huang 2010b, Huang et al. 2010a).

1.5   OUTLINE OF BOOK

Although this book is titled *Polyurethane Shape Memory Polymers*, the focus is actually on the polyurethane SMP invented by Dr. S. Hayashi. The reason for this is twofold. One is because this SMP is one of the most successfully marketed SMPs

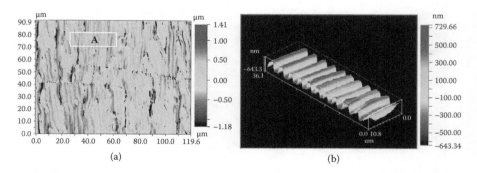

FIGURE 1.20 Typical wrinkles atop polyurethane SMP. (From Huang et al., *Journal of Materials Chemistry*, 20: 3367–3381, 2010. With permission.)

and may be the only commercially available (on a large scale and with a range of products) polyurethane SMP at present. The second reason is that this SMP is the only polyurethane SMP that has been extensively and systematically investigated based on the range of reported applications. By focusing on this particular SMP in this book, we can reveal most of its features and its full potential. Other polyurethane SMPs and other types of SMPs will be discussed briefly as needed. In some cases, more details are presented when necessary for clarity and completeness.

ACKNOWLEDGMENTS

We thank S.Y. Tan, N. Liu, Y. Zhao, Z. Ding, and Dr. C.C. Wang for their help in conducting some of the experiments reported in this chapter. In addition, we thank Dr. C.C. Wang for help in compiling this chapter.

REFERENCES

Aschwanden M and Stemmer A (2006). Polymeric electrically tunable diffraction grating based on artificial muscles. *Optics Letters*, 31, 2610–2612.

Baer GM, Wilson TS, Benett WJ et al. (2007a). Fabrication and *in vitro* deployment of a laser-activated shape memory polymer vascular stent. *Biomedical Engineering Online*, 6, 43.

Baer G, Wilson TS, Matthews DL et al. (2007b). Shape memory behavior of thermally stimulated polyurethane for medical applications. *Journal of Applied Polymer Science*, 103, 3882–3892.

Bajpai AK, Shukla SK, Bhanu S et al. (2008). Responsive polymers in controlled drug delivery. *Progress in Polymer Science*, 33, 1088–1118.

Behl M, Razzaq MY, and Lendlein A (2010). Multifunctional shape-memory polymers. *Advanced Materials*, 22, 3388–3410.

Bellin I, Kelch S, Langer R et al. (2006). Polymeric triple-shape materials. *Proceedings of the National Academy of Sciences of the United States of America*, 103, 18043–18047.

Bellin I, Kelch S, and Lendlein A (2007). Dual-shape properties of triple-shape polymer networks with crystallisable network segments and grafted side chains. *Journal of Materials Chemistry*, 17, 2885–2891.

Buckley CP, Prisacariu C, and Caraculacu A (2007). Novel triol-crosslinked polyurethanes and their thermorheological characterization as shape-memory materials. *Polymer*, 48, 1388–1396.

Buckley PR, Wilson TS, Benett WJ, Hartman J, Saloner D, Maitland DJ (2007). Shape memory polymer stent with expandable foam: a new concept for endovascular embolization of fusiform aneurysms. *IEEE Transactions on Biomedical Engineering*, 54 (6), 1157–1160.

Buckley PR, McKinley GH, Wilson TS et al. (2006). Inductively heated shape memory polymer for the magnetic actuation of medical devices. *IEEE Transactions on Biomedical Engineering*, 53, 2075–2083.

Cao F and Jana SC (2007). Nanoclay-tethered shape memory polyurethane nanocomposites. *Polymer*, 48, 3790–3800.

Caruso MM, Davis DA, Shen Q et al. (2009). Mechanically induced chemical changes in polymeric materials. *Chemical Reviews*, 109, 5755–5798.

Chaterji S, Kwon IK, and Park K (2007). Smart polymeric gels: redefining the limits of bio-medical devices. *Progress in Polymer Science*, 32, 1083–1122.

Chen MC, Tsai HW, Chang Y et al. (2007a). Rapidly self-expandable polymeric stents with a shape-memory property. *Biomacromolecules*, 8, 2774–2780.

Chen S, Hu J, Liu Y et al. (2007b). Effect of SSL and HSC on morphology and properties of PHA-based SMPU synthesized bulk polymerization method. *Journal of Polymer Science,* 45, 444–454.

Chen S, Hu J, Yuen C et al. (2009a). Novel moisture-sensitive shape memory polyurethanes containing pyridine moieties. *Polymer*, 50, 4424–4428.

Chen MC, Chang Y, Liu CT et al. (2009b). The characteristics and *in vivo* suppression of neointimal formation with sirolimus-eluting polymeric stents. *Biomaterials*, 30, 79–99.

Chen S, Hu J, Zhuo H et al. (2010). Study on the thermal-induced shape memory effect of pyridine containing supramolecular polyurethane. *Polymer,* 51, 240–248.

Chiodo JD and Boks C (2002). Assessment of end-of-life strategies with active disassembly using smart materials. *Journal of Sustainable Product Design*, 2, 69–82.

Chiodo JD, Harrison DJ, and Billett EH (2001). An initial investigation into active disassembly using shape memory polymers. *Proceedings of Institute of Mechanical Engineers*, 215, 733–741.

Chiodo JD, Jones N, Billett EH et al. (2002). Shape memory alloy actuators for active disassembly using "smart" materials of consumer electronic products. *Materials and Design*, 23, 471–478.

Cho JW, Kim JW, Jung YC et al. (2005). Electroactive shape-memory polyurethane composites incorporating carbon nanotubes. *Macromolecular Rapid Communications*, 26, 412–416.

Choi N and Lendlein A (2007). Degradable shape-memory polymer networks from oligo[(L-lactide)-ran-glycolide]dimethacrylates. *Soft Matter*, 3, 901–909.

Chung SE, Park CH (2010). The thermoresponsive shape memory characteristics of polyurethane foam. *Journal of Applied Polymer Science*, 117, 2265–2271.

Deka H, Karak N, Kalita RD et al. (2010). Biocompatible hyperbranched polyurethane/multi-walled carbon nanotube composites as shape memory materials. *Carbon*, 48, 2013–2022.

Dietsch B and Tong T (2007). A review: features and benefits of shape memory polymers (SMPs). *Journal of Advanced Materials*, 39, 3–12.

DiOrio AM, Luo X, Lee KM et al. (2010). A functionally graded shape memory polymer. *Soft Matter*, in press.

Du H and Zhang J (2010). Solvent induced shape recovery of shape memory polymer based on chemically cross-linked poly(vinyl alcohol). *Soft Matter*, 6, 3370–3376.

Everhart MC and Stahl J (2005). Reusable shape memory polymer mandrels. *Proceedings of SPIE*, 5762, 27–34.

Everhart MC, Nickerson DM, and Hreha RD (2006). High-temperature reusable shape memory polymer mandrels. *Proceedings of SPIE*, 6171.

Fan K, Huang WM, Wang CC et al. (2011). Water-responsive shape memory hybrid: design concept and demonstration. *eXPRESS Polymer Letters*, 5, 409–416.

Feninat FE, Laroche G, Fiset M et al. (2002). Shape memory materials for biomedical applications. *Advanced Engineering Materials*, 4, 91–104.

Fulcher JT, Lu YC, Tandon GP et al. (2010). Thermomechanical characterization of shape memory polymers using high temperature nanoindentation. *Polymer Testing*, 29, 544–552.

Funakubo H (1987). *Shape Memory Alloys*. Gordon & Breach, New York.

Gall K, Dunn ML, Liu Y et al. (2002). Shape memory polymer nanocomposites. *Acta Materialia,* 50, 5115–5126.

Gall K, Kreiner P, Turner D et al. (2004). Shape-memory polymers for microelectromechanical systems. *Journal of Microelectromechanical Systems*, 13, 472–483.

Gong JP (2010). Why are double network hydrogels so tough? *Soft Matter*, 6, 2583–2590.

Gunes IS and Jana SC (2008). Shape memory polymers and their nanocomposites: a review of science and technology of new multifunctional materials. *Journal of Nanoscience and Nanotechnology*, 8, 1616–1637.

Gunes IS, Cao F, and Jana SC (2008). Evaluation of nanoparticulate fillers for development of shape memory polyurethane nanocomposites. *Polymer*, 49, 2223–2234.

Gunes IS, Jimenez GA, and Jana SC (2009). Carbonaceous fillers for shape memory actuation of polyurethane composites by resistive heating. *Carbon*, 47, 981–997.

Haertling GH (1999). Ferroelectric ceramics: history and technology. *Journal of the American Ceramic Society*, 82, 797–818.

Han S, Gu BH, Nam KH et al. (2007). Novel copolyester-based ionomer for a shape-memory biodegradable material. *Polymer,* 48, 1830–1834.

Havens E, Snyder EA, and Tong TH (2005). Light-activated shape memory polymers and associated applications. *Proceedings of SPIE*, 5762, 48–54.

Hayashi S (1990). Technical report on preliminary investigation of shape memory polymers. Mitsubishi Heavy Industries Research and Development Center, Nagoya, Japan.

Hayashi S, Kondo S, Kapadia P et al. (1995). Room-temperature-functional shape-memory polymers. *Plastics Engineering*, Feb., 29–31.

Hofmann DC (2010). Shape memory bulk metallic glass composites. *Science*, 329, 1294–1295.

Hornbogen E (1978). Shape change during the 19°C phase transformation of PTFE. *Progress in Colloid and Polymer Science*, 64, 125–131.

Hornbogen E (2006). Comparison of shape memory metals and polymers. *Advanced Engineering Materials*, 8, 101–106.

Hu J (2007). *Shape Memory Polymers and Textiles*. Woodhead Publishing, Cambridge.

Huang W (2002). On the selection of shape memory alloys for actuators. *Materials and Design*, 23, 11–19.

Huang WM (2010). Novel applications and future of shape memory polymers. In *Shape-Memory Polymers and Multifunctional Composites*. Taylor & Francis/CRC, New York, pp. 333–363.

Huang WM and Yang B (2010). Electrical, thermomechanical, and shape-memory properties of the PU shape-memory polymer. In *Shape-Memory Polymers and Multifunctional Composites*. Taylor & Francis/CRC, New York, pp. 109–131.

Huang WM, Yang B, An L et al. (2005). Water-driven programmable polyurethane shape memory polymer: demonstration and mechanism. *Applied Physics Letters*, 86, 1140105.

Huang WM, Lee CW, and Teo HP (2006). Thermomechanical behavior of a polyurethane shape memory polymer foam. *Journal of Intelligent Material Systems and Structures*, 17, 753–760.

Huang WM, Ding Z, Wang CC et al. (2010a). *Shape Memory Materials. Materials Today*, 13, 54–61.

Huang WM, Yang B, Zhao Y et al. (2010b). Thermo-moisture responsive polyurethane shape-memory polymer and composites: a review. *Journal of Materials Chemistry*, 20, 3367–3381.

Huang WM, Fu YQ, and Zhao Y (2010c). High performance polyurethane shape-memory polymer and composites. *PU Magazine*, 7, 155–160.

Huck WTS (2008). Responsive polymers for nanoscale actuation. *Materials Today,* 11, 24–32.

Inoue K, Yamashiro M, and Iji M (2009). Recyclable shape-memory polymer: poly(lactic acid) crosslinked by a thermoreversible Diels-Alder reaction. *Journal of Applied Polymer Science*, 112, 876–885.

Jang MK, Hartwig A, and Kim BK (2009). Shape memory polyurethanes cross-linked by surface modified silica particles. *Journal of Materials Chemistry*, 19, 1166–1172.

Jung YC, So HH, and Cho JW (2006). Water-responsive shape memory polyurethane block copolymer modified with polyhedral oligomeric silsesquioxane. *Journal of Macromolecular Science B*, 45, 453–461.

Jung F, Wischke C, and Lendlein A (2010a). Degradable, multifunctional cardiovascular implants: challenges and hurdles. *MRS Bulletin*, 35, 607–613.

Jung YC, Kim HH, Kim YA et al. (2010b). Optically active multi-walled carbon nanotubes for transparent, conductive memory-shape polyurethane film. *Macromolecules*, 43, 6106–6112.

Jung YC, Yoo HJ, Kim YA et al. (2010c). Electroactive shape memory performance of poly-urethane composite having homogeneously dispersed and covalently crosslinked carbon nanotubes. *Carbon*, 48, 1598–1603.

Karp JM and Langer R (2007). Development and therapeutic applications of advanced biomaterials. *Current Opinions in Biotechnology*, 18, 454–459.

Kelch S, Steuer S, Schmidt AM et al. (2007). Shape-memory polymer networks from oligo [(ε-hydroxycaproate)-co-glycolate] dimethacrylates and butyl acrylate with adjustable hydrolytic degradation rate. *Biomacromolecules*, 8, 1018–1027.

Kim JH, Kang TJ, and Yu WR (2010). Stimulation of mechanical behavior of temperature-responsive braided stents made of shape memory polyurethanes. *Journal of Biomaterials*, 43, 632–643.

Knight PT, Lee KM, Chung T et al. (2009). PLGA-POSS end-linked networks with tailored degradation and shape memory behavior. *Macromolecules*, 42, 6596–6605.

Kostanski LK, Huang R, Filipe CDM et al. (2009). Interpenetrating polymer networks as a route to tunable multi-responsive biomaterials: development of novel concepts. *Journal of Biomaterials Science*, 20, 271–297.

Kundys B, Viret M, Colson D et al. (2010). Light-induced size changes in $BiFeO_3$ crystals. *Nature Materials*, 9, 803–805.

Kunzelman J, Chung T, Mather PT et al. (2008). Shape memory polymers with built-in threshold temperature sensors. *Journal of Materials Chemistry*, 18, 1082–1086.

Kuriyagawa M, Kawamura T, Hayashi S et al. (2010). Reinforcement of polyurethane-based shape memory polymer by hindered phenol compounds and silica particles. *Journal of Applied Polymer Science*, 117, 1695–1702.

Lee BS, Chun BC, Chung YC et al. (2001). Structure and thermomechanical properties of polyurethane block copolymers with shape memory effect. *Macromolecules*, 34, 6431–6437.

Lee SH, Jang MK, Kim SH et al. (2007). Shape memory effects of molded flexible polyurethane foam. *Smart Materials and Structures*, 16, 2486–2491.

Lendlein A, Ed. (2010). *Shape-Memory Polymers*. Springer, Heidelberg.

Lendlein A and Langer R (2002). Biodegradable, elastic shape memory polymers for potential biomedical applications. *Science*, 296, 1673–1676.

Lendlein A, Jiang H, Junger O et al. (2005). Light-induced shape-memory polymer. *Nature*, 434, 879–882.

Leng J and Du S, Eds. (2010). *Shape-Memory Polymers and Multifunctional Composites*. Taylor & Francis, Boca Raton, FL.

Leng JS, Huang WM, Lan X et al. (2008a). Significantly reducing electrical resistivity by forming conductive Ni chains in a polyurethane shape-memory polymer/carbon-black composite. *Applied Physics Letters*, 92, 204101.

Leng J, Lv H, Liu Y et al. (2008b). Synergic effect of carbon black and short carbon fiber on shape memory polymer actuation by electricity. *Journal of Applied Physics*, 104, 104917.

Leng JS, Lan X, Liu YJ et al. (2008c). Electrical conductivity of thermo-responsive shape-memory polymer with embedded micron sized Ni powder chains. *Applied Physics Letters*, 92, 014104.

Li F, Qi L, Yang J et al. (2000). Polyurethane/conducting carbon black composites: structure, electric conductivity, strain recovery behavior, and their relationships. *Journal of Applied Polymer Science*, 75, 68–77.

Liang C, Rogers CA, and Malafeew E (1997). Investigation of shape memory polymers and their hybrid composites. *Journal of Intelligent Material Systems and Structures*, 8, 380–386.

Liu C and Mather PT (2002). Thermomechanical characterization of a tailored series of shape memory polymers. *Journal of Applied Medical Polymers*, 6, 47–52.

Liu Y, Gall K, Dunn ML et al. (2003). Thermomechanical recovery couplings of shape memory polymers in flexure. *Smart Materials and Structures*, 12, 947–954.

Liu G., Ding X, Cao Y et al. (2005). Novel shape-memory polymer with two transition temperatures. *Macromolecular Rapid Communications*, 26, 649–652.

Liu G, Guan C, Xia H et al. (2006). Novel shape-memory polymer based on hydrogen bonding. *Macromolecular Rapid Communications*, 27, 1100–1104.

Liu C, Qin H, and Mather PT (2007a). Review of progress in shape-memory polymers. *Journal of Materials Chemistry*, 17, 1543–1558.

Liu N, Huang WM, Phee SJ et al. (2007b). A generic approach for producing various protrusive shapes on different size scales using shape-memory polymer. *Smart Materials and Structures*, 16, N47–N50.

Lorenzo V, Diaz-Lantada A, Lafont P et al. (2009). Physical aging of a PU-based shape memory polymer: influence on their applicability to the development of medical devices. *Materials and Design*, 30, 2431–2434.

Lu H, Liu Y, Gou J et al. (2010). Synergistic effect of carbon nanofiber and carbon nanopaper on shape memory polymer composite. *Applied Physics Letters*, 96, 084102.

Luo X and Mather PT (2010). Triple-shape polymeric composites (TSPCs). *Advanced Functional Materials*, 20, 2649–2656.

Luo H, Liu Y, Yu Z et al. (2008). Novel biodegradable shape memory material based on partial inclusion complex formation between #-cyclodextrin and poly(#-caprolactone). *Macromolecules*, 9, 2573–2577.

Lv H, Leng J, Liu Y et al. (2008). Shape-memory polymer in response to solution. *Advanced Engineering Materials*, 10, 592–595.

Maitland DJ, Metzger MF, Schumann D et al. (2002). Photothermal properties of shape memory polymer micro-actuators for treating stroke. *Lasers in Surgery and Medicine*, 30, 1–11.

Mather PT, Luo X, and Rousseau IA (2009). Shape memory polymer research. *Annual Review of Materials Research*, 39, 445–471.

Meng Q and Hu J (2008). Self-organizing alignment of carbon nanotube in shape memory segmented fiber prepared by *in situ* polymerization and melt spinning. *Composites A*, 39, 314–321.

Meng Q and Hu J (2009). A review of shape memory polymer composites and blends. *Composites A*, 20, 1661–1672.

Meng H and Hu J (2010). A brief review of stimulus-active polymers responsive to thermal, light, magnetic, electric, and water/solvent stimuli. *Journal of Intelligent Material Systems and Structures*, 21, 589–885.

Metcalfe A, Desfaits AC, Salazkin I et al. (2003). Cold hibernated elastic memory foams for endovascular interventions. *Biomaterials*, 24, 491–497.

Metzger MF, Wilson TS, Maitland DJ (2005). Laser-activated shape memory polymer microactuator for thrombus removal following ischemic stroke: preliminary in vitro analysis. *IEEE Journal of Selected Topics in Quantum Electronics*, 11 (4), 892–900.

Metzger MF, Wilson TS, Schumann D et al. (2002). Mechanical properties of mechanical actuator for treating ischemic stroke. *Biomedical Microdevices*, 4, 89–96.

Mitsumata T, Gong JP, and Osada Y (2001). Shape memory functions and motility of amphiphilic polymer gels. *Polymers for Advanced Technologies*, 12, 136–150.

Miyazaki S, Fu YQ, Huang WM, Eds. (2009). *Thin Film Shape Memory Alloys: Fundamentals and Device Applications*. Cambridge University Press, Cambridge.

Mohr R, Kratz K, Weigel T et al. (2006). Initiation of shape-memory effect by inductive heating of magnetic nanoparticles in thermoplastic polymers. *Proceedings of the National Academy of Sciences of the United States of America*, 103, 3540–3545.

Muller WW and Pretsch T (2010). Hydrolytic aging of crystallizable shape memory poly(ester urethane): Effects on the thermomechanical properties and visco-elastic modeling. *European Polymer Journal*, 46, 1745–1758.

Nardo LD, Alberti R, Cogada A et al. (2009). Shape memory polymer foams for cerebral aneurysm reparation: effects of plasma sterilization on physical properties and cytocompatibility. *Acta Biomaterials*, 5, 1508–1518.

Onaca O, Enea R, Hughes DW et al. (2009). Stimuli-responsive polymersomes as nanocarriers for drug and gene delivery. *Macromolecular Bioscience*, 9, 129–139.

Ortega JM, Wilson TS, Benett WJ et al. (2007). A shape memory polymer dialysis needle adapter for the reduction of hemodynamic stress within arteriovenous grafts. *IEEE Transactions on Biomedical Engineering*, 54, 1722–1724.

Osada Y and Gong J (1998). Soft and wet materials: polymer gels. *Advanced Materials*, 10, 827–837.

Pan GH, Huang WM, Ng ZC et al. (2008). Glass transition temperature of polyurethane shape memory polymer reinforced with treated/non-treated attapulgite (palygorskite) clay in dry and wet conditions. *Smart Materials and Structures*, 17, 045007.

Pandit P, Gupta SM, and Wadhawan VK (2004). Shape-memory effect in PMN-PT(65/35) ceramic. *Solid State Communications*, 131, 665–670.

Ping P, Wang W, Chen X et al. (2005). Poly(ε-caprolactone) polyurethane and its shape-memory property. *Biomacromolecules*, 6, 587–592.

Poilane C, Delobelle P, Lexcellent C et al. (2000). Analysis of the mechanical behavior of shape memory polymer membranes by nanoindentation, bulging and point membrane deflection tests. *Thin Solid Films,* 379, 156–165.

Pretsch T (2010a). Triple-shape properties of a thermoresponsive poly(ester urethane). *Smart Materials and Structures*, 19, 015006.

Pretsch T (2010b). Review on the functional determinants and durability of shape memory polymers. *Polymers*, 2, 120–158.

Pretsch T and Muller WW (2010). Shape memory poly(ester urethane) with improved hydrolytic stability. *Polymer Degradation and Stability*, 95, 880–888.

Pretsch T, Jakob I, and Muller W (2009). Hydrolytic degradation and functional stability of a segmented shape memory poly(ester urethane). *Polymer Degradation and Stability*, 94, 61–73.

Prima MD, Gall K, McDowell DL et al. (2010a). Deformation of epoxy shape memory polymer foam I: experiments and macroscale constitutive modeling. *Mechanics of Materials*, 42, 304–314.

Prima MD, Gall K, McDowell DL et al. (2010b). Deformation of epoxy shape memory polymer foam II: macroscale modeling and simulation. *Mechanics of Materials,* 42, 315–325.

Prima MAD, Gall K, McDowell, DL et al. (2010c). Cyclic compression behavior of epoxy shape memory polymer foam. *Mechanics and Materials,* 42, 405–416.

Qi HJ, Nguyen TD, Castro F et al. (2008). Finite deformation thermomechanical behavior of thermally induced shape memory polymers. *Journal of the Mechanics and Physics of Solids*, 56, 1730–1751.

Ratna D and Karger-Kocsis J (2008). Recent advances in shape memory polymers and composites: a review. *Journal of Materials Science*, 43, 254–269.

Razzaq MY, Anhalt M, Frormann L et al. (2007). Thermal, electrical and magnetic studies of magnetite filled polyurethane shape memory polymers. *Materials Science and Engineering A*, 444, 227–235.

Rezanejad S and Kokabi M (2007). Shape memory and mechanical properties of cross-linked polyethylene/clay nanocomposites. *European Polymer Journal*, 43, 2856–2865.

Sahoo NP, Jung YC, and Cho JW (2007). Electroactive shape memory effect of polyurethane composites filled with carbon nanotubes and conducting polymer. *Materials and Manufacturing Processes*, 22, 419–423.

Schmaljohann D (2006). Thermo- and pH-responsive polymers in drug delivery. *Advanced Drug Delivery Reviews*, 58, 1655–1670.

Sharp AA, Panchawagh HV, Ortega A et al. (2006). Toward a self-deploying shape memory polymer neuronal electrode. *Journal of Neural Engineering*, 3, L23–L30.

Snyder R, Rauscher M, Vining B et al. (2010). Shape memory polymer sensors for tracking cumulative environmental exposure. *Proceedings of SPIE*, 7645, 76450C.

Sokolowski W and Chmielewski A (1999). Cold hibernated elastic memory (CHEM) expandable structures. *NASA Technical Brief*, 56–57.

Sokolowski W and Tan SC (2007). Advanced self-deployable structures for space applications. *Journal of Spacecraft and Rockets*, 44, 750–754.

Sokolowski W, Metcalfe A, Hayashi S et al. (2007). Medical applications of shape memory polymers. *Biomedical Materials*, 2, S23–S27.

Song L, Hu W, Zhang H, Wang G et al. (2010). *In vitro* evaluation of chemically cross-linked shape-memory acrylate-methacrylate copolymer networks as ocular implants. *Journal of Physics and Chemistry B*, 114, 7172–7178.

Srivastava V, Chester SA, and Anand L (2010). Thermally actuated shape-memory polymers: experiments, theory, and numerical simulation. *Journal of the Mechanics and Physics of Solids*, 58, 1100–1124.

Stratakis E, Mateescu A, Barberoglou M et al. (2010). From superhydrophobicity and water repellency to superhydrophilicity. *Chemical Communications,* 46, 4136–4138.

Stuart MAC, Huck WTS, Genzer J et al. (2010). Emerging applications of stimuli-responsive polymer materials. *Nature Materials*, 9, 101–113.

Sun L and Huang WM (2010a). Mechanisms of the multi-shape memory effect and temperature memory effect in shape memory polymers. *Soft Matter*, 6, 4403–4406.

Sun L and Huang WM (2010b). Thermo/moisture responsive shape-memory polymer for possible surgery/operation inside living cells in future. *Materials and Design*, 31, 2684–2689.

Sun L, Zhao Y, Huang WM et al. (2009). Formation of combined surface features of protrusion array and wrinkles atop shape-memory polymer. *Surface Review and Letters*, 16, 929–933.

Swain MV (1986). Shape memory behavior in partially stabilized zirconia ceramics. *Nature*, 322, 234–236.

Tey SJ, Huang WM, and Sokolowski W (2001). Influence of long-term storage in cold hibernation on strain recovery and recovery stress of polyurethane shape memory polymer foam. *Smart Materials and Structures*, 10, 321–325.

Thornton AJ, Alsberg E, Albertelli M et al. (2004). Shape-defining scaffolds for minimally invasive tissue engineering. *Transplantation*, 77, 1798–1803.

Tobushi H, Hara H, Yamada E et al. (1996). Thermomechanical properties in a thin film of shape memory polymer of polyurethane series. *Smart Materials and Structures*, 5, 483–491.

Tobushi H, Hashimoto T, Hayashi S et al. (1997). Thermomechanical constitutive modeling in shape memory polymer of polyurethane series. *Journal of Intelligent Material Systems and Structures*, 8, 711–718.

Tobushi H, Hashimoto T, Ito N et al. (1998). Shape fixity and shape recovery in a thin film of shape memory polymer of polyurethane series. *Journal of Intelligent Material Systems and Structures*, 9, 127–136.

Tobushi H, Okumura K, Endo M et al. (2000). Thermomechanical properties of polyurethane-shape memory polymer foam. *Journal of Advanced Science*, 12, 281–286.

Tobushi H, Okumura K, Hayashi S et al. (2001a). Thermomechanical constitutive model of shape memory polymer. *Mechanics of Materials*, 33, 545–554.

Tobushi H, Okumura K, Endo M et al. (2001b). Strain fixity and recovery of polyurethane-shape memory polymer foam. *Transactions of the Materials Research Society of Japan*, 26, 351–354.

Tobushi H, Okumura K, Endo M et al. (2001c). Thermomechanical properties of polyurethane-shape memory polymer foam. *Journal of Intelligent Material Systems and Structures*, 12, 283–287.

Tobushi H, Matsui R, Hayashi S et al. (2004). The influence of shape-holding conditions on shape recovery of polyurethane-shape memory polymer foams. *Smart Materials and Structures*, 13, 881–887.

Tobushi H, Hayashi S, Hoshio K et al. (2006). Influence of strain-holding conditions on shape recovery and secondary-shape forming in polyurethane-shape memory polymer. *Smart Materials and Structures*, 15, 1033–1038.

Tobushi H, Pieczyska E, Ejiri Y et al. (2009). Thermomechanical properties of shape-memory alloy and polymer for their composites. *Mechanics of Advanced Materials and Structures*, 16, 236–247.

Wache HM, Tartakowska DJ, Hentrich A et al. (2003). Development of a polymer stent with shape memory effect as a drug delivery system. *Journal of Materials Science: Materials in Medicine*, 14, 109–112.

Wei ZG, Sandstrom R, and Miyazaki S (1998). Shape-memory materials and hybrid composites for smart systems I: shape-memory materials. *Journal of Materials Science*, 33, 3743–3762.

Willett JL (2008). Humidity-responsive starch-poly(methyl acrylate) films. *Macromolecular Chemistry and Physics*, 209, 764–772.

Wills HH (1977). A near perfect shape-memory ceramic material. *Nature*, 266, 706–707.

Wilson TS, Buckley PR, Benett WJ et al. (2007a). Prototype fabrication and preliminary *in vitro* testing of a shape memory endovascular thrombectomy device. *IEEE Transactions on Biomedical Engineering*, 54, 1657–1666.

Wilson TS, Benett WJ, Loge JM et al. (2005). Laser-activated shape memory polymer intravascular thrombectomy device. *Optics Express*, 13, 8204–8213.

Wischke C, Neffe AT, Steuer S et al. (2009). Evaluation of a degradable shape-memory polymer network as matrix for controlled drug release. *Journal of Controlled Release*, 138, 243–250.

Xie T (2010). Tunable polymer multi-shape memory effect. *Nature*, 464, 267–270.

Xie T and Rousseau IA (2009). Facile tailoring of thermal transition temperatures of epoxy shape memory polymers. *Polymer*, 50, 1852–1856.

Xu W and Li Q (2010). Constitutive modeling of shape memory polymer base self-healing syntactic foam. *International Journal of Solids and Structures*, 47, 1306–1316.

Xu J, Shi W, and Pang W (2006). Synthesis and shape memory effects of Si-O-Si cross-linked hybrid polyurethanes. *Polymer*, 47, 457–465.

Xu B, Huang WM, Pei YT et al. (2009). Mechanical properties of attapulgite clay reinforced polyurethane shape memory nanocomposites. *European Polymer Journal*, 45, 1904–1911.

Xu B, Fu YQ, Ahmad M et al. (2010). Thermomechanical properties of polystyrene-based shape memory nanocomposites. *Journal of Materials Chemistry*, 20, 3442–3448.

Yakacki CM, Shandas R, Lanning C et al. (2007). Unconstrained recovery characterization of shape-memory polymer networks for cardiovascular applications. *Biomaterials*, 28, 2255–2263.

Yakacki CM, Shandas R, Safranski D et al. (2008). Strong, tailored, biocompatible shape-memory polymer. *Advanced Functional Materials*, 18, 2428–2435.

Yang B (2007). Influence of moisture in polyurethane shape memory polymers and their electrically conductive composites. PhD dissertation, Nanyang Technological University, Singapore.

Yang B, Huang WM, Li C et al. (2006). Effect of moisture on the thermomechanical properties of a polyurethane shape memory polymer. *Polymer*, 47, 1348–1356.

Yang B, Huang WM, Li C et al. (2005a). Qualitative separation of the effects of carbon nano-powder and moisture on the glass transition temperature of polyurethane shape memory polymer. *Scripta Materialia*, 53, 105–107.

Yang B, Huang WM, Li C et al. (2005b). Effects of moisture on the glass transition temperature of polyurethane shape memory polymer filled with nano carbon powder. *European Polymer Journal*, 41, 1123–1128.

Yang B, Huang WM, Li C et al. (2004). On the effects of moisture in a polyurethane shape memory polymer. *Smart Materials and Structures*, 13, 191–195.

Yu Y, Nakano M, and Ikeda T (2003). Directed bending of a polymer film by light. *Nature*, 425, 145.

Zhang S, Feng Y, Zhang L et al. (2010). Biodegradable polyester urethane networks for controlled release of aspirin. *Journal of Applied Polymer Science*, 116, 861–867.

Zhao Y, Huang WM, and Fu YQ (2010). Micro/nano-scale wrinkling atop shape memory polymers. *Journal of Micromechanics and Micromachining*, in revision.

Zhao Y, Thorkelsson K, Mastroianni AJ et al. (2009). Small-molecule-directed nanoparticle assembly towards stimuli-responsive nanocomposites. *Nature Materials*, 8, 979–985.

2 Thermomechanical Behavior of Polyurethane Shape Memory Polymer

2.1 INTRODUCTION

A piece of shape memory polymer (SMP) can be processed into a pre-determined (original) shape by molding, casting, or coating, etc. The shape memory effect (SME) of a thermo-responsive SMP is illustrated in Figure 2.1 and includes the following steps:

1. SMPs are easily deformed at a temperature above their glass transition temperature T_g. Only a small force is required to maintain the deformed shape.
2. The constraint is removed after cooling below T_g, resulting in a very small elastic shape recovery; deformation is largely maintained. This is the temporary shape.
3. This temporary shape holds permanently unless the material is heated above its T_g, which triggers full recovery to its original shape.
4. The SME cycle may be repeated again and again.

These steps constitute a typical approach for evaluating the properties of a thermo-responsive SMP. The standard procedure for the thermomechanical characterization of the stress–strain–temperature (σ-δ-T) relationship of an SMP follows the following four steps (Tobushi et al. 1992, Tobushi et al. 2001); see Figure 2.2.

1. At a high temperature T_h ($> T_g$), the SMP specimen is loaded to a pre-determined maximum strain (ε_m) at a constant strain rate ($\dot{\varepsilon}$).
2. The SMP sample is held at ε_m and cooled to a low temperature T_l ($< T_g$).
3. After full unloading, only a very small amount of elastic strain is recovered. ε_f is the residual strain.
4. The free-standing sample is heated from T_l to T_h at a constant heating rate ($\dot{T}$). The pre-strain is almost fully recovered with only a very small amount of strain ε_i left at T_h.

In the course of this study, an ether-based thermoplastic polyurethane SMP invented by Dr. Shunichi Hayashi (formerly of Mitsubishi Heavy Industries, now with SMP Technologies, Japan) was used as a typical example to reveal the thermomechanical and shape memory properties of polyurethane SMPs. This SMP was expected to be

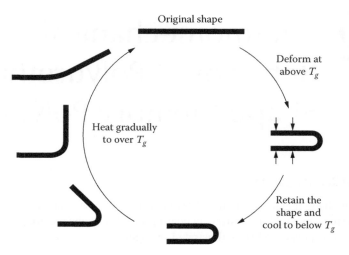

FIGURE 2.1   Shape memory effect (SME).

thermo-responsive because the stimulus for triggering shape recovery is heat. The as-received raw material was in pellet form. According to SMP Technologies (http://www.smptechno.com/index_en.html), the T_g of this SMP can be tailored to meet the requirement of a particular application. Note that instead of glass transition the SME is based on melting in some SMPs.

The specific polyurethane SMP discussed in this chapter is designated SMP MM3520. Its nominal T_g according to SMP Technologies is 35°C. Prof. Hisaaki Tobushi and his collaborators (including Dr. Hayashi) previously conducted a

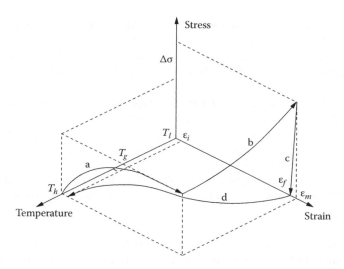

FIGURE 2.2   Three-dimensional stress–strain–temperature diagram showing loading path of thermomechanical test. (Reprinted from Tobushi H et al. *Mechanics of Materials*, 33, 545, 2001. With permission.)

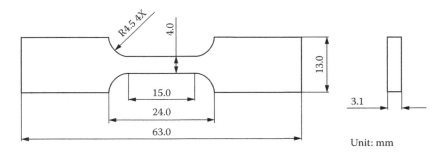

FIGURE 2.3 Dimensions of sample. (From Yang B et al., *Smart Mater. Struct.*, 13: 191, 2004. With permission.)

number of experiments on various types of polyurethane SMPs from the company (Tobushi et al. 1992, 1996, 1997, 1998, 2001). We present here a more systematic study of SMP MM3520 to give readers a full picture of this particular material. The experimental results provide basic knowledge and necessary references for the later chapters of this book.

After drying in a vacuum oven at 80°C for 12 hours, the raw materials were molded into the shape shown in Figure 2.3 by a laboratory injection-molding machine (Manumold 77/30) following the processing procedure suggested by SMP Technologies. Each type of test was repeated a few times on different samples. Only the representative results are presented in this chapter.

2.2 GLASS TRANSITION TEMPERATURE AND THERMAL STABILITY

T_g represents both the glass transition temperature and also the shape recovery temperature at which the SME is activated for this thermo-responsive SMP. At a temperature below T_g, this SMP is in the glass state and is stiff and hard to deform (with high modulus). In this state, the soft segment in the SMP is frozen in place. It may be able to vibrate slightly, but no significant segmental motion occurs. In this state, the bulk material is relatively difficult to deform.

If the SMP is heated gradually, it will enter a glass transition region. The transition does not start or finish instantaneously, but takes place gradually over a temperature range. Within this range, soft segments wiggle around and the heat capacity of SMP increases by an order of magnitude. As one of the conventional approaches, differential scanning calorimeter (DSC) testing can be used to detect the step increases in heat capacity in order to determine the T_g, defined as the median point of the glass transition range in the heating ramp. Likewise, in shape memory alloys (An and Huang 2006), the heating rate may have remarkable influence on the exact T_g determined. A higher heating speed results in a higher T_g. A constant heating and cooling rate of 20°C/min was applied in our experiment. The specimens used for the DSC test (PerkinElmer DSC 7) were cut from the samples shown in Figure 2.3. Each sample weighed ~10 mg. Each specimen was subjected to one test.

Figure 2.4 shows a plot of the DSC results for SMP MM3520 over one heating and cooling cycle. T_g is determined to be 36.7°C—very close to the given nominal

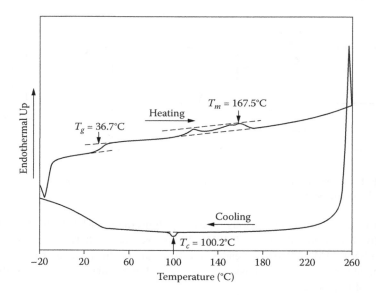

FIGURE 2.4 DSC result for polyurethane SMP MM3520. (From Yang B, 2007. Influence of moisture in polyurethane shape memory polymers and their electrically conductive composites. PhD dissertation. Nanyang Technological University, Singapore.)

value of 35°C. We can see that the glass transition occurs within a temperature range of 30 to 40°C. Furthermore, the DSC revealed that this polyurethane SMP is a semi-crystalline polymer that experiences both glass transition and melting in the heating ramp. The SMP starts to melt at 110°C. At 180°C, the polymer is almost fully melted in the heating ramp; it crystallizes at about 100.2°C in the cooling ramp. The melting temperature (T_m) and crystallization temperature (T_c) are defined as the peak points in the melting range and crystallization range, respectively.

Thermogravimetric analysis (TGA) was conducted to investigate the thermal stability of the polyurethane SMP. TGA was carried out with a specimen weighing around 20 mg at a constant heating rate of 10°C/min, purged with nitrogen gas on a PerkinElmer TGA 7. Figure 2.5 reveals only a slight weight loss of SMP MM3520 before heating to 260°C, which is largely attributed to the evaporation of moisture in the material; 260°C is defined as the decomposition temperature of this polyurethane SMP.

2.3 DYNAMIC MECHANICAL PROPERTIES

Dynamic mechanical analysis (DMA) was used to investigate the dynamic mechanical properties of SMP MM3520. In this test, a small load is applied on a sample in a sinusoidal fashion so that the sample is always deformed elastically. For a perfectly elastic material, the stress and strain curves are perfectly in phase. On the other hand, a perfectly viscous material flows under loading, instead of deforming

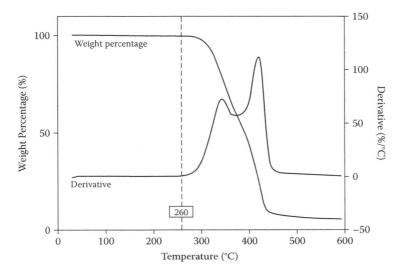

FIGURE 2.5 TGA result for polyurethane SMP MM3520. (From Yang B, 2007. Influence of moisture in polyurethane shape memory polymers and their electrically conductive composites. PhD dissertation. Nanyang Technological University, Singapore.)

reversibly. Under dynamic loading, its stress–strain curves are out of phase by 90° since the strain is proportional to the changing rate of the stress. The polyurethane SMP behaves in a way combining both factors, i.e., it simultaneously reacts elastically and flows to some extent. Therefore, its stress and strain curves are out of phase by a phase angle (δ) that is less than 90°. DMA is used to measure the amplitudes of stress and strain as well as δ.

The storage modulus, loss modulus, and tangent delta can be recorded against the temperature. The storage modulus is the modulus of the elastic portion of material, and the loss modulus is the modulus of the viscous portion. Tangent delta is defined as the ratio of the loss modulus over the storage modulus that indicates the damping capability of a material.

DMA (PerkinElmer DMA 7) was carried out in the three-point bending mode with a 10 mm span. The specimens were heated in a hot chamber at a constant rate of 5°C/min. The viscoelastic behavior of a polymer implies a time dependence of its properties. Thus, the mechanical vibration frequency used for DMA tests has a remarkable effect on the results (Merzlyakov et al. 1999, Zheng and Wong 2003, Zhou et al. 2005). A constant frequency of 1 Hz recommended in the instrument manual was used. The rectangular specimens with dimensions of 15 × 4.0 × 3.1 mm were cut from the samples as shown in Figure 2.3. The result for SMP MM3520 is shown in Figure 2.6. It reveals that the storage modulus that corresponds to the stiffness of the material decreases sharply in the glass transition region. The ratio of storage modulus in the glass state to that in the rubber state was up to 200 to 300. Furthermore, tangent delta reached its maximum of about 1.45 during glass transition; the tangent delta of a typical rubber is 0.2 to 0.4. Thus,

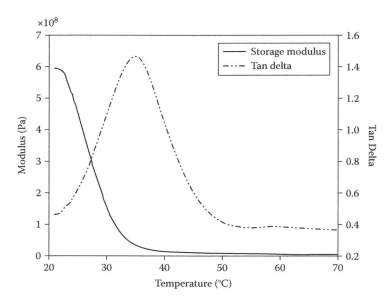

FIGURE 2.6   DMA result for polyurethane SMP MM3520. (From Yang B, 2007. Influence of moisture in polyurethane shape memory polymers and their electrically conductive composites. PhD dissertation. Nanyang Technological University, Singapore.)

in the glass transition region, the polyurethane SMP can be used as a good damping material for energy dissipation.

The temperature corresponding to the peak of tangent delta is an alternative definition for T_g. The T_g of this SMP defined in this way is about 35°C—slightly lower than the result obtained with DSC. The small difference can be ascribed to the higher heating rate applied in the DSC test that produced a slight lag in the transition. Note that the quantity of the sample used for the DMA was much larger than that used for the DSC. To achieve reliable test results, a lower heating rate was applied in the DMA to ensure that the heat was distributed more evenly in the sample.

2.4   UNIAXIAL TENSION IN GLASS STATE

Uniaxial tensile tests were conducted to investigate the behavior of the polyurethane SMP under uniaxial tension. SMP samples were stretched uniaxially at a constant strain rate at room temperature (about 22°C) with an Instron 5569 device. Typical results at three different strain rates, namely 10^{-2}/s, 10^{-3}/s, and 10^{-4}/s, are shown in Figure 2.7. Note that unless otherwise stated, all strains and stresses are engineering types of strain and stress. According to the thermal tests (DSC and DMA), the polyurethane SMP is in a glass state at room temperature.

Figure 2.7 reveals that after a small linear elastic deformation, this SMP experiences a distinct upper yield point, followed by an apparent plateau, and then hardening at all tested strain rates, similar to what occurs in mild steels. The yielding–plateau phenomenon is similar to the well known Luder band phenomenon, in

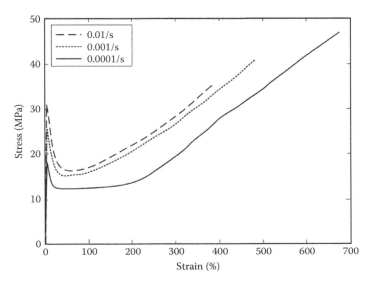

FIGURE 2.7 Strain-versus-stress relationships of polyurethane SMP MM3520 at room temperature. (From Yang B, 2007. Influence of moisture in polyurethane shape memory polymers and their electrically conductive composites. PhD dissertation. Nanyang Technological University, Singapore.)

which necking and propagation may be observed upon stretching. Figure 2.8 depicts the propagation. The hardening is a result of the reorientation of polyurethane SMP molecular chains that induces crystallization. The elongation limit of this material is over 300% in terms of engineering strain and increases with the decrease of the strain rate. Furthermore, the yield strength increases upon increase of the strain rate while the ultimate strength follows an opposite trend.

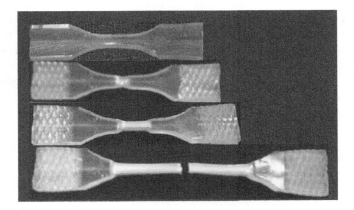

FIGURE 2.8 Samples after uniaxial stretching at different strains (until fracture) at room temperature. (From Yang B, 2007. Influence of moisture in polyurethane shape memory polymers and their electrically conductive composites. PhD dissertation. Nanyang Technological University, Singapore.)

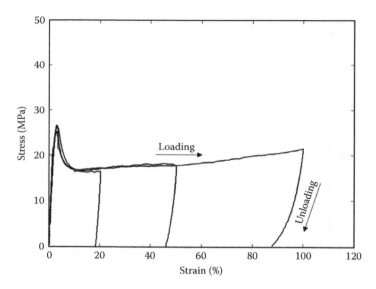

FIGURE 2.9 Strain-versus-stress relationships of polyurethane SMP MM3520 at different strains in loading and unloading test. (From Yang B, 2007. Influence of moisture in polyurethane shape memory polymers and their electrically conductive composites. PhD dissertation. Nanyang Technological University, Singapore.)

Figure 2.9 presents the strain-versus-stress relationships of SMP MM3520 upon loading to different maximum strains, namely, 20, 50, and 100%, followed by unloading to zero stress. The applied constant loading/unloading strain rate was 10^{-3}/sec. Upon unloading, recovery is largely attributed to elastic recovery, in particular at the early unloading stages in the small maximum strain cases.

To check whether the residual strain after unloading is fully recoverable, the SMP samples were heated gradually using a digital hot plate. Figure 2.10 shows that a sample can virtually fully recover its original shape upon heating over its T_g after pre-stretching to 100% strain. It demonstrates the virtual lack of apparent permanent deformation in this polyurethane SMP even after pre-stretching to 100% strain in its glass state. The recovery is rather uniform everywhere. No propagation phenomena can be observed.

2.5 UNIAXIAL TENSION IN RUBBER STATE

Uniaxial tensile tests were carried out at different constant strain rates and at $T_g + 15°C$ where the polyurethane SMP is in the rubber state. The T_g value was 36.7°C based on the previous DSC test. The obtained stress-versus-strain responses are plotted in Figure 2.11 and demonstrate the typical viscoelastic properties of the polyurethane SMP MM3520 at this temperature. At all strain

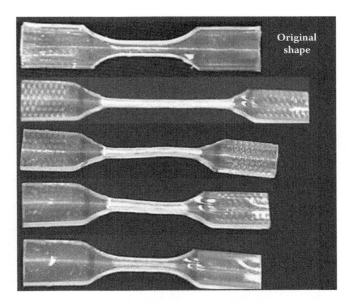

FIGURE 2.10 Recovery sequence (top to bottom) of a piece of polyurethane SMP MM3520 after pre-stretching 100% at room temperature. The original shape is shown for comparison purposes. (From Yang B, 2007. Influence of moisture in polyurethane shape memory polymers and their electrically conductive composites. PhD dissertation. Nanyang Technological University, Singapore.)

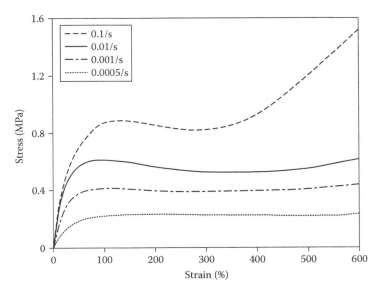

FIGURE 2.11 Strain-versus-stress relationships of polyurethane SMP MM3520 at $T_g + 15°C$. (From Yang B, 2007. Influence of moisture in polyurethane shape memory polymers and their electrically conductive composites. PhD dissertation. Nanyang Technological University, Singapore.)

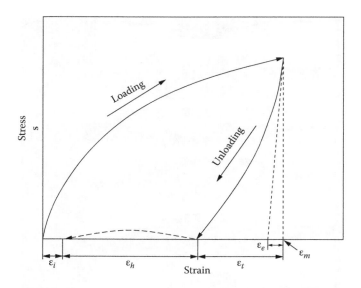

FIGURE 2.12 Illustration of loading and unloading uniaxial tensile test. (From Yang B, 2007. Influence of moisture in polyurethane shape memory polymers and their electrically conductive composites. PhD dissertation. Nanyang Technological University, Singapore.)

rates, the elongation limit was above 600% strain, but the exact stress-versus-strain response was highly strain rate dependent. At a high strain rate of 10^{-1}/s, the curve closely resembles that in the glass state. At low strain rates, the curve is dominated by slant; the extension increases continuously with very little variation in the applied force.

To investigate the strain recovery of SMP MM3520 at a high temperature, a series of loading and unloading uniaxial tensile tests were conducted following the procedure described below (refer to Figure 2.12).

1. Stretch the sample uniaxially to a prescribed maximum strain (ε_m), e.g., 50%, at a constant strain rate ($\dot{\varepsilon}$) and a temperature above T_g.
2. Unload to zero stress at the same constant strain rate ($\dot{\varepsilon}$) and temperature.
3. Hold the sample at this temperature for a given period, e.g., 40 minutes, without applying external load.

As illustrated in Figure 2.12, ε_e, ε_t, ε_h, and ε_i denote the elastic strain, total instant recovery strain upon unloading, free recovery strain during holding (without external load), and final residual strain, respectively. In the deformation of a viscoelastic polymer, ε_m is defined (Clegg and Collyer 1986) as:

$$\varepsilon_m = \varepsilon_e + \varepsilon_r + \varepsilon_i \qquad (2.1)$$

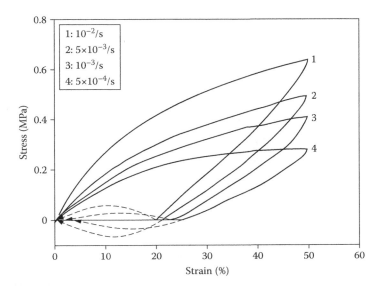

FIGURE 2.13 Results of loading and unloading tensile tests at four different strain rates and at $T_g + 15°C$. (From Yang B, 2007. Influence of moisture in polyurethane shape memory polymers and their electrically conductive composites. PhD dissertation. Nanyang Technological University, Singapore.)

where ε_r is the rubber elastic strain. Practically, the elastic strain (ε_e) is part of the total instant recovery strain (ε_t). Hence,

$$\varepsilon_m = \varepsilon_t + \varepsilon_h + \varepsilon_i \qquad (2.2)$$

Results of loading and unloading tensile tests carried out at four different strain rates at $T_g + 15°C$ following the above procedure are plotted in Figure 2.13. In general, recovery is almost complete. The residual strain (ε_i) is a very small portion of ε_m and decreases continuously upon further holding. Thus, ε_i can be attributed largely to the viscous flow and the contribution of reorientation in polymer chains should be limited.

The result also reveals that a higher strain rate results in more instant recovery strain. It is known that, if $t \ll \tau$, the strain energy in a polymer can be stored by quick mechanical deformation at a temperature above T_g (Wineman and Rajagopal 2000). Here, t is the deformation time and τ is a characteristic relaxation time for a polymer at this temperature. Recovery happens upon unloading, indicating the return of the stretched polymer chains to more equilibrium and coiled conformations. Thus, quick deformation is preferred for more recoverable strain.

Figure 2.13 shows that a large portion of ε_m (up to 20%) is ε_h, i.e., the polyurethane SMP may only partially recover upon mechanical unloading. A significant amount of strain may be recovered, but only gradually with the elapse of time. The strain recovery, especially ε_t, is highly dependent on the strain rate. With the increase of strain rate, ε_t becomes smaller while ε_h increases. Because ε_i is mainly ascribed to the viscous flow of a material that is highly dependent on the strain

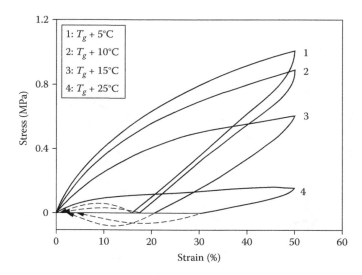

FIGURE 2.14 Results of loading and unloading tensile tests at a constant strain rate of 10^{-2}/s at different temperatures. (From Yang B, 2007. Influence of moisture in polyurethane shape memory polymers and their electrically conductive composites. PhD dissertation. Nanyang Technological University, Singapore.)

rate, ε_i increases with the strain rate. Therefore, in the applications of this polyurethane SMP, an instant load is preferred, so that the pre-strain in a polyurethane SMP can be fully recovered upon heating above T_g.

To investigate the dependence of strain recovery on temperature and deformation, the SMP samples were loaded to a maximum strain of 50% and then unloaded to zero stress at a constant strain rate of 10^{-2}/s at four different temperatures above T_g. The resultant stress-versus-strain curves are plotted in Figure 2.14. It reveals that ε_t deceases while ε_i increases with the increase of temperature. At a low temperature, the polyurethane SMP can instantly recover more upon unloading than at a high temperature. Because the material is easier to deform at a high temperature, less stress is required to reach the same strain. Above T_g, the polyurethane SMP, as a viscoelastic material, shows more viscous properties at a high temperature. The strain induced by viscous flow cannot be fully recovered immediately. Hence, ε_i increases with temperature as shown in Figure 2.14 and the working temperature of this SMP is suggested to range from T_g + 5°C to T_g + 10°C for better instant recovery.

The dependence of strain recovery upon the maximum pre-strain was investigated by a series of cyclical tensile tests. The samples were loaded to different maximum strains and then unloaded to zero stress at a strain rate of 10^{-2}/s and at T_g + 5°C. The maximum pre-strains were from 50 to 600% at a 50% interval. The relationships of stress versus strain in these tests are plotted in Figure 2.15. The envelope formed by these cycles coincides very well with the stress-versus-strain curve of the single extension test. The SMP samples experienced obvious "creeping" in the strain range from 100 to 300%, then hardening due to reorientation and crystallization. The result

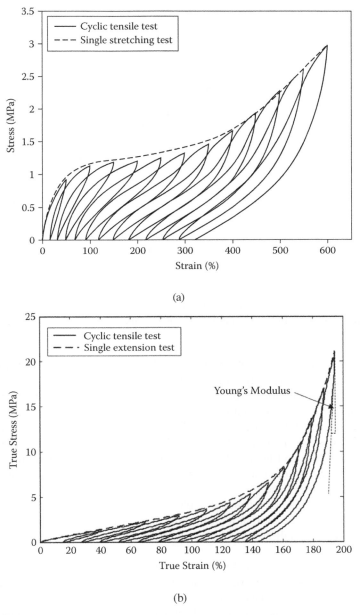

(a)

(b)

FIGURE 2.15 Results of cyclic tensile test and single stretching test at a constant strain rate of 0.01/s at $T_g T_g + 5°C$. (a) Engineering stress versus engineering strain. (b) True stress versus true strain. (From Yang B, 2007. Influence of moisture in polyurethane shape memory polymers and their electrically conductive composites. PhD dissertation. Nanyang Technological University, Singapore.)

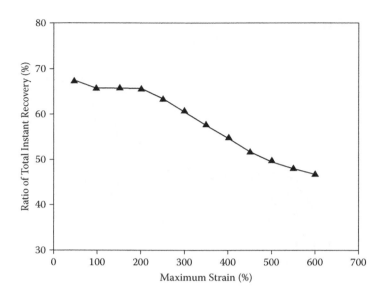

FIGURE 2.16 Ratios of total instant recovery at different maximum strains. (From Yang B, 2007. Influence of moisture in polyurethane shape memory polymers and their electrically conductive composites. PhD dissertation. Nanyang Technological University, Singapore.)

in loading and unloading cyclic test shares the same general trend in the stress-versus-strain behavior as in the single stretching test.

In Figure 2.16, the ratio of total instant recovery is plotted against the maximum strain. The ratio of total instant recovery is calculated as the total instant recovery strain over the maximum strain in a cycle. Generally speaking, the ratio decreases with the increase of maximum strain. The decrease is more obvious at a maximum strain above 200% and negligible at a maximum strain range from 100 to 200%. Based on the results, the material is hardened above 200% due to the reorientation and crystallization of the polymer chains. Thus, the decrease in the total instant recovery ratio can also be attributed to reorientation and crystallization. A one-step sudden decrease in the ratio of total instant recovery may be a result of decoupling in the imperfect crystalline part of the polymer in the first two cycles (Tobushi et al. 1996).

The evolution of Young's modulus with the changes in maximum strain is plotted in Figure 2.17. Young's modulus was calculated from the linear portion of the true stress-versus-true strain curve at the beginning of unloading in each cycle (see Figure 2.15 for illustration). The modulus decreases in the first two cycles because the polymer chains with imperfect crystalline parts are decoupled in these two cycles at the beginning of the test (Tobushi et al. 1996). After that, Young's modulus increases gradually until 250% strain is reached. Beyond that point, it increases almost linearly due to hardening caused by the reorientation and crystallization of the polymer chains. At this point, we may conclude that the MM3520 polyurethane SMP is better for applications with maximum strains below 200% for a high ratio of total instant recovery exceeding 60%

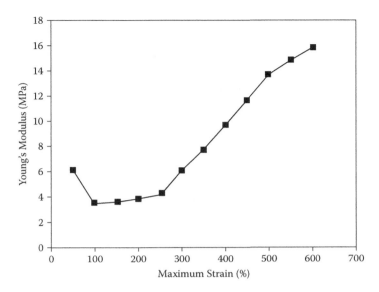

FIGURE 2.17 Evolution of Young's modulus with maximum strain. (From Yang B, 2007. Influence of moisture in polyurethane shape memory polymers and their electrically conductive composites. PhD dissertation. Nanyang Technological University, Singapore.)

at $T_g + 5°C$. Above that maximum strain, hardening of the SMP results in an increase of Young's modulus and a decrease of the ratio of total instant recovery.

The stability of shape recovery properties of the polyurethane SMP was investigated experimentally by a cyclic tensile test at a constant maximum strain of 100%, a constant strain rate of $10^{-2}/$ s and $T_g + 5°C$. One hundred cycles were carried out on a piece of polyurethane SMP MM3520. Figure 2.18 plots results only at selected cycle numbers. It is apparent that stress decreases with the increase of cycle number. The decrease is sharp at the beginning and then becomes gradual. This occurs because loading distorts the polymer chains to trigger viscous flow. This distortion causes the reorientation of polymer chains and thus results in a decrease of stress. The reorientation is somewhat limited by the physical links among the polymer chains. Thus, after a sufficient number of cycles and over a long period, the reorientation occurs and viscous flow gradually becomes stable (Wineman and Rajagopal 2000).

Our recent experimental results on a polystyrene SMP from Cornerstone Research Group revealed a possible breakdown of some elastic segments (Liu et al. 2007) that may represent another cause of stress decrease upon cyclic loading. For a better view, the time- or cycle-versus-strain relationship is plotted in Figure 2.19. The strain at the bottom point of each cycle corresponds to the residual strain after unloading. The residual strain, corresponding to zero stress, increases substantially in the first 20 cycles and then becomes gradual. After about 40 cycles, a stabilized instant recoverable strain of about 30% can be obtained. This can be ascribed to the

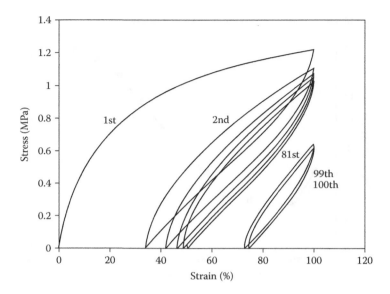

FIGURE 2.18 Stress-versus-strain relationship of polyurethane SMP MM3520 in cyclic tensile test at constant maximum strain of 100% and $T_g + 5°C$. (From Yang B, 2007. Influence of moisture in polyurethane shape memory polymers and their electrically conductive composites. PhD dissertation. Nanyang Technological University, Singapore.)

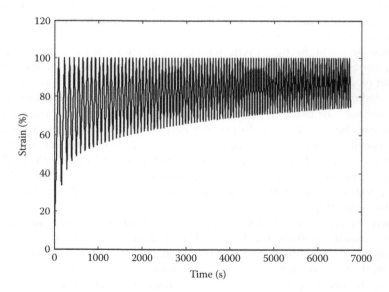

FIGURE 2.19 Time- or cycle-versus-strain relationship of polyurethane SMP MM3520 in cyclic tensile test with constant maximum strain of 100% at $T_g + 5°C$. (From Yang B, 2007. Influence of moisture in polyurethane shape memory polymers and their electrically conductive composites. PhD dissertation. Nanyang Technological University, Singapore.)

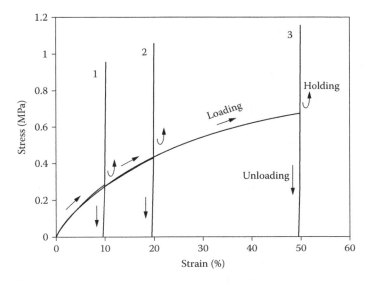

FIGURE 2.20 Typical stress-versus-strain curves upon loading (at 50°C) and unloading (at room temperature). (From Yang B, 2007. Influence of moisture in polyurethane shape memory polymers and their electrically conductive composites. PhD dissertation. Nanyang Technological University, Singapore.)

slow-down or cessation of the viscous flow of the polymer after many cycles over a long time period.

2.6 RECOVERY TESTS

The unique feature of SMP materials is the shape memory effect (SME). From a thermomechanical point of view, recovery stress and recoverable strain are essential concerns in real engineering applications of the SME. Two types of tests, namely the constrained recovery test and the free recovery test, were carried out for characterizing the SME of MM3520.

Polyurethane SMP wires of 1.0 mm diameter were prepared by extrusion.* The wires were uniaxially stretched at 50°C by an Instron 5565 instrument with a 100 N load cell to three different maximum strains, namely 10, 20, and 50%. The initial gauge length was 40.0 mm in all experiments. The wires were then rapidly cooled to room temperature in 3 minutes with their maximum strain held, followed by unloading to zero stress. A constant strain rate of 5×10^{-3}/s was applied during both loading and unloading. Figure 2.20 presents three typical stress-versus-strain curves in these processes. The unloading process is virtually linearly elastic.

Subsequently, the pre-strained wires were divided into two groups for two types of recovery tests. The wires in the first group were heated with their lengths fixed.

---

* Refer to http://www.smptechno.com/index_en.html for instruction on extrusion procedure.

This is the so-called constrained recovery test. The recovery stress was measured. The wires in the second group were heated without constraints so that they could deform freely (free recovery test). In both types of tests, the wires were heated in a hot chamber at a constant rate of 2°C/min. For all pre-strained wires in the free recovery test, the recovery ratio (ratio of measured recovery strain to pre-strain; 10, 20, or 50%) was applied as a measure of the recovery. Figure 2.21 presents the evolution of the recovery stress and recovery ratio against temperature.

In Figure 2.21(a), the recovery stress reaches a peak at ~30°C and then falls continuously upon further heating, in particular at temperatures over 40°C. The recovery stress almost vanishes at 60°C. Greater pre-strain produced greater recovery stress. Upon heating to 60°C, the recovery (without constraint) was ~100% as shown in Figure 2.21(b). In other words, the wires fully regained their original shapes.

2.7 SUMMARY

During the glass transition, the dynamic mechanical properties of the MM3520 polyurethane SMP changed abruptly. Its storage modulus slumped by 200 to 300 times while its damping ratio increased to 1.45. MM3520 appears to be a good damping material for efficient energy dissipation. It is thermally stable; it loses only a little weight due to the evaporation of moisture before decomposition starts at 260°C in a nitrogen environment. MM3520 has an elongation limit over 300% and experiences yielding and hardening from uniaxial extension in the glass state. On the other hand, it has an elongation limit exceeding 600% and behaves like a typical viscoelastic material upon loading in the rubber state. Necking and propagation, i.e., the Luder band phenomenon, are observed during uniaxial tension in the low temperature glass (stiff) state but not in the high temperature rubber (soft) state. This is opposite to the behaviors of shape memory alloys that exhibit the Luder band in the high temperature stiffer phase (Huang 2005).

The thermomechanical behavior and shape recovery ability of the polyurethane SMP are highly dependent on strain rate, temperature, and maximum strain. This material can almost fully recover a pre-strain loaded below and above its T_g. For more strain recovery, a higher strain rate is preferred during loading. Furthermore, it is preferable for this polyurethane SMP to deform at a temperature ranging from $T_g + 5°C$ to $T_g + 10°C$ for more recoverable strain, especially for faster recoverable strain. Because severe hardening at over 200% strain results in some irreversible strain, the polyurethane SMP should be used at a strain rate below 200%.

The stability of shape recovery properties of the MM3520 polyurethane SMP is highly dependent on the cycle number. Upon cycling, the total instant recoverable strain deteriorates and the stress corresponding to the maximum strain decreases. However, the total instant strain recovery almost stabilizes after enough loading and unloading cycles. Upon heating, the SMP can fully recover its original shape in a free recovery test or generate high recovery stress in a constrained recovery test. The former can be utilized for shape recovery, while the latter can be used for actuation. These results serve as a platform for further investigation of this polyurethane SMP covered in subsequent chapters.

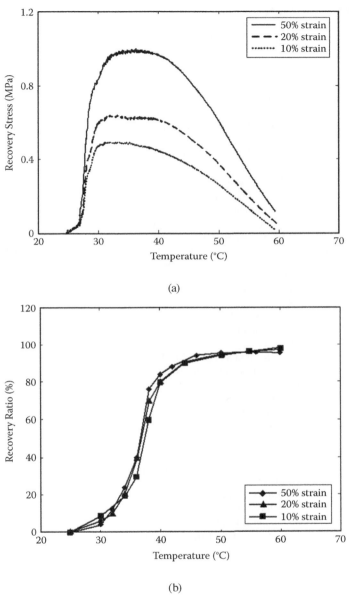

FIGURE 2.21 Recovery upon heating. (a) Recovery stress as function of temperature. (b) Shape recovery ratio as function of temperature. (From Yang B, 2007. Influence of moisture in polyurethane shape memory polymers and their electrically conductive composites. PhD dissertation. Nanyang Technological University, Singapore.)

ACKNOWLEDGMENT

We wish to thank N.W. Khun for his help in compiling this chapter.

REFERENCES

An L and Huang WM (2006). Transformation characteristics of shape memory alloys upon thermal cycling. *Materials Science and Engineering A*, 420, 220–227.

Clegg DW and Collyer AA (1986). *Mechanical Properties of Reinforced Thermoplastics.* Elsevier, Amsterdam.

Huang WM (2005). Transformation front in shape memory alloys. *Materials Science and Engineering A*, 392, 121–129.

Liu N, Huang WM, and Phee SJ (2007). A secret garden of micro butterflies: phenomenon and mechanism. *Surface Review and Letters*, 14, 1187–1190.

Merzlyakov M, Wurm A, Zorzut M et al. (1999). Frequency and temperature amplitude dependence of complex heat capacity in the melting region of polymers. *Journal of Macromolecular Science B*, 38, 1045–1054.

Tobushi H, Hara H, Yamada E et al. (1996). Thermomechanical properties in a thin film of shape memory polymer of polyurethane series. *Smart Materials and Structures*, 5, 483–491.

Tobushi H, Hashimoto T, Hayashi S et al. (1997). Thermomechanical constitutive modeling in shape memory polymer of polyurethane series. *Journal of Intelligent Material Systems and Structures*, 8, 711–718.

Tobushi H, Hashimoto T, Ito N et al. (1998). Shape fixity and shape recovery in a film of shape memory polymer of polyurethane series. *Journal of Intelligent Material Systems and Structures*, 9, 127–136.

Tobushi H, Hayashi S, and Kojima S (1992). Mechanical properties of shape memory polymer of polyurethane series: basic characteristics of stress–strain–temperature relationship). *JSME International Journal*, 35, 296–302.

Tobushi H, Okumura K, Hayashi S et al. (2001). Thermomechanical constitutive model of shape memory polymer. *Mechanics of Materials*, 33, 545–554.

Wineman AS and Rajagopal KR (2000). *Mechanical Response of Polymers*, Cambridge University Press, Cambridge.

Yang B (2007). Influence of moisture in polyurethane shape memory polymers and their electrically conductive composites. PhD dissertation. Nanyang Technological University, Singapore.

Yang B, Huang WM, Li C et al. (2004). On the effects of moisture in a polyurethane shape memory polymer. *Smart Materials and Structures*, 13, 191–195.

Zheng W and Wong SC (2003). Electrical conductivity and dielectric properties of PMMA/ expanded graphite composites. *Composites Science and Technology*, 63, 225–235.

Zhou JS, Yan FY, Tian N et al. (2005). Effect of temperature on the tribological and dynamic mechanical properties of liquid crystalline polymer. *Polymer Testing*, 24, 270–274.

3 Effects of Moisture on Glass Transition Temperature and Applications

3.1 INTRODUCTION

The formation of hydrogen bonding in polyurethane is generally known to have a major effect on its morphology and overall properties (Yen et al. 1999, Luo et al. 1997, Yoon and Han 2000, Heintz et al. 2002, Teo et al. 1997). The primary bands in polyurethane, the N-H stretching as a proton donor, and the carbonyl stretching as a proton acceptor are sensitive to hydrogen bonding. Therefore, the formation of hydrogen bonds is characterized by the shift of the Fourier transform infrared (FTIR) absorbance peak to a lower frequency. The strength of hydrogen bonding can be quantified by the magnitude of the shift (Yen et al. 1999, Luo et al. 1997, Yoon and Han 2000, Chen et al. 2000, Brunette et al. 1982).

Moisture is one of the most important environmental factors that may significantly affect the properties of polyurethane shape memory polymers (SMPs). From a real engineering application view, it is necessary to know the reliabilities of SMPs in different environments, for instance, a humid environment. This chapter discusses the effects of moisture on the glass transition temperature (T_g) of the MM3520 polyurethane SMP (SMP Technologies, Japan). For this SMP, T_g is also the shape recovery temperature—a key parameter in thermo-responsive SMP applications. The mechanism behind the effects of moisture on the T_g is investigated and after detailing the effects of moisture on the T_g, two new features are proposed for potential applications of this SMP.

3.2 MOISTURE ABSORPTION IN ROOM-TEMPERATURE WATER

To investigate the moisture absorption of the MM3520 polyurethane SMP studied in Chapter 2, hot-pressed SMP sheets ~1.0 mm thick (refer to http://www.smptechno.com/index_en.html for the recommended processing procedure from SMP Technologies) were immersed into room-temperature water. After different durations of immersion, the sheets were removed from the water, blown with an air gun to remove water from their surfaces, then cut into small pieces for thermogravimetric analysis (TGA) to determine their moisture fractions. Thin samples weighing ~20 mg and sliced from SMP sheets were heated in a TGA 2950 (TA Instruments) from 30 to 330°C at a rate of 20°C/minute. Because the TGA results are used together with DSC results for analysis, the heating rate should be the same in both tests.

Figure 3.1 presents TGA results after different immersion lengths. Note that all samples start to decompose at ~260°C as seen by the rapid decrease in weight fraction. As expected, weight loss becomes more significant with increased immersion time. In the samples immersed in water for more than 12 hours, significant weight loss gradually occurred between 100 and 180°C before decomposition. This loss may be attributed to the evaporation of absorbed water in the polymer. For convenience, a weight fraction at 240°C was chosen as the reference for comparisons of results in subsequent studies. This means that the total loss of weight at 240°C is roughly taken as the total amount of water absorbed in the samples, although some water may still evaporate above 240°C (not significant in the current study) before decomposition occurs. Figure 3.2 further summarizes the weight ratio of water (R_e) to SMP versus immersion time. R_e is obtained by the following equation:

$$R_e = w/(1-w) \times 100\% \tag{3.1}$$

where w is the SMP loss of weight at 240°C in the TGA test. Figure 3.2 reveals that the moisture content in the material increases at the beginning and gradually becomes almost constant after 240 hours of immersion. Thus, the sample may be considered fully saturated after 240 hours of immersion. In the saturated state, after

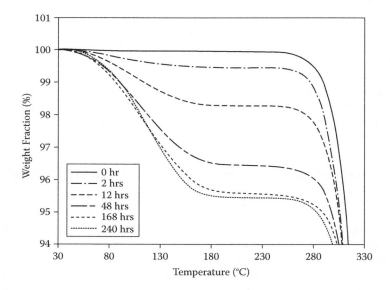

FIGURE 3.1 TGA results of MM3520 polyurethane SMP after various times of immersion in water. (Reprinted from Yang B, Huang WM, Li C et al. *Polymer*, 47, 1348–1356, 2006. With permission.)

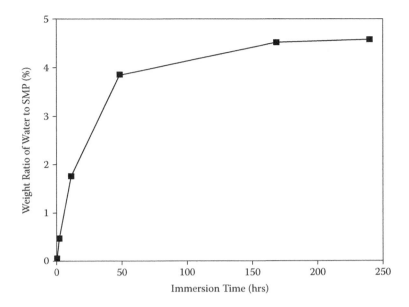

FIGURE 3.2 Ratio of water to SMP in weight versus immersion time. (From Yang B, 2007. Influence of moisture in polyurethane shape memory polymers and their electrically conductive composites. PhD dissertation, Nanyang Technological University, Singapore.)

immersion into room temperature water (about 22°C), water constituted ~4.8% of the weight.

3.3 GLASS TRANSITION TEMPERATURE AFTER IMMERSION

Differential scanning calorimeter (DSC) tests were carried out on MM3520 specimens using a DSC 2920 (TA Instruments) to examine changes in T_g after immersion in water for varying lengths of time. The specimens for testing weighed around 10 mg and the constant heating/cooling rate was 20°C/minute. The T_g was taken at the median point in the range of glass transition during heating. Figure 3.3 plots the DSC results and reveals that the T_g decreased remarkably with the increase of immersion time.

For a better illustration, the T_g values spanning the onset and end of glass transition against immersion time are summarized in Figure 3.4. Note that the onset and end of glass transition were obtained by drawing a tangent to the inclining portions of the DSC curve in the glass transition region at the start and finish parts, respectively. With the increase of immersion time T_g decreased significantly, it dropped ~35°C after 240 immersion hours. The decrease is rapid at the beginning and becomes moderate as the immersion time is prolonged. For immersion time exceeding 168 hours, the samples are close to a saturated state and the change in T_g with immersion time is minor. Moreover, the temperature range of glass transition widened from ~10 to 40°C with the increase of immersion time.

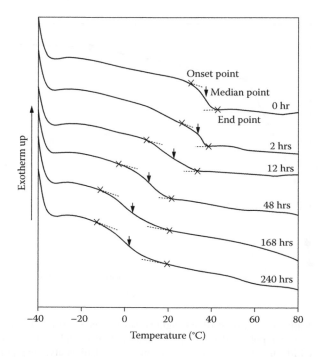

FIGURE 3.3 DSC results after immersion in water for different hours. (Reprinted from Yang B, Huang WM, Li C et al. *Polymer*, 47, 1348–1356, 2006. With permission.)

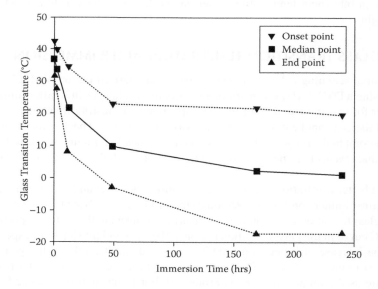

FIGURE 3.4 Evolution of T_g and onset and end point of glass transition against immersion time. (Reprinted from Yang B, Huang WM, Li C et al. *Polymer*, 47, 1348–1356, 2006. With permission.)

3.4   EVOLUTION OF GLASS TRANSITION TEMPERATURE UPON THERMAL CYCLING

The results from the previous sections show that absorbed water or moisture can significantly decrease the T_g of this SMP. A very important issue in materials processing and applications is whether the T_g can be restored to its original value via dehydration because most engineering applications require reliable and stable materials. To answer this question, cyclic DSC tests were carried out on the MM3520 SMP samples after various hours of immersion. For each cycle, a sample was first heated from $-20°C$ to a selected temperature and then cooled back to $-20°C$ at a constant heating or cooling rate of $20°C/minute$. In each test of a sample, six heat temperatures (80, 100, 140, 160, 180, and $240°C$) were used in increasing order.

Figure 3.5 plots one typical cyclic DSC result (a) and a zoom-in view (b) of the saturated SMP after 240 hours of immersion. The T_g changed significantly upon heating within the temperature range of 140 to $180°C$. The evolution of T_g against heating temperature is plotted in Figure 3.6. As the heating temperature increased, the T_g gradually approached its original value of $35°C$. However, upon heating below $100°C$, the increase in T_g was minor. The increase was more significant upon heating over $100°C$ up to $180°C$—almost coinciding with the temperature range at which the evaporation of water takes place, as shown in Figure 3.1. Further heating to $240°C$ can virtually bring the T_g almost fully back to its original value.

As we can see in Figure 3.1, a significant part of the moisture is removed from the sample at a temperature above $100°C$. Thus, the restoration of the T_g can be attributed directly to the evaporation of moisture. Because the moisture is removed only through evaporation, the whole process is reversible. Consequently, the T_g can be reduced if samples are immersed into water again, but after that the original T_g can also be restored by heating the samples to $180°C$ or above. From an engineering application view, $180°C$ appears to be the critical temperature for refreshing this SMP.

3.5   INTERACTION OF WATER AND POLYURETHANE SMP

We have demonstrated that moisture exerts a strong influence on the glass transition of the MM3520 polyurethane SMP. However, the mechanism behind this phenomenon remains unknown. The following sections aim to identify the exact mechanism causing the effects of moisture on the T_g of this SMP, in particular the interaction between water and the SMP.

In this study, FTIR spectroscopy was used to identify the interactions of water with the SMP. The specimens for FTIR tests were 1.0 mm thick polyurethane SMP sheets. FTIR spectra were collected by averaging 70 scans at a resolution of 4 cm^{-1} in a reflection mode from a spectrometer (Nicolet Magna IR-560).

The full FTIR spectrum of the polyurethane SMP at room temperature without immersion in water is presented in Figure 3.7. Some peaks of interest in this study are marked according to references (e.g., Yen et al. 1999, Luo et al. 1997, Yoon and Han 2000). Strong hydrogen bonding is evidenced in the polyurethane SMP where the infrared band of the bonded N-H stretching occurs at 3289 cm^{-1} while that of

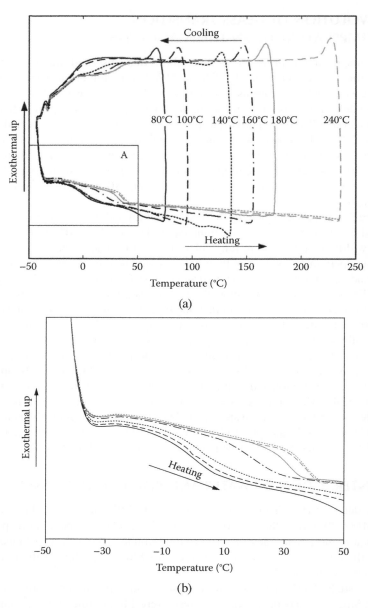

FIGURE 3.5 Cyclic DSC curves of a saturated SMP. (a) Overall view. (b) Zoom-in of A. (From Yang B, 2007. Influence of moisture in polyurethane shape memory polymers and their electrically conductive composites. PhD dissertation, Nanyang Technological University, Singapore.)

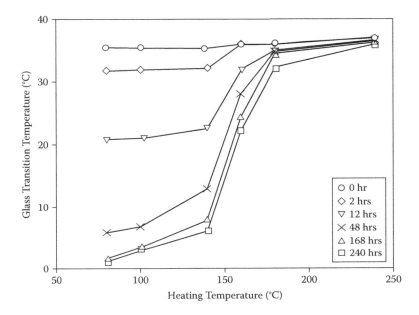

FIGURE 3.6 Evolution of T_g upon thermal cycling. (From Yang B, 2007. Influence of moisture in polyurethane shape memory polymers and their electrically conductive composites. PhD dissertation, Nanyang Technological University, Singapore.)

free N-H stretching occurs at 3498 cm⁻¹. Conversely, the infrared band of free C=O stretching at 1724 cm⁻¹ shifts to that of the bonded one at 1701 cm⁻¹.

Figure 3.8 presents the FTIR spectra of samples after different hours of immersion in the N-H and C=O stretching regions, respectively. According to Figure 3.8a, the infrared band intensity of hydrogen-bonded N-H stretching shows no significant

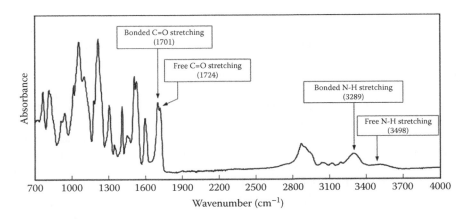

FIGURE 3.7 FTIR spectra of polyurethane SMP without immersion. (Reprinted from Yang B, Huang WM, Li C et al. *Polymer*, 47, 1348–1356, 2006. With permission.)

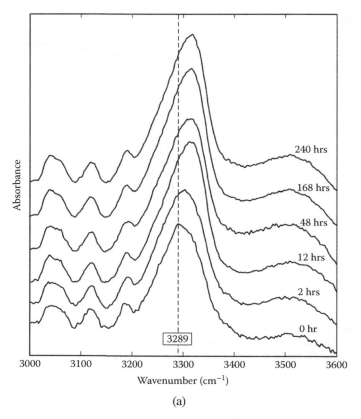

(a)

FIGURE 3.8 FTIR spectra of polyurethane SMPs after different immersion hours. (a) N-H stretching region. (b) C=O stretching region. (Reprinted from Yang B, Huang WM, Li C et al. *Polymer*, 47, 1348–1356, 2006. With permission.)

change as compared with that of the free N-H stretching. However, with the increase of immersion time, the infrared band of hydrogen-bonded N-H stretching shifts to a higher frequency. The shift is more significant in samples with shorter immersion times and becomes stable in samples with immersion times longer than 48 hours. On the other hand, Figure 3.8b indicates that after immersion, the infrared band of hydrogen-bonded C=O stretching shifts slightly to a lower frequency. Furthermore, with the increase of immersion time, the infrared band intensity of bonded C=O stretching becomes more striking as compared with that of free C=O stretching, indicating that a longer immersion time results in more C=O groups involved in hydrogen bonding.

The heating process mentioned previously directly influences the absorbed water in the polymer. The FTIR spectra of the saturated polyurethane SMP sample after 240 hours of immersion under different heating temperatures in the heating process are plotted in Figure 3.9. As revealed in Figure 3.9a, after increases of heating temperature, both the infrared band position and the intensity of the hydrogen-bonded N-H stretching shifted back to the original values of a dry sample as shown in

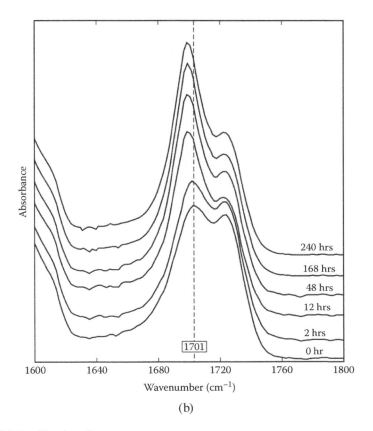

FIGURE 3.8   (Continued)

Figure 3.8a. Meanwhile, as shown in Figure 3.9b, the infrared band of the bonded C=O stretching not only shifted to a higher frequency almost identical to that of the dry samples but also regained its dominance among all intensities (Figure 3.8b).

3.6   CORRELATION OF MOISTURE, GLASS TRANSITION TEMPERATURE, AND HYDROGEN BONDING

Utilizing the data obtained from previous tests, namely TGA and DSC, the relation between the T_g and ratio of moisture to SMP in weight percentage for all SMP samples in immersion and heating processes can be found. In Figure 3.10, samples immersed for different numbers of hours and then heated to different temperatures were found to have similar slanted, L-shaped curves between T_g and water content. The change in T_g is clearly divided into two stages. At a lower heating temperature, the transition temperature is kept almost constant despite the continuous reduction of water content. However, beyond a critical temperature, the T_g starts to increase linearly with further decreases in water content. Note a turning point in the L-shaped curve during the heating process in Figure 3.10. Referring to the water ratio in weight

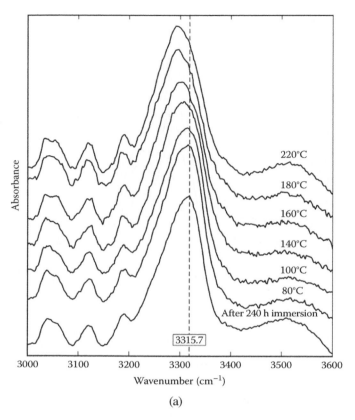

(a)

FIGURE 3.9 FTIR spectra of saturated samples (240 hours of immersion) after heating to different temperatures. (a) N-H stretching region. (b) C=O stretching region. (Reprinted from Yang B, Huang WM, Li C et al. *Polymer*, 47, 1348–1356, 2006. With permission.)

percentage at this turning point in the TGA heating curve in Figure 3.1, the critical temperature can be identified as 120°C.

The heating temperature of 120°C demonstrated in Figure 3.10 and Figure 3.1 has a clear physical meaning. The total absorbed water in the polyurethane SMP can be divided into free water and bound water (Herrera-Gómez et al. 2001). Bound water can be removed from the polymer only at a higher heating temperature. In this case, the critical heating temperature is around 120°C and the bound water moves out of the polymer in an approximate linear fashion with increases of heating temperature. As for the free water, all horizontal segments in Figure 3.10 indicate that it exerts a negligible effect on the T_g. A similar phenomenon between the T_g and ratio of water to SMP in weight percentage has been found in another polyurethane SMP from SMP Technologies, namely MM5520 (Huang et al. 2005).

As shown in Figure 3.10, the lines for the immersion and heating processes intersect in the minus zone of moisture content. This point corresponds to the real dry state of the polyurethane SMP. Thus, at 240°C used as a reference in the earlier

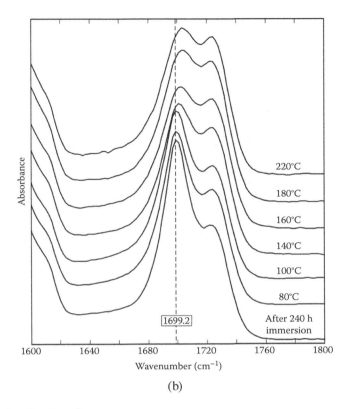

(b)

FIGURE 3.9   (Continued)

calculation, some moisture remains trapped in the material. The real ratio of moisture to SMP in weight R can be obtained according to Yang (2007):

$$R = (R_t + R_e)/(1 - R_t) \tag{3.2}$$

where R_t is the positive valued ratio of moisture to SMP composite in weight at the intersection point (Figure 3.1) and R_e is the measured ratio of moisture to SMP in weight based on the weight of the sample at 240°C.

The amounts of free and bound water during the immersion process now can be further identified by Equation (3.1). Using the two segments in the slanted L-shaped curves, the ratios of free, bound, and total absorbed water in the polymer can be determined as functions of immersion time, as shown in Figure 3.11. Moisture absorption increases dramatically in the first 48 hours of immersion, and at any point the free water is more predominant than the bound water.

FTIR results provide a clearer picture of the hydrogen-bonding mechanism behind the change of T_g in the polyurethane SMP (Figures 3.8 and 3.9). The mechanism is largely dominated by the hydrogen bonding between the N-H and C=O groups. For better illustration, based on Figure 3.8, Figure 3.12 plots the infrared bands of the

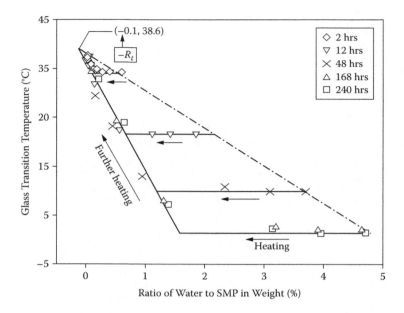

FIGURE 3.10 T_g versus ratio of water to SMP in weight percentage. (Reprinted from Huang WM, Yang B, An L et al. *Applied Physics Letters*, 86, 114105, 2005. With permission.)

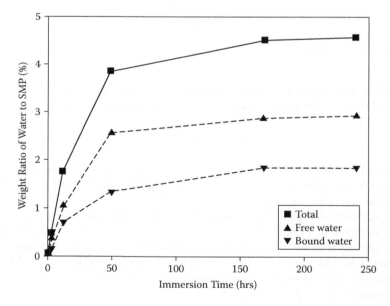

FIGURE 3.11   Ratio of water to SMP in weight versus immersion time. (Reprinted from Yang B, Huang WM, Li C et al. *Polymer*, 47, 1348–1356, 2006. With permission.)

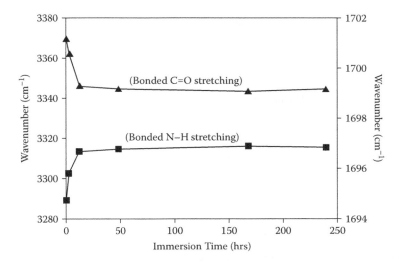

FIGURE 3.12 Infrared band of bonded C=O and N-H stretchings versus immersion time in water. (From Yang B, 2007. Influence of moisture in polyurethane shape memory polymers and their electrically conductive composites. PhD dissertation, Nanyang Technological University, Singapore.)

hydrogen-bonded N-H and C=O stretchings as functions of immersion time in water. As we can see, the shift of infrared bands in the hydrogen-bonded N-H and C=O groups is significant in the first 48 hours of immersion and then flattens out. The water content in the polyurethane SMP also increases with the increase of immersion time in a similar manner (Figure 3.11). This reveals that water has direct effects on the hydrogen bonding in the polyurethane SMP, which can be explained based on the model in Figure 3.13, modified from Lim et al. (1999).

As illustrated in Figure 3.13, some water molecules absorbed in the polyurethane SMP upon immersion act as bridges between the hydrogen-bonded N-H and C=O groups (site *a* in Figure 3.13). In this model, only one interaction is possible between the water and bonded N=O (site *a*) that directly relates to the hydrogen bonding in the SMP. Thus, the change of the bonded N-H infrared band induced by water may be interpreted as the effect of water on this hydrogen bonding. The loosely bound water directly weakens the hydrogen bonding, as shown by the shift of the infrared band of the hydrogen-bonded N-H to a higher frequency. Based on the function of water as a plasticizer, the T_g is reduced.

On the other hand, some absorbed water molecules can form double hydrogen bonds with two already hydrogen-bonded C=O groups (site *b* in Figure 3.13). Due to the hydrogen bonding in site *a*, the infrared band of bonded C=O stretching shifts up to a higher frequency while the hydrogen bond in site *b* brings it down to a lower frequency. These two hydrogen bonds may work together and counteract. According to Puffr and Sebenda (1967), water in site *b* is more firmly bound than that in site *a* (Lim et al. 1999). Consequently, the infrared band of hydrogen-bonded C=O decreases (Figure 3.12).

Figure 3.14 illustrates the infrared bands of hydrogen-bonded N-H and C=O stretchings against temperature during heating. Both N-H and C=O infrared bands

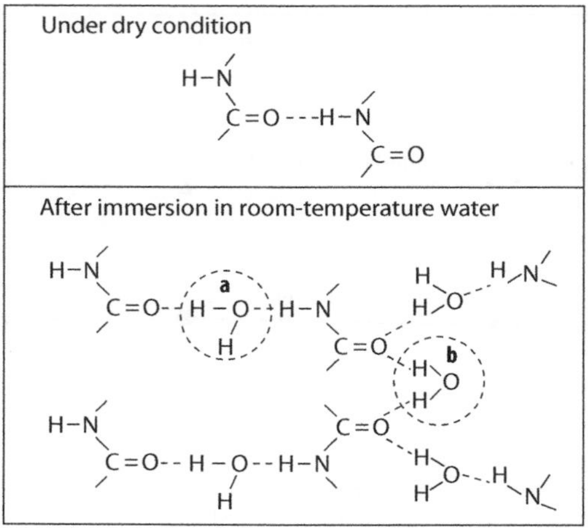

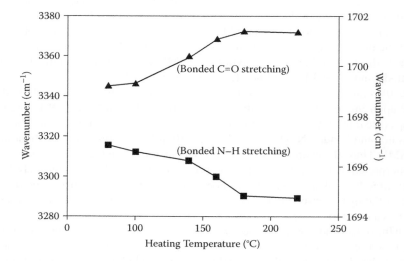

FIGURE 3.13 Effects of water on hydrogen bonding in MM3520 polyurethane SMP. (Reprinted from Yang B, Huang WM, Li C et al. *Polymer*, 47, 1348–1356, 2006. With permission.)

FIGURE 3.14 Infrared band of bonded C=O and N-H stretchings versus heating temperature. (From Yang B, 2007. Influence of moisture in polyurethane shape memory polymers and their electrically conductive composites. PhD dissertation, Nanyang Technological University, Singapore.)

change remarkably in the range of 100 to 180°C, coinciding with the temperature range of significant water loss shown in Figure 3.1. With the evaporation of water, especially bound water (Figure 3.11), in the heating process, the interaction between water and polymer is removed. Therefore, the hydrogen bonding between N-H and C=O gradually reverts to its original state. Water that acts as a plasticizer is removed upon heating and the SMP finally recovers its original T_g.

3.7   NEW FEATURES BASED ON EFFECTS OF MOISTURE

Because the T_g of the MM3520 polyurethane SMP may be reduced by immersion into room-temperature water, it is possible to utilize this phenomenon to design moisture-responsive functionally gradient SMPs. The basic idea is to lower the T_g of the material by immersing it into water. To verify these features—(1) moisture response and (2) functionally gradient T_g (both can dramatically widen the applications of the material)—three tests were performed on wires prepared by extrusion at 200°C to diameters of 1.5 mm.

The first test was water-driven recovery to verify the moisture-responsive feature. A piece of straight SMP wire was bent into a circular shape at 40°C and the shape was retained during cooling back to room temperature (about 22°C). No apparent shape recovery was found after the deformed wire was kept in a dry cabinet at 30% relative humidity (RH) and room temperature for 1 week. However, after immersion into room-temperature water for about 30 minutes, the material starts to recover gradually (Figure 3.15). This experiment confirms the moisture-responsive feature of this SMP.

Another piece of SMP wire was used for the gradient T_g test. The wire was divided into three segments of identical length from top to bottom. The segments were immersed into room-temperature water for 0, 30 minutes, and 5 hours, respectively. This resulted in three different T_g values: 36, 28, and 15°C, respectively, from top to bottom segments as measured by a DSC 2920 (TA Instruments) at a heating rate of 20°C/minute. Subsequently, the wires were deformed at 40°C into *M* shapes and cooled to 10°C; shapes were retained. Upon exposure to air, the bottom segment recovered in about 30 seconds (Figure 3.16). After heating to about 30°C with a hot plate, the middle segment became straight in about 1 minute. Upon further heating to 40°C and holding for ~1 minute, the top segment regained its original shape. This experiment successfully demonstrates recovery in sequence—a useful feature for programmable recovery.

Combining both features, the recovery of a piece of polyurethane SMP can be actuated by water in a programmable manner. Another SMP wire was used for this test. The top half of the wire was immersed in room-temperature water ~20 minutes, while the bottom half was kept dry. Subsequently, the wire was deformed into a Z shape at a high temperature and then cooled to room temperature with the deformed shape retained. Figure 3.17 shows the sequence of recovery in water. Note that the top half with a lower T_g recovers first, followed by the recovery of the bottom half. After about 75 minutes, the wire almost fully regains its original shape.

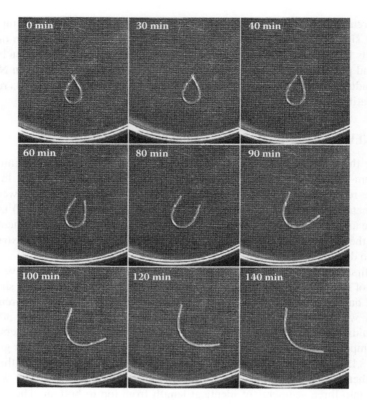

FIGURE 3.15 Water-driven recovery. (Reprinted from Huang WM, Yang B, An L et al. *Applied Physics Letters*, 86, 114105, 2005. With permission.)

3.8 RECOVERY TESTS

To examine the shape memory effect described in Section 2.6, we conducted free recovery and constrained recovery tests on 1 mm diameter MM3520 wires produced by extrusion. However, instead of heating to test thermal response, we immersed the pre-stretched wires (with 10, 20, and 50% maximum strain as discussed in Section 2.6) into room-temperature water (moisture response) for recovery. Figure 3.18 presents the evolution of the recovery stress and recovery ratio against the immersion time.

Figure 3.18a reveals that higher pre-strain produced higher recovery stress in wires immersed in water. The recovery stress started to increase dramatically after ~2.5 hours of immersion and reached a maximum in ~4 hours. Thereafter, the stress reduced but only very slightly. In contrast, the recovery ratio after a 10-hour immersion (Figure 3.18b) was lower in the highly pre-strained wires, and significant recovery started after ~4 hours of immersion.

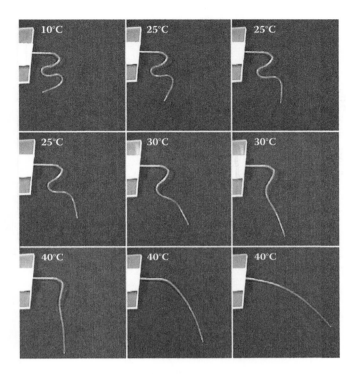

FIGURE 3.16   Recovery of SMP wire in programmable manner upon heating. (Reprinted from Huang WM, Yang B, An L et al. *Applied Physics Letters*, 86, 114105, 2005. With permission.)

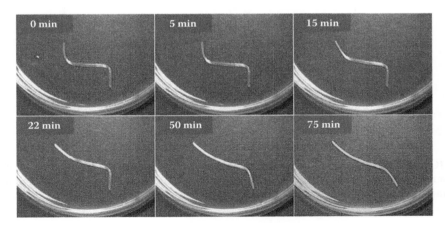

FIGURE 3.17   Recovery of a piece of functionally gradient SMP wire actuated by water in a sequence. (Reprinted from Huang WM, Yang B, An L et al. *Applied Physics Letters*, 86, 114105, 2005. With permission.)

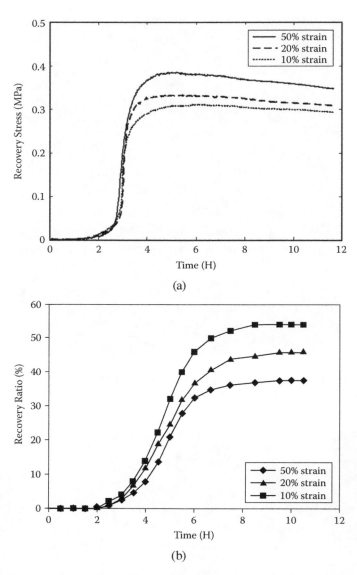

FIGURE 3.18 Recovery in room-temperature water. (a) Recovery stress as function of immersion time. (b) Shape recovery ratio as function of immersion time. (Reprinted from Yang B, Huang WM, Li C et al. *Polymer*, 47, 1348–1356, 2006. With permission.)

3.9   SUMMARY

This chapter systematically investigated the effects of moisture on the T_g values of a polyurethane SMP. Experimental results reveal that the T_g may be reduced substantially by immersion of the material into room-temperature water. The reduction in T_g continues until the SMP is saturated. Moreover, the change in T_g is reversible by heating or dehydrating.

The mechanism behind the influence of moisture on the T_g is identified. Water absorbed in the polyurethane SMP interacts with the polymer chains and weakens the hydrogen bonding between N-H and C=O groups. With water as a plasticizer, the T_g of the polyurethane SMP is reduced significantly. Heating or dehydrating can cancel these effects so that the SMP recovers its original T_g.

Water absorbed by the polyurethane SMP can be classified as free water and bound water and both are quantitatively identified in the immersion and heating processes. The free water can be fully removed at ~120°C by evaporation. Free water absorbed in the polyurethane SMP has a negligible effect on the T_g, while bound water significantly reduces the T_g in an almost linear manner. Furthermore, a temperature above 120°C is required to remove the bound water. A similar phenomenon involving T_g value and the ratio of water to SMP in weight percentage was also found in the MM5520 polyurethane SMP of SMP Technologies (Huang et al. 2005).

Based on these findings, two new features of the polyurethane SMP are demonstrated. One is the actuation triggered by water, and the other is the recovery in a programmable manner after introducing a gradient T_g by immersing different parts into water for different periods of time. The MM3520 polyurethane SMP is both thermo-responsive and moisture responsive. A number of its novel applications are presented in subsequent chapters.

ACKNOWLEDGMENT

We wish to thank N.W. Khun for his help in compiling this chapter.

REFERENCES

Brunette CM, Hsu SL, and MacKnight WJ (1982). Hydrogen-bonding properties of hard-segment model compounds in polyurethane block copolymers. *Macromolecules*, 15, 71–77.

Chen TK, Tien YI, and Wei KH (2000). Synthesis and characterization of novel segmented polyurethane/clay nanocomposites. *Polymer,* 41, 1345–1353.

Heintz AM, McKiernan RL, Gido SP et al. (2002). Crystallization behavior of strongly interacting chains. *Macromolecules,* 35, 3117–3125.

Herrera-Gómez A, Velázquez-Cruz G, and Martín-Polo MO (2001). Analysis of the water bound to a polymer matrix by infrared spectroscopy. *Journal of Applied Physics*, 89, 5431–5437.

Huang WM, Yang B, An L et al. (2005). Water-driven programmable polyurethane shape memory polymer: demonstration and mechanism. *Applied Physics Letters*, 86, 114105.

Lim LT, Britt IJ, and Tung MA (1999). Sorption and transport of water vapor in nylon 6,6 film. *Journal of Applied Polymer Science*, 71, 197–206.

Luo N, Wang DL, and Ying SK (1997). Hydrogen-bonding properties of segmented polyether poly(urethane urea) copolymer. *Macromolecules*, 30, 4405–4409.

Puffr R and Sebenda J (1967). On the structure and properties of polyamides 27, 28. *Journal of Polymer Science C*, 16, 79–93.

Teo HS, Chen CY, and Kuo JF (1997). Fourier transform infrared spectroscopy study on effects of temperature on hydrogen bonding in amine-containing polyurethanes and poly (urethane-urea)s. *Macromolecules*, 30, 1793–1799.

Yang B, Huang WM, Li C et al. (2006). Effect of moisture on the thermomechanical properties of a polyurethane shape memory polymer. *Polymer*, 47, 1348–1356.

Yang B (2007). Influence of moisture in polyurethane shape memory polymers and their electrically conductive composites. PhD dissertation, Nanyang Technological University, Singapore.

Yen FS, Lin LL, and Hong JL (1999). Hydrogen-bond interactions between urethane–urethane and urethane–ester linkages in a liquid crystalline poly(ester-urethane). *Macromolecules*, 32, 3068–3079.

Yoon JP and Han CD (2000). Effect of thermal history on the rheological behavior of thermoplastic polyurethanes. *Macromolecules*, 33, 2171–2183.

4 Electrically Conductive Polyurethane Shape Memory Polymers

4.1 INTRODUCTION

Conductive polymers can be realized in two ways. One is to produce a polymer that is intrinsically conductive. Heeger et al. received the Nobel Prize in Chemistry in 2000 for developing these intrinsically conductive polymers (ICPs). However, commonly used ICPs such as polyacetylene, polyaniline, and polypyrrole have typical conductivity values around 10^{-10} to 10^{5}S/m. They do not have stable mechanical properties and are expensive and difficult to prepare, particularly during polymerization (Cotts and Reyes 1986, Nalwa 1997, Morgan and Foot 2001). The second approach is to dope or load conductive fillers into an insulator. Graphite, carbon powders, and metallic particles are widely used. Due to very low cost and convenient fabrication (only conventional processes are required to prepare polymers), the latter approach is more popular in engineering practice (Jäger et al. 2001, Thommerel et al. 2002, Zheng and Wong 2003).

The percolation theory (Stauffer and Aharony 1994, Wu and Mclachlan 1997, Jäger et al. 2001) may be used to describe the dependence of electrical resistivity on the volume fractions of conductive fillers in conductive composites. The electrical resistivity (ρ) for randomly distributed conductors in an insulator matrix of insulator–conductor binary composites is expressed as:

$$\rho = A(\phi_c - \phi_f)^s \quad \text{for} \quad \phi_f < \phi_c \tag{4.1}$$

$$\rho = B(\phi_f - \phi_c)^{-r} \quad \text{for} \quad \phi_f > \phi_c \tag{4.2}$$

where ϕ_f is the volume fraction of the conductive filler, ϕ_c is the percolation threshold, and s and r are the critical exponents close to 0.7 and 2.0, respectively, and are affected by distribution, geometry of conductive fillers, and other factors (Sahimi 1994, Das et al. 2002). A and B are critical coefficients relating to the resistivity of the polymer matrix and the conductive filler.

Since native shape memory polymers (SMPs) are intrinsically electrically nonconductive, they cannot be conveniently heated to trigger the shape memory effect (SME) by means of Joule heating, as shape memory alloys (SMAs) can (Huang 2002). Li et al. (2000) reported their preliminary experimental results on conductive SMPs following the second approach. They used polyurethane SMP–carbon black

(CB) composites prepared by a solution–precipitation process, followed by compressive molding. Their investigation focused on structures, electrical conductivities, strain recoveries, and the relationships of these composites.

The percolation threshold was obtained at 20% weight fraction of carbon black. However, a highly loaded composite revealed a very low strain recovery speed and recoverable strain ratio, although stiffness and strength improved. For example, an SMP filled with 25% weight fraction of carbon had an electrical conductivity of 10^{-1} Ωm, but only recovered less than 70% of pre-loaded strain. Subsequent intensive research efforts in this area involved different polymer systems (Leng et al. 2007, 2008, Beloshenko et al. 2005, Sahoo et al. 2007, Jung et al. 2010a, 2010b, Koerner et al. 2004, Luo and Mather 2010, Cho et al. 2005). Readers may refer to the relevant sections in Chapter 7 for more details.

This chapter reports a systematic study of a CB-filled polyurethane SMP product (MM5520, SMP Technologies, Japan) that has a nominal glass transition temperature (T_g) of 55°C, a melting temperature around 200°C, and a bulk density of 1.25 g/cm^3 (refer to http://www.smptechno.com/index_en.html). In Chapter 3, we saw that the T_g of this type of polyurethane SMP can drop below room temperature due to the influence of moisture, and special instruments for sub-ambient temperature tests were not available during this study. Therefore, SMP MM5520, which has a higher T_g than SMP MM3520, was chosen for this chapter to avoid sub-ambient temperature tests.

4.2   PREPARATION OF ELECTRICALLY CONDUCTIVE POLYURETHANE SMP

Carbon nano powders bought from Degussa were used as conductive fillers. The technical data for carbon nano powders are listed in Table 4.1. The data indicate that these carbon powders are highly structured with Brunauer, Emmett, and Teller (BET) surface areas up to 1000 m^2/g—a measure of the total area of carbon powder including the external and internal surface areas.[*] Furthermore, these carbon powders with dibutylphthalate (DBP) absorption rates of 420 mL/100 g are more highly aggregated than ordinary carbon powders with DBP absorptions below 90 mL/100 g. Note that DBP absorption is an indirect measure for determining carbon powder structure by measuring the void volume between the individual carbon aggregates and agglomerates.

Before processing, the polyurethane SMP and carbon powders were dried in a vacuum oven at 80°C for 12 hours to remove moisture. After the pellets of polyurethane SMP were melted at 200°C in the mixing head of a Haake Rheocord 90 for 1 minute, carbon powders were added slowly and blended with the melted SMP at 200°C. The rotation speed of the mixer was 60 rpm. The whole process lasted ~20 minutes. SMP composites with five different volume fractions of carbon powder (4, 7, 10, 13, and 15%) were fabricated. The volume fraction of carbon powder ($_f$) was calculated by:

---

[*] Degussa Corporation (2002). *Technical Bulletin: Pigments.* Parsippany, NJ.

TABLE 4.1
Conductive Carbon Power Technical Data

Material	Average Powder Size	Specific Gravity	Purity	DBP Absorption	BET Surface Area
Carbon powder	~30 nm	~1.85	98.4%	420 mL/100 g	1000 m²/g

Source: Degussa Corporation, Parsippany, NJ.

$$f = \frac{1}{1+(M_m/\rho_m)\times(\rho_f/M_f)} \times 100\% \qquad (4.3)$$

where M_m, ρ_m, M_f, and ρ_f denote the mass of SMP, the bulk density of SMP stated by SMP Technologies, the mass of carbon powders, and the bulk density of carbon powders stated by Degussa, respectively.

CBX denotes the polyurethane SMP composites with X% volume fractions of carbon powders. Hence, CB0 is the pure SMP without carbon powder. After blending, the mixture was processed into desired shapes at 200°C for testing. Three kinds of samples were prepared: (1) thin sheets with thicknesses of 2.0 mm, 1.0 mm, and 0.5 mm were fabricated using a laboratory hot press, (2) wires with individual diameters of 2.0 mm were prepared by extrusion, and (3) cylindrical samples with diameters of 15.0 mm and lengths of 20.0 mm were molded. Before testing, all samples were kept in a dry cabinet at a relative humidity (RH) below 30%.

4.3 SHAPE RECOVERY BY PASSING ELECTRICAL CURRENT

The SMP composite CB13 was hot-pressed into a plate shape with a thickness of 2.0 mm and then cut into an *U* shape for demonstrating the SME upon heating by passing an electrical current through it. The *U*-shaped sample was connected to a 15 volt DC power supply through two electrodes. An infrared camera (Thermovision 900, AGEMA) was used to observe the temperature distribution in the sample. Figure 4.1a reveals the temperature distribution in the sample heated by an electrical current for 45 seconds. The sample was then easily bent at 60°C—above the T_g. After switching off the electrical power and cooling back to room temperature (about 22°C) with the deformed shape held still, a *permanently* bent shape was formed. When reheated above the T_g by means of passing an electrical current (Joule heating), the sample recovered its original shape as shown in Figure 4.1b.

4.4 DISTRIBUTION OF CARBON POWDER IN POLYURETHANE SMP

The dispersion of carbon powders within the conductive SMP was studied by investigating the cryofracture surfaces of samples without coating using a scanning electron microscope (SEM, Leica Cambridge S360). Figure 4.2 illustrates the formation

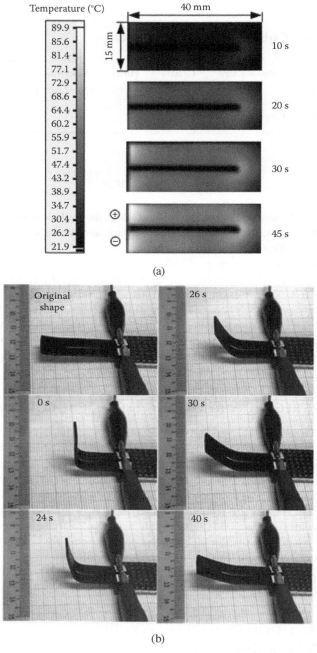

(a)

(b)

FIGURE 4.1 Shape recovery in SMP composite (with 13% nano carbon powder in volume). (a) Temperature distribution shown by infrared camera. (b) Shape recovery upon passing an electrical current. (Reprinted from Yang B, Huang WM, Li C et al. *European Polymer Journal*, 41, 1123–1128, 2005. With permission.)

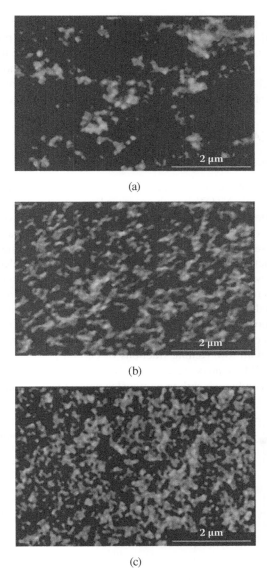

(a)

(b)

(c)

FIGURE 4.2   SEM images of cryofracture surfaces. (a) CB4. (b) CB7. (c) CB13. (From Yang B, 2007. Influence of moisture in polyurethane shape memory polymers and their electrically conductive composites. PhD dissertation, Nanyang Technological University, Singapore.)

of carbon networks in the SMP composites filled with different volume fractions of carbon powders. Note that the white areas are carbon powders. The figure shows that carbon powders distribute within the polyurethane SMP matrix randomly, aggregating as clusters instead of separating from each other. The average size of these clusters is around 100 to 200 nm in diameter. The aggregation of carbon powders in other carbon fine-powder-filled polymers has also been reported (Flandin et al. 2001, Knite et al. 2002, Carmona and Ravier 2002). Carbon powders distribute randomly in CB4 as isolated agglomerates (Figure 4.2a). With the increase of carbon powder content, carbon agglomerates connect with each other and form continuous carbon networks (Figure 4.2b). With the continuous increase of carbon loading up to 13%, the number of carbon networks increases dramatically (Figure 4.2c).

4.5 ELECTRICAL RESISTIVITY

4.5.1 DEPENDENCE ON LOADING OF CARBON POWDER

Electrical resistivity of the conductive SMPs was measured by a four-point resistivity probe system (SINGNTONE) with an upper limit of $10^{10}\,\Omega$m. The resistivities of the SMP composites with smaller volume fractions of carbon powder were over this limit. Therefore, a digital high resistivity determiner (RP2680) was used instead.

Figure 4.3 presents the electrical resistivity of SMP composites filled with different volume fractions of carbon powders. It shows that the electrical resistivity of the composite with less than 4% volume fraction of carbon powder was almost constant. A sharp transition occurred between the 4 to 7% volume fraction, known as the percolation threshold range (Ishigure et al. 1999, Zheng and Wong 2003). The percolation threshold is a critical value that indicates the transition of a material from insulative to conductive (Ishigure et al. 1999, El-Tantawy et al. 2002). For the SMP composites with low volume fractions of carbon powders (<4%), the carbon aggregates in the SMP matrix were relatively more separated (Figure 4.2a). Large gaps between the conductive carbon aggregates present significant physical barriers

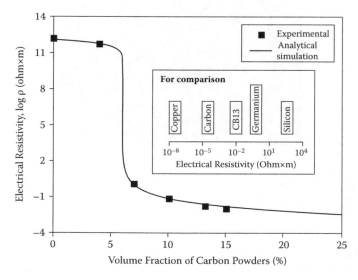

FIGURE 4.3 Electrical resistivities versus volume fractions of carbon powders. (From Yang B, 2007. Influence of moisture in polyurethane shape memory polymers and their electrically conductive composites. PhD dissertation, Nanyang Technological University, Singapore.)

to electron flow, so that almost no conductive channel can be formed in the material and the electrical resistivity is very high. As the volume fraction of conductive filler reaches the percolation threshold, the gaps between conductive carbon aggregates are reduced, as shown in Figure 4.2b. Some carbon aggregates even directly contact each other. Thus, electrons can jump more easily. The three-dimensional conducting network is constructed and the composite becomes much more conductive (Sheng et al. 1978, Azulay et al. 2003), resulting in a sharp decrease of electrical resistivity with the increase of carbon powder content. A further increase of the volume fraction of carbon powders reduced the gap slightly, and only a few more conductive channels formed (Figure 4.2c). Hence, reduction in resistivity is gradual.

From data fitting of the experimental results, two equations were obtained as follows:

$$\rho = 1.0 \times 10^{13} (6\% - _f)^{0.7} \quad \text{for} \quad _f < 6\% \tag{4.4}$$

$$\rho = 1.6 \times 10^{-4} (_f - 6\%)^{-1.9} \quad \text{for} \quad _f > 6\% \tag{4.5}$$

Hence, the percolation threshold for this carbon-powder-filled polyurethane SMP is 6% (volume fraction) as shown in Figure 4.3.

4.5.2 EFFECTS OF TEMPERATURE AND UNIAXIAL MECHANICAL STRAIN

Normally, the temperature range in a real engineering application of the polyurethane SMP and its conductive composites is from room temperature to about 100°C as reported in the literature (Chiodo et al. 1999, Cadogan et al. 2002, Wache et al. 2003, Gall et al. 2004). In our study, each conductive SMP was heated from room temperature to ~100°C and the electrical resistivity of each composite was recorded against temperature using the four-point resistivity probe system. The results were plotted as shown in Figure 4.4 and show no apparent change in electrical resistivity in all conductive SMP composites within this temperature range.

In real applications, the conductive SMPs are deformed and then heated for recovery. Therefore, the understanding of the effects of strain on the electrical resistivity of a conductive SMP is very important. A testing setup as illustrated in Figure 4.5 was designed and fabricated for this purpose. An Instron 5569 device with a hot chamber was used to stretch and compress the conductive SMP samples at 60°C and at a constant crosshead speed yield a fixed strain rate of 10^{-3}/s. Conductive SMP wires with diameters of 2.0 mm and gauge lengths of 30.0 mm were used for tensile testing. Cylindrical samples with diameters of 15.0 mm and gauge lengths of 20.0 mm were used for compressive testing. Two copper electrodes were attached to the ends of the samples. A multimeter was coupled with the Instron to measure the resistance (R') of samples. The electrical volume resistivity (ρ) was calculated:

$$\rho = R' \times S'/L \tag{4.6}$$

where S' is the cross-sectional area of the sample and L is the gauge length.

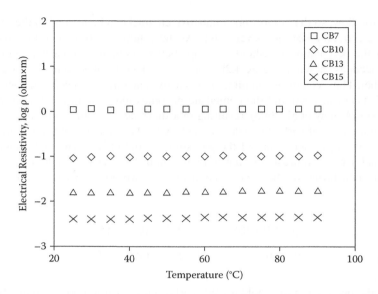

FIGURE 4.4 Electrical resistivity as a function of temperature. (From Yang B, 2007. Influence of moisture in polyurethane shape memory polymers and their electrically conductive composites. PhD dissertation, Nanyang Technological University, Singapore.)

Figure 4.6a plots the tensile strain against electrical resistivity. It shows that, generally speaking, with an increase of strain the electrical resistivity increases in a nonlinear fashion. The increase of electrical resistivity is more remarkable at the lower strain range, i.e., below 10%. With the increase of loading of carbon powders, the increase of electrical resistivity in the lower strain range becomes more significant.

When a conductive composite is subjected to a tensile strain, it is expected that two simultaneous processes will operate within the material: (1) the breakdown of existing conductive networks due to an increase in the gap between the carbon powder aggregates and (2) reformation of new conductive networks due to reorientation of carbon aggregates (Aneli et al. 1999, Flandin et al. 2001). At a lower tensile strain, the breakdown process is more prominent than the formation process. Thus, the net result is a reduction in the number of conductive networks, indicating an increase of resistivity. However, at a higher tensile strain, rather than formation of holes and destruction of the conductive networks, the extension produces new conductive pathways and/or improves the existing pathways by reorientation of carbon powders. For highly structured carbon powders, rotation, translation, and possible shape changes of the asymmetric aggregates may preserve the number of contacts and conductive pathways. These processes somewhat counterbalance the effects of breakdowns of conductive networks, resulting in a slower rate of change in resistivity against the extension at a higher strain (Kost et al. 1984, Pramanik et al. 1993, Das et al. 2002).

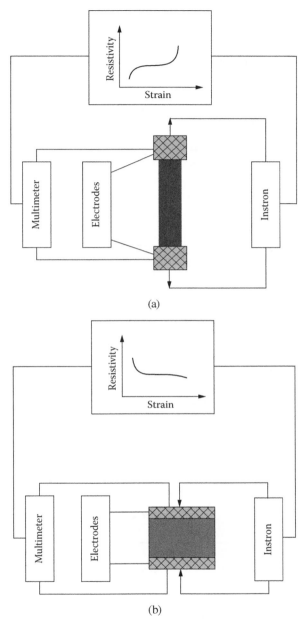

FIGURE 4.5 Experimental setup. (a) Tensile test. (b) Compressive test. (From Yang B, 2007. Influence of moisture in polyurethane shape memory polymers and their electrically conductive composites. PhD dissertation, Nanyang Technological University, Singapore.)

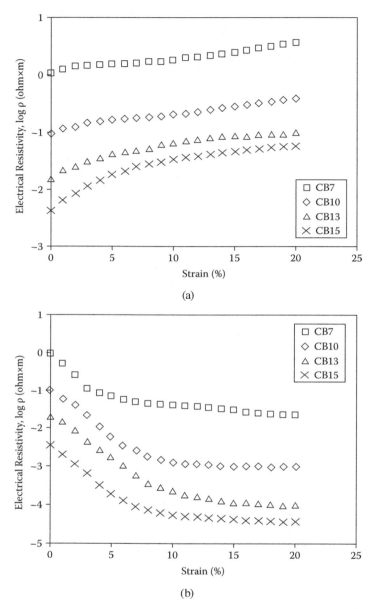

FIGURE 4.6 Electrical resistivity versus strain relationship. (a) In tension. (b) In compression. (From Yang B, 2007. Influence of moisture in polyurethane shape memory polymers and their electrically conductive composites. PhD dissertation, Nanyang Technological University, Singapore.)

However, a compressive strain has a different effect on electrical resistivity. The testing results are plotted in Figure 4.6b. With an increase of compressive strain, the electrical resistivity decreases instead. The most significant decrease happens at a lower strain range, i.e., below 10% (same result as with tension). Under a lower compressive strain, the gap between carbon aggregates in the SMP is reduced so that the tunneling conduction becomes possible between some aggregates. Some aggregates may even contact each other physically. Thus, more conductive pathways are constructed, resulting in a decrease in resistivity. These effects are less significant at a higher strain range because all the possible conductive pathways are formed upon compression in the lower strain range.

4.6 THERMAL STABILITY

The thermal stability of carbon-powder-filled polyurethane SMP composites was tested using a TGA 2950 (TA Instruments). The samples weighing 20 to 25 mg were heated at a constant rate of 20°C/minute in nitrogen atmosphere. The results are plotted in Figure 4.7. The decomposition of the pure polyurethane SMP (CB0) started at ~260°C. In general, with the addition of carbon powders, the onset of decomposition temperature increased slightly and the full decomposition occurred at a higher temperature.

It has been reported that the incorporation of nano fillers may improve the thermal stabilities of some polymer composites (Gilman et al. 2000, Zeng et al. 2004). The increase in the degradation temperature may arise for two possible reasons. One

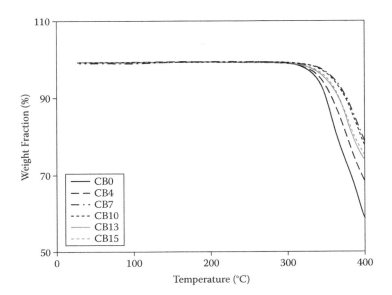

FIGURE 4.7 TGA results of dry SMP composites with different volume fractions of carbon powders. (Reprinted from Yang B, Huang WM, Li C et al. *European Polymer Journal*, 41, 1123–1128, 2005. With permission.)

is the restriction on the mobility of the macromolecules imposed by, in our case, the carbon nano powders, causing a reduction when tension is induced by thermal excitation in the carbon–carbon bond (Bryk 1991, Ahmad et al. 1995). The other probable explanation may relate to the capacity of carbon powders for heat absorption. According to the black body principle, black carbon nano powders are excellent absorbers of heat. Thermal degradation in a polymer matrix occurs only after a certain amount of heat energy has been absorbed by the material. Heat initiates the degradation process and breakdown of the matrix structure by causing ruptures or scissions in the molecular chains. With an increase of carbon nano powder content in an SMP composite, more heat is absorbed by the carbon powders. Thus, higher temperature is required to overcome the required threshold energy for commencement of degradation.

4.7 UNIAXIAL TENSILE TESTING AT ROOM TEMPERATURE

Uniaxial tensile tests were carried out using the Instron 5569 with a 1 kN load cell to investigate the properties of SMP composites under uniaxial tensile at room temperature. Figure 4.8 illustrates the dimensions of the testing sample. The gauge length (distance between two clamps) was set at 20 mm. During testing, a constant strain rate of 5×10^{-3}/s was applied. Engineering strain and engineering stress were used, and the strain calculated from the displacement of the crosshead over the original gauge length. The original cross-sectional area of sample, namely 4.0×1.0 mm^2, was used to calculate the stress.

Figure 4.9 plots the stress–strain curves of the uniaxial tensile tests at room temperature. In general, upon loading, the SMP composites behaved more or less similarly. They experienced yielding, cold drawing, and then failure. With the increase of carbon powders, the yielding strength of the SMP significantly increased.

The relationship of Young's modulus and the elongation limit versus volume fraction of carbon powder ($_f$) is plotted as shown in Figure 4.10. Note that Young's modulus is calculated from the early loading part of the stress–strain curve. Carbon powders reinforced the conductive SMP so that the Young's modulus of CB15 was

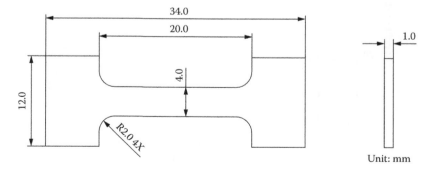

FIGURE 4.8 Dimensions of sample. (From Yang B, 2007. Influence of moisture in polyurethane shape memory polymers and their electrically conductive composites. PhD dissertation, Nanyang Technological University, Singapore.)

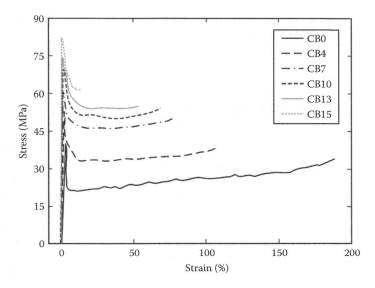

FIGURE 4.9 Stress–strain curves at room temperature. (From Yang B, 2007. Influence of moisture in polyurethane shape memory polymers and their electrically conductive composites. PhD dissertation, Nanyang Technological University, Singapore.)

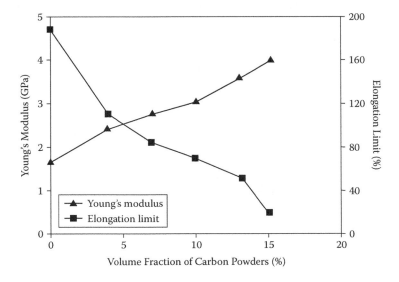

FIGURE 4.10 Relationship of Young's modulus and elongation limit versus $_f$. (From Yang B, 2007. Influence of moisture in polyurethane shape memory polymers and their electrically conductive composites. PhD dissertation, Nanyang Technological University, Singapore.)

about twice that of pure SMP. However, the presence of carbon powders reduces the elongation limit significantly. Even a mere 4% volume fraction of carbon powders results in a serious decrease in the elongation limit from 185% in CB0 to 110% in CB4.

4.8 SHAPE MEMORY PROPERTIES UPON HEATING

Thermomechanical tests were carried out to investigate the shape memory properties of SMP composites. Samples with dimensions as shown in Figure 4.8 were used. The testing procedure is illustrated in Figure 4.11. In step a, the sample is stretched uniaxially to a maximum strain (ε_m) of 20% at a constant strain rate of 5×10^{-3}/s by an Instron Microforce materials test system with a 100 N load cell. The sample was tested at a constant temperature (T_h) of 65°C inside a hot chamber, then held at ε_m and cooled to room temperature (T_r) in 4 minutes (step b), during which the stress relaxes at the very beginning of holding due to viscous flow as the material is in the rubber state. In step c, the sample is unloaded to zero stress, yielding a small amount of strain recovery. The fixed strain (ε_f) corresponds to the strain at the end of unloading.

From a thermomechanical point of view, the recovery stress and recoverable strain are essential concerns in applications of SMPs. Thus, two types of tests (constrained recovery and free recovery) were performed after step c to measure the maximum recoverable strain and stress during recovery. The samples were separated into two groups for different tests. The samples from one group were heated on a digital hot plate at a constant rate of 2°C/minute and their lengths were allowed to vary. The length of a sample was measured using a microscope

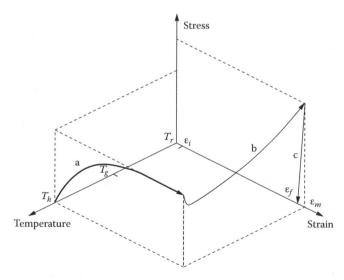

FIGURE 4.11 Thermomechanical test procedure. (Modified from Tobushi H, Okumura K, Hayashi S et al. *Mechanics of Materials*, 33, 545–554, 2001. With permission.)

and then the recovered strain was calculated and recorded along with the temperature. This is the so-called free recovery test. The second group of samples was tested under the same conditions except that the lengths of these samples were fixed. The recovery stress was recorded by the Instron against temperature. This is the so-called constrained recovery test.

To evaluate the shape memory properties of the SMP composites, the ratio of fixable strain (R_f) and the ratio of recoverable strain (R_r) were calculated using the following equations:

$$R_f = \varepsilon_f / \varepsilon_m \times 100\% \qquad\qquad (4.7)$$

$$R_r = (\varepsilon_m - \varepsilon_f + \varepsilon_r)/\varepsilon_m \times 100\% \qquad\qquad (4.8)$$

where ε_r is the total recovered strain at the end of the free recovery test.

4.8.1 Fixable Strain

The stress-versus-strain relationships in the thermomechanical tests are plotted in Figure 4.12. Note that with the increase in volume fractions of carbon powders, the SMP composite gains more resistance to deformation. The change is more remarkable from CB4 to CB7. At the beginning of holding and cooling, the stress relaxes—attributed to the viscous flow of SMPs in the rubber state. The stress then increases due to the increase of Young's modulus of materials in transition from the rubber to the glass state and thermal contraction upon cooling. Only a small linear elastic strain recovery results from unloading. The linear elastic recovery at the end of unloading in all SMP composites was below 3%.

Figure 4.13 plots the ratio of fixable strain against the volume fraction of carbon powder. It reveals that the presence of carbon powders in the SMP matrix deteriorates the shape fixity. At a lower loading of carbon powder below 4% volume fraction, this effect is almost negligible while the ratio of fixable strain gains significant reduction when the content of carbon powder reaches 7%. Further increases of carbon powder reduce the ratio of fixable strain gradually. In general, although carbon powders decrease shape fixity, all SMP composites still have reasonable shape fixity so that over 95% of preload strain is fixable after cooling and subsequent removal of the external constraint.

4.8.2 Recoverable Strain

The free recovery strain as a function of temperature in the free recovery test is plotted in Figure 4.14. It reveals that all samples recover the preloaded strain abruptly between 50 and 60°C. With the increase of carbon powder, the start temperature for recovery becomes lower, while the finish temperature is higher, so that the temperature range for strain recovery is widened. Such effects may be attributed to the change of glass transition kinetics due to the loading of carbon powders.

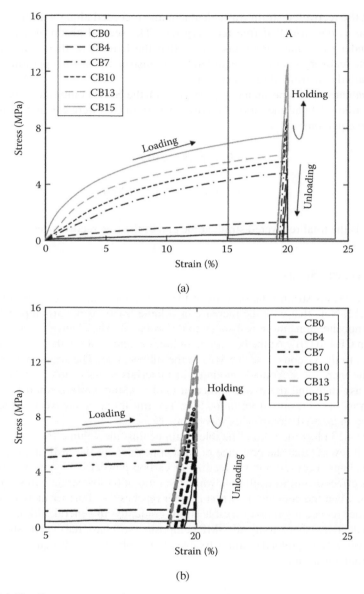

FIGURE 4.12   Stress-versus-strain curves at $T_g + 10°C$. (a) Overall view. (b) Zoom-in view. (From Yang B, 2007. Influence of moisture in polyurethane shape memory polymers and their electrically conductive composites. PhD dissertation, Nanyang Technological University, Singapore.)

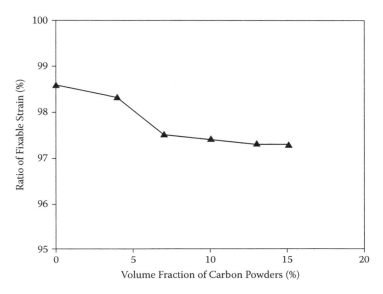

FIGURE 4.13　Ratio of fixable strain versus volume fractions of carbon powders. (From Yang B, 2007. Influence of moisture in polyurethane shape memory polymers and their electrically conductive composites. PhD dissertation, Nanyang Technological University, Singapore.)

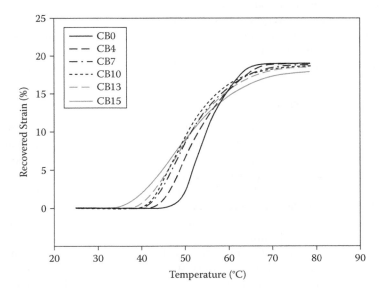

FIGURE 4.14　Recovered strain as function of temperature. (From Yang B, 2007. Influence of moisture in polyurethane shape memory polymers and their electrically conductive composites. PhD dissertation, Nanyang Technological University, Singapore.)

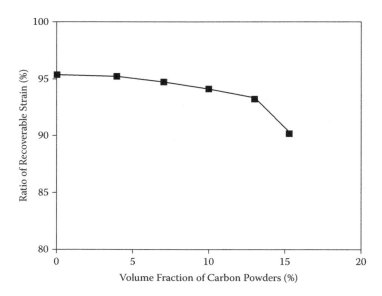

FIGURE 4.15 Ratio of recoverable strain against volume fractions of carbon powders. (From Yang B, 2007. Influence of moisture in polyurethane shape memory polymers and their electrically conductive composites. PhD dissertation, Nanyang Technological University, Singapore.)

For a better illustration, Figure 4.15 plots the ratio of recoverable strain against the volume fraction of carbon powder. It reveals that carbon powders exert limited effects on the shape recovery abilities of SMPs at a lower $_f$ values. Even when the volume fraction reaches 10%, the SMP composite is still capable of recovering over 90% of the pre-strain. However, with the increase of carbon powder over 13% in $_f$, the trend of decreasing the ratio of recoverable strain becomes significant.

4.8.3 RECOVERY STRESS

Figure 4.16 plots the recovery stress against temperature in the constrained recovery test. Generally, little change in recovery stress occurred below 35°C. The change was largely negative and may be attributed to the thermal expansion of SMP composites. With further increases in temperature, the recovery stress increased quickly because of the SME and then reached a peak. After that, it decreased gradually and even approached zero. The decrease of recovery stress is a result of the relaxation of SMPs attributed to the viscous flow of polymer chains at higher temperatures. Furthermore, the results reveal that the loading of carbon powders significantly increased the maximum recovery stress of SMP composites. With the increase in volume fraction of carbon powders, the decrease of recovery stress upon further heating slowed after maximum stress was reached. Because carbon powders restrict the viscous flow in SMPs, slower relaxation in heavily carbon-loaded SMP composites is expected.

Maximum recovery stress upon heating was plotted against volume fractions of carbon powders as shown in Figure 4.17. It shows that the loading of carbon powders

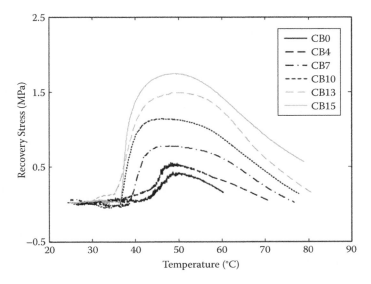

FIGURE 4.16 Recovery stress as a function of temperature. (From Yang B, 2007. Influence of moisture in polyurethane shape memory polymers and their electrically conductive composites. PhD dissertation, Nanyang Technological University, Singapore.)

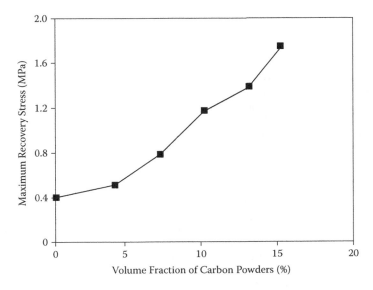

FIGURE 4.17 Maximum recovery stress against volume fractions of carbon powders. (From Yang B, 2007. Influence of moisture in polyurethane shape memory polymers and their electrically conductive composites. PhD dissertation, Nanyang Technological University, Singapore.)

almost led to an exponential increase in maximum recovery stress and that loading of carbon powders is an efficient way to increase recovery stress.

4.9 SUMMARY

Carbon nano powders distributed within the MM5520 polyurethane SMP matrix randomly with an agglomerate size of 100 to 200 nm in diameter. A 6% volume fraction of carbon powder is the percolation threshold. Over this threshold, the conductive network is virtually formed in the SMP composites, resulting in a sharp transition from insulative to electrically conductive status. The electrical resistivity of the SMP composite with a 13% volume fraction of carbon powders was about 10^{-2} Ωm. Furthermore, carbon powders reinforced the SMP composites, resulting in an increase of Young's modulus. The elongation limit, however, was reduced.

These conductive SMPs show good shape memory properties. The ratio of fixable strain is above 90% when the maximum strain is 20%. The presence of carbon powders significantly affects shape recovery ability. Based on a 20% maximum strain, more than 90% strain is recoverable in low $_f$ (<15%) SMP composites, but the freely recoverable strain decreases below 90% for high $_f$ ($\geq$15%). The carbon-powder-filled SMP composite (CB13) exhibited excellent electrical conductivity and good shape memory properties. It can be heated directly and efficiently by passing an electrical current to achieve recovery.

ACKNOWLEDGMENT

We wish to thank N.W. Khun for his help in compiling this chapter.

REFERENCES

Ahmad MY, Mustafah J, Mansor MS et al. (1995). Thermal properties of polypropylene/rice husk ash composites. *Polymer International*, 38, 33–43.
Aneli JN, Zaikov GE, and Khananashvili LM (1999). Effects of mechanical deformations on the structurization and electric conductivity of electric conducting polymer composites. *Journal of Applied Polymer Science*, 74, 601–621.
Azulay D, Eylon M, Eshkenazi O et al. (2003). Electrical–thermal switching in carbon-black–polymer composites as a local effect. *Physical Review Letters,* 90, 236601–236604.
Beloshenko VA, Varyukhin VN, and Voznyak YV (2005). Electrical properties of carbon-containing epoxy compositions under shape memory effect realization. *Composites A*, 36, 65–70.
Bryk MT (1991). *Degradation of Filled Polymers: High Temperature and Thermo-Oxidative Processes*. Ellis Horwood, Chichester.
Cadogan DP, Scarborough SE, Lin JK et al. (2002). Shape memory composite development for use in gossamer space inflatable structures. *AIAA/ASME/ASCE/AHS/ASC Structures, Structural Dynamics, and Materials Conference*, 2, 1294–1304.
Carmona F and Ravier J (2002). Electrical properties and mesostructure of carbon black-filled polymers. *Carbon*, 40, 151–156.
Chiodo JD, Billett EH, and Harrison DJ (1999). Active disassembly using shape memory polymers for the mobile phone industry. *Proceedings of IEEE International Symposium on Electronics and the Environment* (Cat. No. 99CH36357), pp. 151–156.

Cho JW, Kim JW, Jung YC et al. (2005). Electroactive shape-memory polyurethane composites incorporating carbon nanotubes. *Macromolecular Rapid Communications*, 26, 412–416.

Cotts DB and Reyes Z (1986). *Electrically Conductive Organic Polymers for Advanced Applications*. Noyes Data Corporation, Park Ridge, NJ.

Das NC, Chaki TK, and Khastgir D (2002). Effect of processing parameters, applied pressure and temperature on the electrical resistivity of rubber-based conductive composites. *Carbon*, 40, 807–816.

Das NC, Chaki TK, and Khastgir D (2002). Effect of axial stretching on electrical resistivity of short carbon fibre and carbon black-filled conductive rubber composites. *Polymer International*, 51, 156–163.

El-Tantawy F, Kamada K, and Ohnabe H (2002). *In situ* network structure, electrical and thermal properties of conductive epoxy resin–carbon black composites for electrical heater applications. *Materials Letters*, 56, 112–126.

Flandin L, Hiltner A, and Baer E (2001). Interrelationships between electrical and mechanical properties of a carbon black-filled ethylene-octene elastomer. *Polymer*, 42, 827–838.

Gall K, Kreiner P, Turner D et al. (2004). Shape-memory polymers for microelectromechanical systems. *Journal of Microelectromechanical Systems*, 13, 472–483.

Gilman JW, Jackson CL, Morgan AB et al. (2000). Flammability properties of polymer-layered-silicate nanocomposites: polypropylene and polystyrene nanocomposites. *Chemistry of Materials*, 12, 1866–1873.

Huang W (2002). On the selection of shape memory alloys for actuators. *Materials and Design*, 23, 11–19.

Ishigure Y, Iijima S, Ito H et al. (1999). Electrical and elastic properties of conductor–polymer composites. *Journal of Materials Science*, 34, 2979–2985.

Jäger KM, McQueen DH, Tchmutin IA et al. (2001). Electron transport and ac electrical properties of carbon black polymer composites. *Journal of Physics D*, 34, 2699–2707.

Jung YC, Yoo HJ, Kim YA et al. (2010a). Electroactive shape memory performance of polyurethane composite having homogeneously dispersed and covalently crosslinked carbon nanotubes. *Carbon*, 48, 1598–1603.

Jung YC, Kim HH, Kim YA et al. (2010b). Optically active multi-walled carbon nanotubes for transparent, conductive memory-shape polyurethane film. *Macromolecules*, 43, 6106–6112.

Knite M, Teteris V, Polyakov B et al. (2002). Electric and elastic properties of conductive polymeric nanocomposites on macro- and nanoscales. *Materials Science and Engineering C*, 19, 15–19.

Koerner H, Price G, Pearce NA et al. (2004). Remotely actuated polymer nanocomposites: stress recovery of carbon-nanotube-filled thermoplastic elastomers. *Nature Materials*, 3, 115–120.

Kost J, Narkis M, and Foux A (1984). Resistivity behavior of carbon-black-filled silicone rubber in cyclic loading experiments. *Journal of Applied Polymer Science*, 29, 3937–3946.

Leng J, Lv H, Liu Y et al. (2007). Electroactive shape-memory polymer filled with nanocarbon particles and short carbon fibers. *Applied Physics Letters*, 91, 144105.

Leng J, Lv H, Liu Y et al. (2008). Synergic effect of carbon black and short carbon fiber on shape memory polymer actuation by electricity. *Applied Physics Letters*, 104, 104917.

Li FK, Qi L, Yang J et al. (2000). Polyurethane/conducting carbon black composites: structure, electric conductivity, strain recovery behavior, and their relationships. *Journal of Applied Polymer Science*, 75, 68–77.

Luo X and Mather PT (2010). Conductive shape memory nanocomposites for high speed electrical actuation. *Soft Matter*, 6, 2146–2149.

Morgan H and Foot PJS (2001). The effects of composition and processing variables on the properties of thermoplastic polyaniline blends and composites. *Journal of Materials Science*, 36, 5369–5377.

Nalwa HS (1997). *Handbook of Organic Conductive Molecules and Polymers 2*, Wiley, New York.

Pramanik PK, Khastagir D, and Saha TN (1993). Effect of extensional strain on the resistivity of electrically conductive nitrile-rubber composites filled with carbon filler. *Journal of Materials Science*, 28, 3539–3546.

Sahimi M (1994). *Applications of Percolation Theory*, Taylor & Francis, London.

Sahoo NG, Jung YC, and Cho JW (2007). Electroactive shape memory effect of polyurethane composites filled with carbon nanotubes and conductive polymer. *Materials and Manufacturing Processes*, 22, 419–423.

Sheng P, Sichel EK, and Gittleman JI (1978). Fluctuation-induced tunneling conduction in carbon-polyvinylchloride composites. *Physical Review Letters*, 40, 1197–1200.

Stauffer D and Aharony A (1994). *Introduction to Percolation Theory*, Taylor & Francis, London.

Thommerel E, Valmalette JC, Musso J et al. (2002). Relations between microstructure, electrical percolation and corrosion in metal–insulator composites. *Materials Science and Engineering A*, 328, 67–79.

Tobushi H, Okumura K, Hayashi S et al. (2001). Thermomechanical constitutive model of shape memory polymer. *Mechanics of Materials*, 33, 545–554.

Wache HM, Tartakowska DJ, Hentrich A et al. (2003). Development of a polymer stent with shape memory effect as a drug delivery system. *Journal of Materials Science: Materials in Medicine*, 14, 109–112.

Wu J and Mclachlan DS (1997). Percolation exponents and thresholds obtained from the nearly ideal continuum percolation system graphite-boron nitride. *Physical Reviews B*, 56, 1236–1248.

Yang B, Huang WM, Li C et al. (2005), Effects of moisture on the glass transition temperature of polyurethane shape memory polymer filled with nano carbon powder. *European Polymer Journal*, 41, 1123–1128.

Yang B (2007). Influence of moisture in polyurethane shape memory polymers and their electrically conductive composites. PhD dissertation, Nanyang Technological University, Singapore.

Zeng J, Saltysiak B, Johnson WS et al. (2004). Processing and properties of poly (methyl methacrylate)/carbon nano fiber composites. *Composites B*, 35, 173–178.

Zheng W and Wong SC (2003). Electrical conductivity and dielectric properties of PMMA–expanded graphite composites. *Composites Science and Technology*, 63, 225–235.

5 Effects of Moisture on Electrically Conductive Polyurethane Shape Memory Polymers

5.1 INTRODUCTION

As presented in Chapter 3, moisture exerts a significant influence on the glass transition temperature (T_g) of a polyurethane shape memory polymer (SMP) from SMP Technologies, Japan. After blending with carbon nano powders, electrically conductive SMP composites were produced and characterized as detailed Chapter 4. They can be directly heated for recovery by Joule heating (passing an electrical current) so that a wider range of applications can be realized. However, the effects of moisture on the electrically conductive SMP composites are still not known. A good understanding of this issue is significant in engineering applications. Therefore, this chapter aims to address this issue by studying the samples discussed in Chapter 4. Again, CBX denotes the polyurethane SMP (MM5520) composites with X% volume fraction of carbon powders. Hence, CB0 is the pure SMP without carbon powder.

5.2 ABSORPTION OF MOISTURE IN ROOM-TEMPERATURE WATER

To study the absorption of moisture in electrically conductive SMP composites, hot-pressed SMP composite sheets about 1.0 mm thick were immersed in 22°C room-temperature water. After different amounts of immersion time, the sheets were cut into small pieces for thermogravimetric analysis (TGA) to determine their moisture fractions. Samples weighing around 20 mg were heated from 30 to 330°C by a TGA 2950 (TA Instruments) at a rate of 20°C/minute.

Figure 5.1 presents one set of TGA results for sample CB13 after different immersion times. Clearly a major weight loss occurred in the range between 100 and 240°C before decomposition occurred. This weight loss is directly attributed to the evaporation of moisture absorbed in the SMP composite. For simplicity (as in Chapter 3), we set the percentage of weight loss at 240°C as the total weight percentage of moisture absorbed. It is obvious that the moisture content in SMP composites increases with increases of immersion time.

Figure 5.2 further summarizes the moisture fraction in weight percentage versus immersion time of samples with different volume fractions of carbon powders.

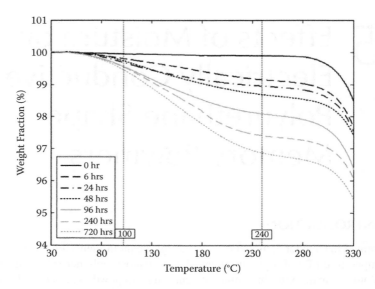

FIGURE 5.1 TGA results for CB13 after different immersion hours in water. (Reprinted from Yang B, Huang WM, Li C et al. *European Polymer Journal*, 41, 1123–1128, 2005. With permission.)

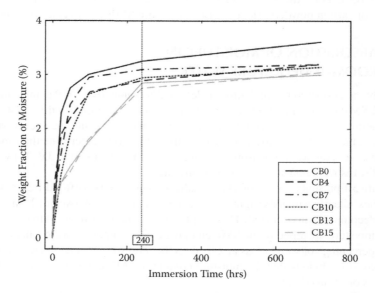

FIGURE 5.2 Moisture fraction in weight percentage versus immersion time. (Reprinted from Yang B, Huang WM, Li C et al. *European Polymer Journal*, 41, 1123–1128, 2005. With permission.)

It reveals that the moisture content increases at the beginning and gradually becomes almost constant after 240 hours. Thus, after 720 immersion hours, a sample may be considered saturated. Furthermore, SMP composites filled with higher contents of carbon powders tend to have less moisture content in the saturated state. The maximum moisture absorption (in weight percentage) of the pure polyurethane SMP is about 3.5%. Furthermore, the loading of carbon powders slows the speed of moisture absorption. This is more obvious at the beginning of immersion.

5.3 ELECTRICAL RESISTIVITY AFTER IMMERSION

Figure 5.3 plots the electrical resistivity of carbon-powder-filled SMPs as a function of immersion time in room-temperature water. It reveals that the electrical resistivity of CB7 and CB10 decreases at the beginning of immersion and then quickly reaches a stable state. However, almost no change in the electrical resistivity of CB13 and CB15 was noted. As the immersion time is directly related to the moisture ratio in SMPs, this improvement of electrical conductivity in CB7 and CB10 may be attributed to the increase of moisture content.

It is believed that the conductive fillers distribute in an insulator in three possible ways: non-contact, close proximity, and physical contact between the conductive fillers or agglomerates (Bhattacharya 1986). The electrons can jump the gap between the conductive fillers in the case of close proximity. The jumping causes the tunneling conductivity in conductive polymer composites (Sheng et al. 1978, Azulay et al. 2003).

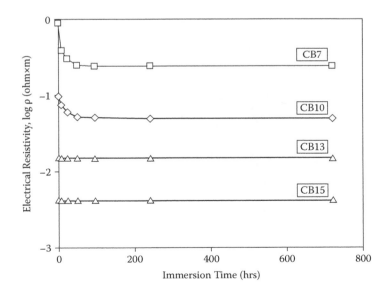

FIGURE 5.3 Electrical resistivity as function of immersion time. (From Yang B, 2007. Influence of moisture in polyurethane shape memory polymers and their electrically conductive composites. PhD dissertation, Nanyang Technological University, Singapore.)

In CB7 and CB10, the tunneling conductivity may be more dominant than that in CB13 and CB15 because of lower loading of carbon powders. It may be reasonable to assume that moisture improves the mobility of electrons so that more tunneling happens in CB7 and CB10. Their subsequent decreases in electrical resistivity are more remarkable than those in CB 13 and CB15 containing higher loadings of carbon powders. Note that in this study, the electrical resistivity was obtained by measuring the surface resistance of SMP samples, where the material absorbs moisture much faster. The material on the surface can quickly reach the saturated state. Thus, the variation in electrical resistivity is more remarkable at the beginning of immersion.

5.4 GLASS TRANSITION TEMPERATURE AFTER IMMERSION

A differential scanning calorimeter (DSC 2920, TA Instruments) was used for testing the original dry samples and immersed in water for varying times to examine the changes in the T_g. The samples weighed ~10 mg and the constant heating or cooling rate was 20°C/minute. The T_g was determined at the median point in the glass transition range during heating. The DSC curves of the dry SMP composites upon heating are plotted as shown in Figure 5.4.

A set of typical DSC results for a wetted sample (CB13) is presented in Figure 5.5. It reveals that the T_g decreases remarkably with the increase of immersion time. Furthermore, the temperature range for glass transition widens. Relationships between the T_g of the SMP composites with different carbon powder contents and

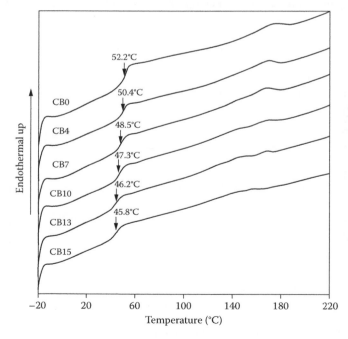

FIGURE 5.4 DSC curves of dry SMP composites upon heating. (Reprinted from Yang B, Huang WM, Li C et al. *European Polymer Journal*, 41, 1123–1128, 2005. With permission.)

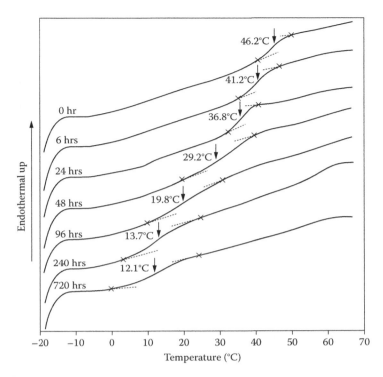

FIGURE 5.5 DSC results for CB13 after different immersion hours in water. (Reprinted from Yang B, Huang WM, Li C et al. *European Polymer Journal*, 41, 1123–1128, 2005. With permission.)

immersion time are summarized in Figure 5.6. The T_g values of all samples fall with the increase of immersion time, following a similar trend. The decrease is significant at the beginning and then becomes mild for immersion times over 240 hours—consistent with the increase of moisture shown in Figure 5.2. With more moisture absorbed into SMP and its composites, the T_g decreases further. This phenomenon persists until samples are saturated. Figure 5.6 also demonstrates that the added carbon powders have the tendency to lower the T_g. This is more obvious if the water content is high. For instance, the T_g of pure polyurethane SMP (CB0) at the saturation point (720 hours) is about 10°C higher than that of all other samples.

5.5 EVOLUTION OF GLASS TRANSITION TEMPERATURE UPON THERMAL CYCLING

The results from the previous sections show that the absorbed moisture can significantly decrease the T_g of electrically conductive SMP composites. To know whether T_g can be restored to its original value by dehydration, cyclic DSC tests were carried out on the SMP composites after immersion in water following the same testing procedure described in Section 3.4.

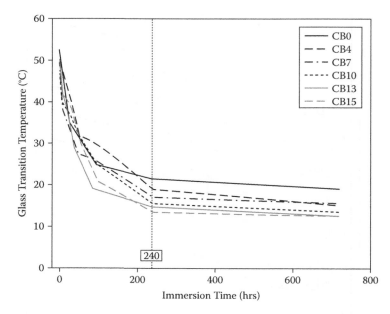

FIGURE 5.6 T_g versus immersion time. (Reprinted from Yang B, Huang WM, Li C et al. *European Polymer Journal*, 41, 1123–1128, 2005. With permission.)

Figure 5.7 plots one typical set of results showing the evolution of T_g upon thermal cycling of CB13. Clearly, after six different thermal cycles all the specimens regain their original T_g values. In general, the regaining process is more significant in the temperature range from 140 to 180°C. Figure 5.8 summarizes the evolution of T_g upon thermal cycling in all saturated SMP composites. Again, it shows that the most significant recovery of T_g occurs between 140 and 180°C. Further heating beyond the melting point (~200°C) can bring the T_g almost fully back to its original value. Recall that a significant part of the moisture is removed at temperatures above 140°C (Figure 5.1). The increase in T_g may be attributed directly to the evaporation of moisture. Because the moisture is removable, the whole process is reversible, and the T_g can be reduced again if samples are immersed into water again.

5.6 CORRELATION OF MOISTURE ABSORPTION, LOADING OF CARBON POWDER, AND GLASS TRANSITION TEMPERATURE

Utilizing the data obtained from previous tests, namely the TGA and DSC, and following the approach described in Section 3.6, we can determine the relation between the T_g and the ratio of moisture to SMP and its composites in weight percentage for all samples in the immersion and heating processes. Figure 5.9 shows two typical results. Similar slanted, L-shaped curves between the transition temperature and water content appeared for all samples upon immersing and subsequent heating. As indicated, the change in T_g can be obviously divided into two stages. The critical water content (water ratio in weight percentage) for dividing the result is ~120°C for

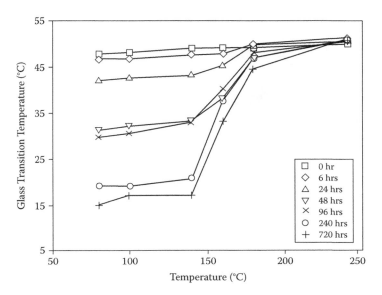

FIGURE 5.7 Evolution of T_g upon thermal cycling in CB13. (From Yang B, 2007. Influence of moisture in polyurethane shape memory polymers and their electrically conductive composites. PhD dissertation, Nanyang Technological University, Singapore.)

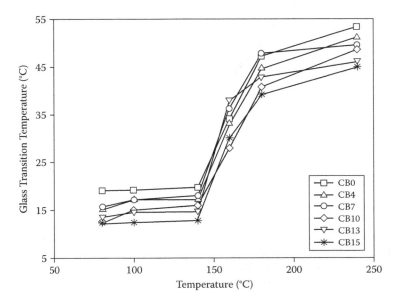

FIGURE 5.8 Evolution of T_g upon thermal cycling in saturated samples. (Reprinted from Yang B, Huang WM, Li C et al. *European Polymer Journal*, 41, 1123–1128, 2005. With permission.)

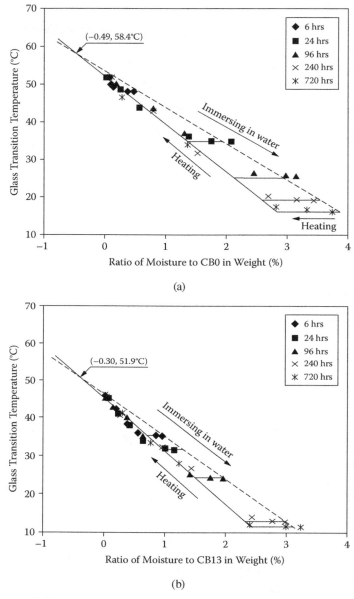

FIGURE 5.9 The T_g versus ratio of moisture to SMP (composite) in weight percentage. (a) CB0 (Reprinted from Huang WM, Yang B, An L et al. *Applied Physics Letters*, 86, 114105, 2005. With permission.) (b) CB13 (Reprinted from Yang B, Huang WM, Li C et al. *Scripta Materialia*, 53, 105–107, 2005. With permission.)

TABLE 5.1

Ratio of Moisture Evaporated over 240°C in SMP Composites

SMPs and Composites	CB0	CB4	CB7	CB10	CB13	CB15
Ratio of moisture evaporated over 240°C (%)	0.49	0.49	0.48	0.38	0.30	0.20

Source: Yang B, 2007. Influence of moisture in polyurethane shape memory polymers and their electrically conductive composites. PhD dissertation, Nanyang Technological University, Singapore.

all cases if the ratios are converted using the water weight fraction at 240°C as the reference. The total absorbed water can be divided into free water and bound water.

Two linear lines were obtained to describe the relationships of T_g versus the ratio of moisture in the immersion and heating processes. These lines intersect in the minus zone of moisture content corresponding to the true dry state of the polyurethane SMP. The ratios of moisture to composites in weight at the positive valued intersection R_t points are shown in Table 5.1 for SMP and its conductive composites. After conversion using Equation (3.2) and the values listed in Table 5.1, the linear relationships of the two ratios (bound water to SMP and total moisture ratio to SMP) are constructed by data fitting and plotted as in Figure 5.10.

Subsequently, the T_g of each polyurethane SMP composite at the true dry and saturated states are obtained. Figure 5.11 plots the T_g of SMP as a function of the volume fraction of carbon powder. It reveals that the T_g levels at the real dry and saturated

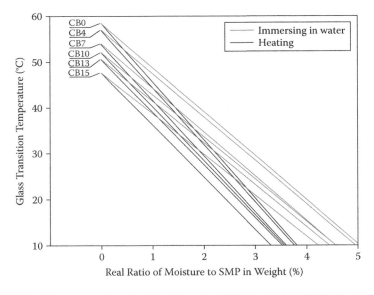

FIGURE 5.10 T_g versus real ratio of moisture to SMP in weight. CB13 (Reprinted from Yang B, Huang WM, Li C et al. *Scripta Materialia*, 53, 105–107, 2005. With permission.)

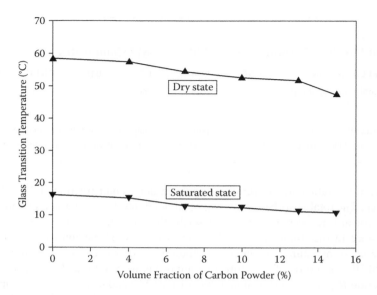

FIGURE 5.11 T_g versus volume faction of carbon powder. (From Yang B, 2007. Influence of moisture in polyurethane shape memory polymers and their electrically conductive composites. PhD dissertation, Nanyang Technological University, Singapore.)

states decrease with the increase of carbon nano powder. However, in general, the loading of carbon powder has almost no effect on the amplitude of decreases in T_g.

Figure 5.12 plots the moisture ratio in the saturated state against the content of carbon powder. The loading of carbon powder had a negligible effect on the ratio of moisture at the saturated state. Generally, SMP and its electrically conductive composites can absorb about 4.3% moisture in weight when saturated in room-temperature water.

Figure 5.13 presents the slopes of the lines in Figure 5.10 as functions of volume fraction of carbon powder for immersion and heating processes, respectively. It reveals that all slopes almost linearly increase with the increase of carbon nano powder. In other words, with the increase of carbon nano powder content, water has less influence on T_g. Furthermore, the difference between the slopes of water immersion and heating becomes smaller with the increase of carbon powder.

5.7   EFFECTS OF MOISTURE ON THERMOMECHANICAL PROPERTIES

5.7.1   Damping Capability

A dynamic mechanical analysis (DMA) was carried out at a constant frequency of 1 Hz on a DMA 2980 (TA Instruments) in film mode. After different lengths of immersion in room-temperature water, strip samples 15.0 mm long, 3.0 mm wide, and 0.5 mm thick were heated at a rate of 4°C/minute.

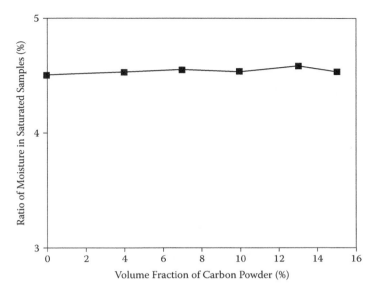

FIGURE 5.12 Ratio of moisture in saturated samples versus volume fraction of carbon powder. (From Yang B, 2007. Influence of moisture in polyurethane shape memory polymers and their electrically conductive composites. PhD dissertation, Nanyang Technological University, Singapore.)

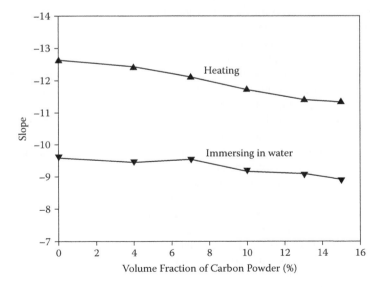

FIGURE 5.13 Slopes of two theoretical lines versus volume fraction of carbon powder. (From Yang B, 2007. Influence of moisture in polyurethane shape memory polymers and their electrically conductive composites. PhD dissertation, Nanyang Technological University, Singapore.)

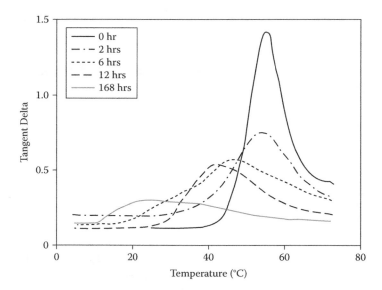

FIGURE 5.14 Tangent delta as function of temperature in CB0. (From Yang B, 2007. Influence of moisture in polyurethane shape memory polymers and their electrically conductive composites. PhD dissertation, Nanyang Technological University, Singapore.)

A typical result of tangent delta as a function of temperature (CB0) is plotted as shown in Figure 5.14. Note that tangent delta is the ratio of the loss modulus to the storage modulus that represents the damping capability of a material under shock or vibration. Figure 5.14 shows that tangent delta generally reaches a maximum near T_g in this polyurethane SMP. The maximum tangent delta of CB0 without water immersion is about 1.4, which is much higher than that of a typical high damping rubber. Therefore, CB0 can be used as a superior damping material near its T_g value. Furthermore, this result reveals that the maximum tangent delta of CB0 decreases significantly with the increase in immersion time. The immersion time and the moisture absorbed by the material are directly related. Therefore, the decrease in damping capability can be directly attributed to the increase of moisture ratio in the material.

Figure 5.15 summarizes the change in maximum tangent delta with the ratio of moisture for the SMP and its composites. Note that the ratio of moisture is determined as described in Section 3.2. Generally, the loading of carbon powder remarkably lowers damping capability. On the other hand, upon gradually absorbing moisture, the maximum tangent delta of SMP and its composites decreases to ~0.2. The maximum tangent delta of CB0 almost linearly decreases with the increase of moisture content. However, with the loading of carbon powders, the decrease in the maximum tangent delta becomes less significant. Especially in CB15, only a minor change occurred in the maximum tangent delta.

To study whether all materials can recover their original mechanical properties after immersion in room-temperature water, two groups of saturated samples were dried in a vacuum oven for 12 hours. According to the previous results, 120°C is the critical point dividing the absorbed water into free and bound water. Thus, two

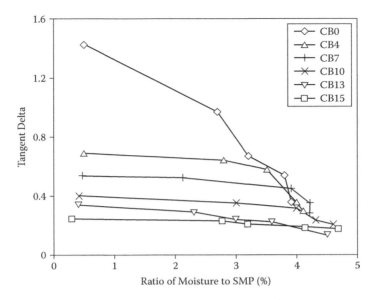

FIGURE 5.15   Change in maximum tangent delta with moisture content. (From Yang B, 2007. Influence of moisture in polyurethane shape memory polymers and their electrically conductive composites. PhD dissertation, Nanyang Technological University, Singapore.)

temperatures, namely, 80 and 120°C, were chosen for drying in a vacuum oven. We expected that upon drying at 80°C for 6 hours, only free water can be evaporated out, while all free water and some bound water can be removed upon drying at 120°C for 6 hours. The typical results of a saturated sample upon drying are shown in Figure 5.16. In general, the saturated SMP and its composite only partially regained their damping capabilities upon the removal of free water. However, after drying at 120°C for 6 hours, their damping capabilities were remarkably regained.

5.7.2 Young's Modulus

Based on the dynamic mechanical analysis (DMA) results, the Young's modulus of the material is obtained as a function of temperature. One typical result (CB0) is presented in Figure 5.17. The difference in the curves of Young's modulus against temperature with different immersion times is apparent.

In this polyurethane SMP, the temperature range for glass transition is about 30°C. It can be assumed that the material is in the glass state below $T_g - 15°C$ and in the rubber state above $T_g + 15°C$. The influence of temperature on Young's modulus in SMPs in a glass or rubber state is minor and it appears that the Young's modulus of the SMP in the glass or rubber state is constant (Tobushi et al. 2001). Normally, in this SMP, $T_g - 15°C$ is considered the Young's modulus in the glass state and $T_g + 15°C$ is considered the Young's modulus in the rubber state. For comparison, the Young's moduli of all samples at $T_g - 15°C$ and $T_g + 15°C$ are summarized against moisture content in Figure 5.18. Note that T_g is the temperature corresponding to the peak of

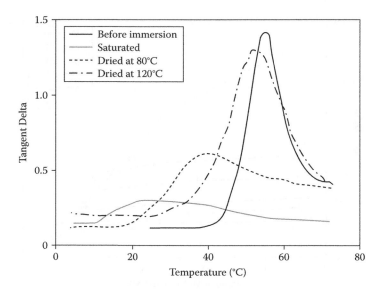

FIGURE 5.16 Tangent delta as a function of temperature in saturated CB0 upon drying. (From Yang B, 2007. Influence of moisture in polyurethane shape memory polymers and their electrically conductive composites. PhD dissertation, Nanyang Technological University, Singapore.)

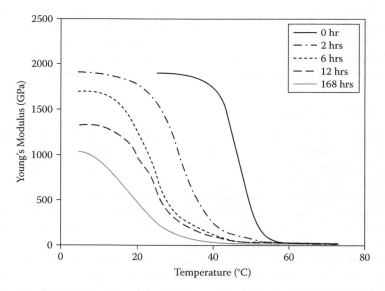

FIGURE 5.17 Young's modulus as a function of temperature in CB0. (From Yang B, 2007. Influence of moisture in polyurethane shape memory polymers and their electrically conductive composites. PhD dissertation, Nanyang Technological University, Singapore.)

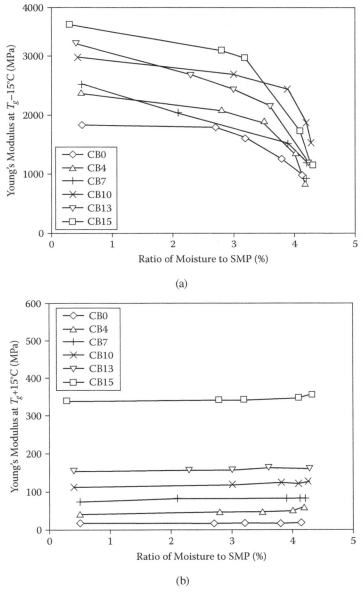

(a)

(b)

FIGURE 5.18 Change in Young's modulus at different moisture contents. (a) At $T_g - 15°C$. (b) At $T_g + 15°C$. (From Yang B, 2007. Influence of moisture in polyurethane shape memory polymers and their electrically conductive composites. PhD dissertation, Nanyang Technological University, Singapore.)

the tangent delta from the DMA testing. The Young's modulus of the SMP and its composites in the glass state decreases with the increase of moisture content. The decrease may be attributed to the moisture that interacts with the polymer chains and plasticizes the polymer. However, negligible change was noted in the Young's modulus in the rubber state.

Figure 5.19 presents one typical result of a saturated sample after drying at 80 and 120°C. It shows that the removal of free water in the sample results in the partial recovery of Young's modulus. Upon drying at 120°C for 6 hours its Young's modulus is remarkably regained.

5.7.3 UNIAXIAL TENSION BEHAVIOR

Uniaxial tensile tests were carried out on SMP wires with 1.0 mm diameters using an Instron 5565 machine with a 100 N load cell at a constant strain rate of 5×10^{-3}/s. The gauge length was 40.0 mm. A hot chamber with a tolerance of ±0.5°C was attached to the Instron for temperature control. Two testing temperatures were chosen: 22°C room temperature and $T_g + 25$°C at which the polyurethane SMP is in the rubber state. In these experiments, the T_g of an individual sample was determined by a DSC.

Figure 5.20a plots one set of stress-versus-strain curves of SMP wires (CB0) tested at room temperature and shows two types of distinct behaviors. When the immersion time is shorter than 12 hours, the SMP behaves like a crystalline material, in which a linear stress-versus-strain relation is found prior to yielding followed by hardening.

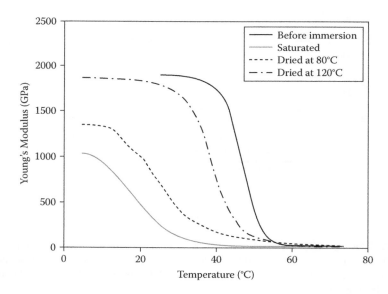

FIGURE 5.19 Young's modulus as function of temperature in saturated CB0 upon drying. (From Yang B, 2007. Influence of moisture in polyurethane shape memory polymers and their electrically conductive composites. PhD dissertation, Nanyang Technological University, Singapore.)

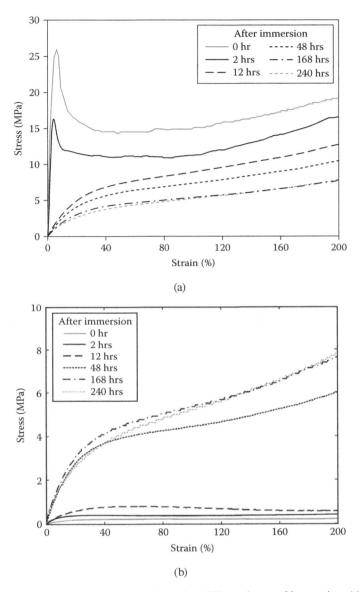

FIGURE 5.20 Stress-versus-strain relation after different hours of immersion. (a) At room temperature. (b) At T_g + 25°C. (From Yang B, 2007. Influence of moisture in polyurethane shape memory polymers and their electrically conductive composites. PhD dissertation, Nanyang Technological University, Singapore.)

When the immersion time exceeds 12 hours, the stress-versus-strain curve is similar to those of rubbers. This curve has no apparent yielding point and shows that a large deformation can be achieved at a low level of stressing. The rubber-like behavior becomes more pronounced over a prolonged immersion period due to the increased mobility of the polymer chains. On the other hand, if the uniaxial tensile test is carried out at 25°C above T_g, the SMP becomes more rigid after a longer immersion period as shown in Figure 5.20b.

To find the individual roles of free water and bound water in stress-versus-strain relations, additional uniaxial tensile tests were conducted. The SMP wires were immersed in water for different numbers of hours, and then heated in air up to 80 and 120°C, respectively, at a constant rate of 10°C/minute. Subsequently, the wires were cooled back down to room temperature in 3 minutes. No apparent change in the stress-versus-strain behavior was found in the SMP wires pre-heated to 80°C. The stress-versus-strain curves of the wires pre-heated to 120°C are presented in Figure 5.21. According to Figure 5.9, 120°C is the critical point for full removal of the free water from the SMP.

Figure 5.21a shows a remarkable change in comparison with the results in Figure 5.20a. All samples exhibit a linear stress-versus-strain relation followed by hardening regardless of the immersion duration. The stiffness of the wires obviously increased after the removal of the free water. Apart from this, a very slight increase in T_g is observed after heating to 120°C (Figure 5.9). However, in a tensile test conducted at T_g + 25°C (Figure 5.21b), the removal of free water has little significant

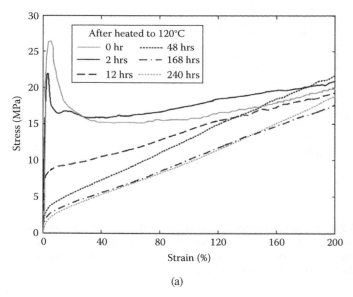

(a)

FIGURE 5.21 Stress-versus-strain relation after different hours of immersion and heating to 120°C. (a) At room temperature. (b) At T_g + 25°C. (From Yang B, 2007. Influence of moisture in polyurethane shape memory polymers and their electrically conductive composites. PhD dissertation, Nanyang Technological University, Singapore.)

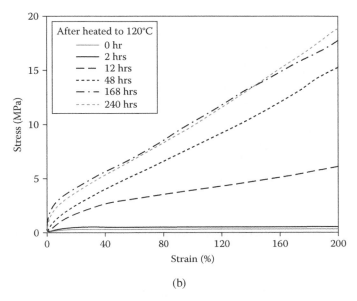

After heated to 120°C
——— 0 hr
——— 2 hrs
— — 12 hrs
······ 48 hrs
—·— 168 hrs
----- 240 hrs

(b)

FIGURE 5.21   (Continued)

influence apart from the stiffening of the wires after immersion exceeding 12 hours. We may conclude that free water exerts only a very limited effect on the tensile behavior of the SMP and bound water has a very strong influence.

5.7.4 MOISTURE-RESPONSIVE SHAPE RECOVERY

Chapter 3 proposed a new feature of polyurethane SMPs: moisture-responsive shape recovery. This feature is also applicable to electrically conductive SMP composites. Figure 5.22 shows the recovery of a CB7 sample actuated by room-temperature water. This section addresses the moisture-responsive shape recovery properties of SMP and its composites that are vital for the practical use of this material.

Moisture–water-actuated free recovery and moisture–water-actuated constrained recovery tests were performed. Two groups of samples were prepared and pre-strained to 20% following the procedure described in Section 4.8. Subsequently, the samples were clamped inside an Instron 5569 with a 100 N load cell. A water tank was attached to the Instron with proper sealing and the samples were immersed completely in room-temperature water for recovery without heating, as shown in Figure 5.23. One group of samples was unconstrained and could vary their lengths. The samples were taken out of water and their changes in length measured after a certain period using a microscope. The recovered strain was calculated and recorded along with immersion times; this is the water-actuated free recovery test. The second group of samples was tested under the same conditions except that their lengths were fixed. The recovery stress was recorded against immersion time via the Instron 5569; this is the water-actuated constrained recovery test.

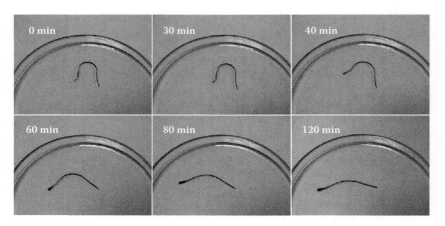

FIGURE 5.22 Recovery of a CB7 sample actuated by room-temperature water. (Reprinted from Yang B, Huang WM, Li C et al. *European Polymer Journal*, 41, 1123–1128, 2005. With permission.)

In Figure 5.24, the recovered strain in the free recovery test is plotted against immersion time. It shows that after immersion in room-temperature water for 60 minutes, all SMPs began to recover. CB0 recovered about 18% strain—about 90% of the preloaded maximum strain. However, with the increase of carbon powder content, the recovered strain decreased and the recovery slowed.

Figure 5.25 presents the relationship between recovery stress and immersion time in the constrained recovery test. In general, recovery stress begins to increase after 60 minutes of immersion and then gradually reaches a maximum. After that, it decreases slowly. About 1.8 MPa stress can be generated in CB0 with constrained recovery. The loading of carbon powders has a negligible effect on maximum recovery stress. However, the time needed for the SMP composite to reach the maximum recovery stress becomes longer with the increase of carbon powder content. After

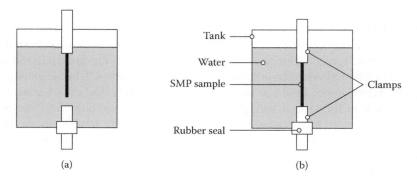

FIGURE 5.23 Experimental setup of water-actuated recovery tests. (a) Free recovery. (b) Constrained recovery. (From Yang B, 2007. Influence of moisture in polyurethane shape memory polymers and their electrically conductive composites. PhD dissertation, Nanyang Technological University, Singapore.)

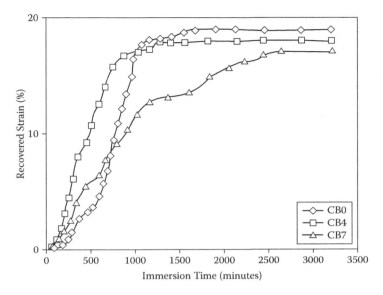

FIGURE 5.24   Recovered strain versus immersion time. (From Yang B, 2007. Influence of moisture in polyurethane shape memory polymers and their electrically conductive composites. PhD dissertation, Nanyang Technological University, Singapore.)

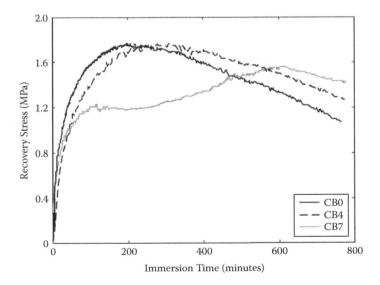

FIGURE 5.25   Recovery stress versus immersion time. (From Yang B, 2007. Influence of moisture in polyurethane shape memory polymers and their electrically conductive composites. PhD dissertation, Nanyang Technological University, Singapore.)

reaching its maximum, recovery stress decreases less remarkably with the increase of carbon powder content. This may be attributed to the constraint of carbon powder to the relaxation of polymer.

5.8 SUMMARY

A series of experiments were conducted to study the effects of moisture on electrically conductive SMPs. Experimental results indicate that the electrical conductivity of conductive SMPs is improved by moisture. This phenomenon is more obvious in SMPs filled with lower contents of carbon powder. The T_g values of conductive SMP composites can be reduced effectively by immersion in water. The T_g continuously drops until the polymer is saturated. Furthermore, the change in T_g is reversible upon heating or dehydration because the removal and absorption of moisture are physical processes.

The effects of carbon nano powder and moisture on the T_g of conductive SMP composites were separated and qualitatively investigated by a series of TGA and cyclic DSC tests. The relationship of T_g versus the real ratio of moisture to SMP in weight of each SMP composite was obtained. The added carbon powders tended to reduce the T_g. The bound water in the SMP material reduced the T_g dramatically in a linear fashion, while the influence of free water on the T_g was negligible. The T_g in the real dry state decreased with the increase of carbon nano powder in the material. With more carbon nano powder, the T_g of SMP became less sensitive to water. Overall, the T_g is affected by the presence of bound water more than the loading of carbon nano powder.

The damping capabilities of conductive SMP composites worsen upon immersion in water. Young's moduli of SMP composites in the glass state decrease with increases in water content. However, little change in Young's modulus occurred in the rubber state upon immersion in water. Upon dehydration, the conductive SMP composites remarkably regained their original damping capabilities and Young's moduli.

Conductive SMP composites can recover pre-strain by immersion in room-temperature water based on a decrease of their T_g values. The water-actuated recoverable strain decreases with the increase of carbon powder content. However, the carbon powder has negligible influence on the maximum stress generated by water-actuated recovery.

ACKNOWLEDGMENT

We wish to thank N.W. Khun for his help in compiling this chapter.

REFERENCES

Azulay D, Eylon M, Eshkenazi O et al. (2003). Electrical–thermal switching in carbon black–polymer composites as a local effect. *Physical Review Letters*, 90, 236601/1–236601/4.
Bhattacharya SK (1986). *Metal-Filled Polymers: Properties and Applications*, Marcel Dekker, New York.

Huang WM, Yang B, An L et al. (2005). Water-driven programmable polyurethane shape memory polymer: demonstration and mechanism. *Applied Physics Letters*, 86, 114105.

Sheng P, Sichel and EK, Gittleman JI (1978). Fluctuation-induced tunneling conduction in carbon-polyvinylchloride composites. *Physical Review Letters*, 40, 1197–1200.

Tobushi H, Okumura K, Hayashi S et al. (2001). Thermomechanical constitutive model of shape memory polymer. *Mechanics of Materials*, 33, 545–554.

Yang B (2007). Influence of moisture in polyurethane shape memory polymers and their electrically conductive composites. PhD dissertation, Nanyang Technological University, Singapore.

Yang B, Huang WM, Li C et al. (2005a). Qualitative separation of the effects of carbon nano powder and moisture on the glass transition temperature of polyurethane shape memory polymer. *Scripta Materialia*, 53, 105–107.

Yang B, Huang WM, Li C et al. (2005b). Effects of moisture on the glass transition temperature of polyurethane shape memory polymer filled with nano carbon powder. *European Polymer Journal*, 41, 1123–1128.

6 Magnetic and Conductive Polyurethane Shape Memory Polymers

6.1 INTRODUCTION

It is well known that heat can be generated when an alternative magnetic field is applied to magnetic particles. This heating technique is called inductive heating. According to Buckley et al. (2006), the Curie temperature T_c of a magnetic particle may be adjusted by changing its composition. Magnetic implant materials treated with this heating technique have been used for interstitial hyperthermia (Shimizu and Matsui 2003). In addition, functionalized magnetite (Fe_3O_4) nano particles have great potential for many bioapplications (Majewski and Thierry 2007). However, we should bear in mind that according to Keblinski et al. (2006), a certain size limit applies to such nano particles for effective inductive heating.

By blending thermo-responsive shape memory polymer (SMP) with magnetic particles, shape recovery can be triggered by inductive heating (Mohr et al. 2006, Razzaq et al. 2007). An additional advantage of this heating technique is that the maximum temperature can be controlled. This is important because magnetic particles lose their magnetic properties above T_c and the inductive heating process stops (Aphesteguy et al. 2009).

This heating technique provides an alternative for achieving real remote and wireless actuation of thermo-responsive SMPs that have a wide range of applications, in particular in biomedical devices for minimally invasive surgery (Buckley et al. 2006). This chapter presents a systematic study of the influences of magnetic particles (Fe_3O_4 and Ni) on the thermomechanical properties of SMP–magnetic particle composites. Furthermore, we explore some unconventional applications of magnetic particles, for example, to alter the surface roughness and morphology and enhance the electrical conductivity of SMPs. Unless otherwise stated, the polyurethane SMP discussed in this chapter is from SMP Technologies, Japan.

6.2 IRON OXIDE MICRO PARTICLES

We first studied the influence of moisture on SMPs filled with Fe_3O_4 micro particles, then showed how to form Fe_3O_4 micro particle chains inside SMPs. Finally, the feasibility of using Fe_3O_4 micro particles for surface roughness and morphology modification was demonstrated. The iron oxide powder (Fe_3O_4, Sigma-Aldrich) used had a specific density of 5.0 g/cm^3, average size <5 μm, and purity of 98%.

6.2.1 Influence of Moisture on T_g

We used MM5520 pellets (nominal T_g: 55°C) in this study. The specific density of this SMP is 1.25 g/cm^3. A Haake Rheomix 600 mixer controlled by a Haake Rheocord 90 was used to mix SMP and iron oxide powder at 200°C at a rotor speed of 3 rpm for 20 minutes. Subsequently, a hot-press machine (Caver) was used to compress the mixture at 200°C into 0.5 mm thickness films. Samples of 10 × 20 mm were cut from the thin films for various experiments.

Figure 6.1 shows cryofracture surfaces of samples with 30 vol.% and 5 vol.% of iron oxide powder. At 5 vol.% of Fe_3O_4, some tiny bright areas are visible. They are more or less single Fe_3O_4 particles that are well distributed. However, at 30 vol.% of Fe_3O_4, instead of single particles, Fe_3O_4 appears as large aggregates. We can even see many holes indicating poor dispersion of Fe_3O_4 within an SMP with a high content of Fe_3O_4.

For simplicity, we assume that all particles are of the same size, well separated, and uniformly distributed. Consider a unit cell containing particles in four different types of configurations as illustrated in Figure 6.2 (top). We can convert them into a standard morphology for easy comparison as shown. The $2t$ indicates the minimum distance between particles, and D is the dimension of a unit cell. Figure 6.2 (bottom) shows the t/D versus volume fraction of particle relationships. We can clearly see that in all configurations, at about 30 vol.% of particles, t/D is about 1/10, indicating that the particles are very close to each other and thus aggregates are very likely to form due to surface tension at the surfaces of the particles, in particular the micron- and nano-sized particles. At 5 vol.%, the particles are distant from each other. They may be well separated and distributed within a matrix.

The experimental procedure and characterization results are as follows. Samples were immersed in room-temperature water (about 22°C) for specific durations of time, ranging from 0 to 720 hours. Cyclic differential scanning calorimeter (DSC) testing (TA DSC 2920) was conducted at a heating and cooling rate of 20°C/minute. Figure 6.3 shows a typical DSC result. Thermogravimetric analysis (TGA, Perkin-Elmer TGA-7) was also conducted at the same 20°C/minute heating rate until 600°C was reached.

Following the same approach as in Chapter 3, we calculated the T_gs and weight fractions (WFs) of absorbed moisture in samples after different hours of immersion into water as seen in the left column of Figure 6.4. The results are not so consistent, which should largely be attributed to the possible nonuniformity of Fe_3O_4, particularly in heavily loaded samples. The general trend is that with the increase of Fe_3O_4 content, the T_g of the composite decreases. On the other hand, as expected, with an increase in immersion time, more water is absorbed, causing a decrease in T_g. In the right column of Figure 6.4, the evolution of T_g during thermal cycling is plotted for all samples. Despite fluctuations, we can see clearly the general trend that the T_g increases gradually upon thermal cycling around 125°C and above. Beyond ~180°C, the T_g returns to the initial value (right before immersion). In comparison with carbon

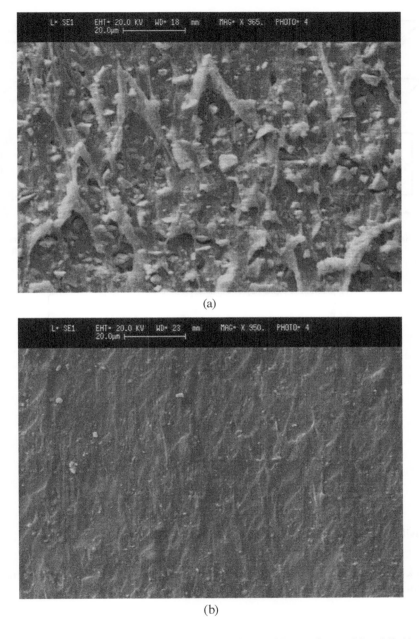

(a)

(b)

FIGURE 6.1 Morphologies of SMPs filled with iron oxide powders at 30 vol.% (a) and 5 vol.% (b). The scale bar length is 20 μm.

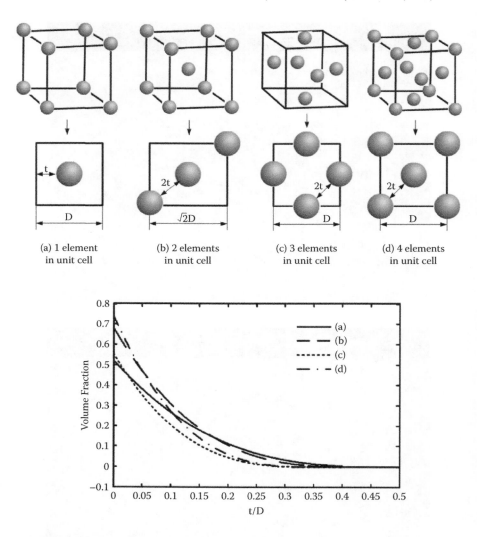

FIGURE 6.2 Relationship of *t/D* versus volume fractions of particles.

black–filled samples, it appears that the moisture absorption ratio in Fe_3O_4 micro-particle-filled samples is far less.

6.2.2 ALIGNMENT OF IRON OXIDE MICRO PARTICLES

To avoid the dispersion problem in mixing as discussed in Section 6.2.1, we used a polyurethane (PU) resin solution (MS 5510 with 30% solids content and 70% dimethylformamide [DMF] solution) to investigate the feasibility of forming Fe_3O_4 micro particle chains. In addition, a low volume fraction of Fe_3O_4 micro particles (5 vol.% or less) was applied to eliminate the aggregation problem.

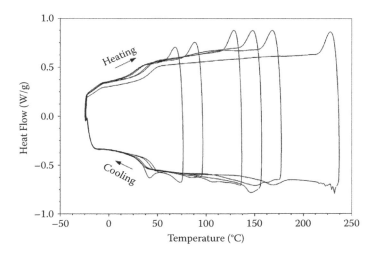

FIGURE 6.3   Cyclic DSC result from 5 vol.% micro Fe_3O_4 particles after 240-hour immersion.

A glass slide was placed on the top of two pieces of square magnetic rods that were spaced ~5 mm apart to generate a uniform magnetic field. SMP–DMF solution was mixed with different volume fractions of Fe_3O_4 micro particles, ultrasonically stirred for uniform scattering. The liquid mixture was then spread on the top surface of the glass slide, where the magnetic field was uniform. After different periods of holding time, the glass slide was heated to 80°C to further dry the thin film of SMP mixture. The thicknesses of the thin films were about 50 μm as determined by a Wyko interferometer.

Figure 6.5 reveals the progress in the formation of magnetic particle chains at four different volume fractions of Fe_3O_4 micro particles. At a low content of magnetic particles, i.e., 0.2 vol.%, the chains did not form as quickly as in other samples, and they are discontinuous even after 2 days (maximum length is ~150 μm, width is ~1 μm). In all cases, the chains were short and narrow at the beginning and gradually become longer and wider. Long, continuous chains were seen in samples with 1 vol.% or more of magnetic particles. Higher magnetic particle content requires less time to form continuous chains and achieves wider chains.

Instead of forming magnetic chains before a solution is dried, an alternative is to dry and cure the solution first and then form magnetic chains at high temperatures (above the melting temperature of SMP) and apply a magnetic field. This procedure was also used to prepare SMP thin films with different magnetic particle contents atop glass slides. Because no magnetic field was applied during curing, the magnetic particles were more or less evenly distributed in all films. Subsequently, the glass slides were placed atop two magnets and heated to 200°C, which is above the melting temperature of this SMP. As we can see in Figure 6.6, magnetic chains formed gradually more or less similarly to the manner discussed above. Because a melted SMP is more viscous than an SMP-DMF solution, it takes more time for chains to form.

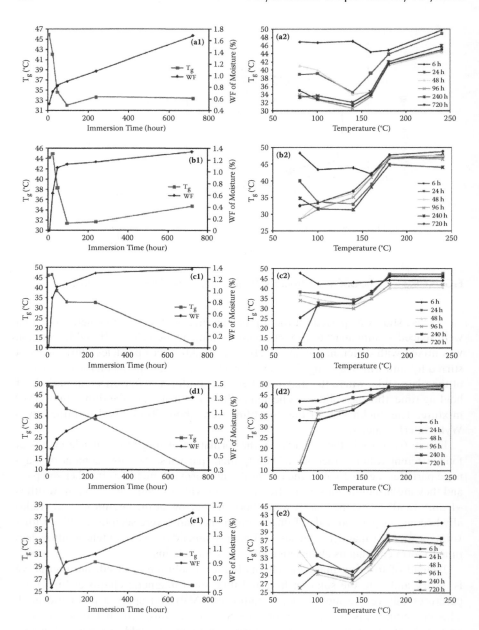

FIGURE 6.4 T_g and its evolution upon thermal cycling: (a) 5 vol.%; (b) 10 vol.%; (c) 18 vol.%; (d) 25 vol.%; (e) 35 vol.%. Left: immersion time versus T_g and weight fraction (WF) of moisture (%). Right: evolution of T_g upon thermal cycling.

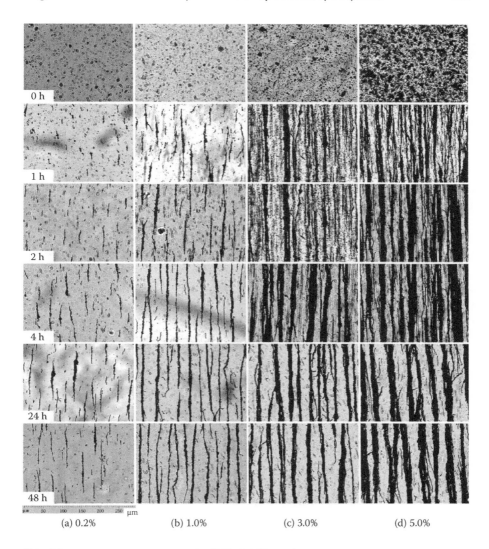

(a) 0.2%   (b) 1.0%   (c) 3.0%   (d) 5.0%

FIGURE 6.5 Formation of chains in SMP–DMF solution.

Inspired by the above study, we heated the thin film samples produced in the above test (samples containing magnetic chains) to 200°C again but the magnetic field direction was switched by 90 degrees. Figure 6.7 shows that the magnetic chains also switch by 90 degrees following a sequence of break-ups of existing chains and regrouping into new chains. A closer look reveals that the higher the content of magnetic particles, the longer the time required for new chains to form. In addition, the new chains are shorter and of lower quality than the original chains. This is because the original chains break up at certain intervals and form new ones. The availability of magnetic particles from the broken chains restricts the quality of new chains.

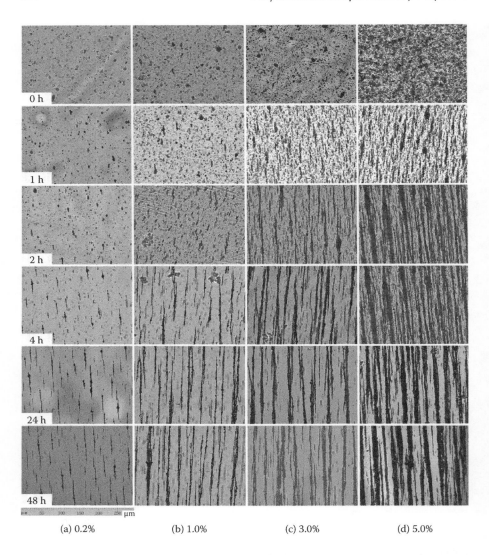

| | (a) 0.2% | (b) 1.0% | (c) 3.0% | (d) 5.0% |

FIGURE 6.6 Formation of chains by heating to 200°C.

6.2.3 Altering Surface Roughness and Morphology

We have seen the mobility of magnetic particles above the melting temperature of SMP material when a magnetic field is applied. We now demonstrate a simple approach to alter the surface roughness and morphology of thin film SMPs using magnetic particles.

In the first test, pure (MS5510) thin SMP films ~5 μm thick were fabricated atop a glass slide. After curing, the surface was scanned using a Wyko interferometer.

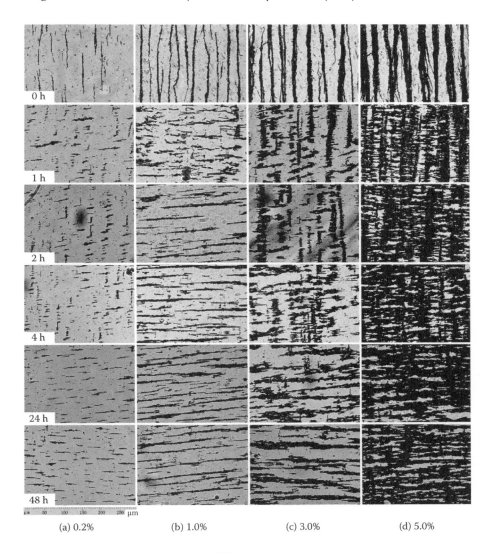

(a) 0.2% (b) 1.0% (c) 3.0% (d) 5.0%

FIGURE 6.7 Switching of chains at 200°C.

Figure 6.8a reveals the typical three-dimensional surface morphology of a thin film. The average surface roughness (R_a) and root mean square surface roughness (R_q) were 12.48 nm and 17.23 nm, respectively. Subsequently, a thin layer of magnetic particles was placed atop the thin film. After heating to 80°C (above the T_g of MS5510), a magnet was placed beneath the glass slide. After 1 hour of heating, the thin film and the magnet were cooled back to room temperature. Figure 6.8b reveals the surface after modification. As we can see, both R_a and R_q almost doubled.

In the second test, SMP thin films (about 5 μm thick) with 0.2 wt.% of magnetic particles were prepared atop a glass slide. The top surface of the SMP thin film was

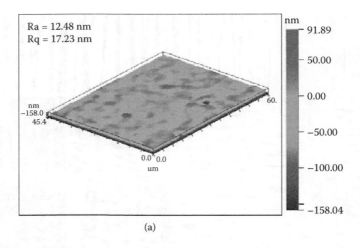

(a)

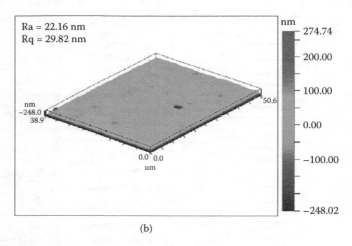

(b)

FIGURE 6.8 Surface morphologies. (a) Original thin film. (b) Film after surface modification.

coated with a thin layer of gold (about 20 nm) to prevent the pull-out of the magnetic particles. Figure 6.9a shows the morphology of one peak and indicates a group of magnetic particles beneath the peak revealed by the Wyko interferometer. After heating to 80°C, a magnetic field was applied to the glass slide along the vertical direction. After 10 minutes, the glass slide was cooled back to room temperature. Figure 6.9 reveals that the height of the peak increased to ~0.55 μm from the original height of ~0.46 μm. The increase in height results from the magnetic force that pulls the magnetic particles upward.

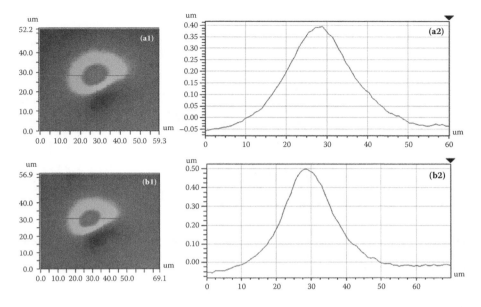

FIGURE 6.9 Morphology of peak. (a) Original shape. (b) Shape after modification. Left: top view. Right: cross-sectional view.

6.3 NICKEL MICRO AND NANO POWDERS

Fe_3O_4 particles are fully biocompatible and are thus ideal for use in biomedical applications. Although Fe_3O_4 is magnetic, it has extremely poor electrical conductivity. Nickel, however, is both magnetic and electrically conductive and can be used to achieve shape recovery of thermo-responsive SMPs by Joule heating.

6.3.1 ALIGNMENT OF NI POWDER

The same polyurethane SMP solution (MS5510) was used. Both micro- and nano-sized nickel particles (Sigma-Aldrich) were used. The average sizes of the micro Ni particles (99.8% purity) ranged from 3 to 7 µm; the average size of the nano Ni particles (99.0% purity) was 50 nm.

Each sample was prepared according to a specific weight ratio of Ni particles to SMP solution. Micro-sized Ni particles (weight ratios of 1, 3, 6, and 15%) and nano-sized Ni particles (weight ratios of 1, 5, and 15%) were mixed with SMP solution to form thin films. Instead of spreading the SMP solution mixed with Ni particles on a glass slide, aluminum molds with central cavities measuring 30 × 30 × 5 mm were used as containers. SMP solution (mixed with Ni powder) was poured into the cavities, which allowed good dimensional control of samples.

Permanent magnets 20 × 20 × 60 mm in size (GSN-35, Alniff Industries, Singapore) were used. A pair of magnets were placed 20 mm apart. A mold was

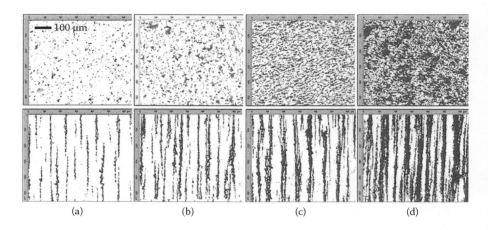

FIGURE 6.10 Magnetic chains formed using different weight ratios of micro-sized Ni powder: (a) 1 wt.% of Ni; (b) 3 wt.% of Ni; (c) 6 wt.% of Ni; (d) 15 wt.% of Ni. Top row: non-chained samples for comparison.

placed atop each pair of magnets. The magnetic flux was measured at 1.35×10^{-4} Wb at the cavity. The experimental steps were:

1. The well-stirred SMP solution–Ni powder mixture is poured into the mold.
2. The mold and magnets are heated to 80°C for 4 hours to cure the SMP. No magnets are used for non-chained samples.
3. A transmitted light microscope with a 5× magnification lens is used to study the chains in the SMP thin films.

Figure 6.10 presents microscope images of chained and non-chained samples with four different weight percentages of micro-sized Ni powder. Figure 6.11 shows microscope images of chained and non-chained samples with three different weight percentages of nano-sized Ni powder.

For the samples filled with micro-sized particles, most chains were 300 μm long and 10 μm wide. As with the samples filled with Fe_3O_4 powder, the lengths and widths of the Ni chains also increased with the weight percentage of Ni powder. As for the samples with nano-sized particles, at low Ni content (1 wt.%), the chain width was generally thinner than that of samples filled with micro-sized Ni. However, at high content of nano-sized Ni powder, the chains were wider. The gap between chains was smaller in the sample with 1 wt.% of nano-sized Ni in comparison with the sample of the same Ni content in micro size. At 10 wt.% of nano-sized Ni powder, the maximum width of Ni chains was ~100 μm—more than double the width of the sample with 15 wt.% of micro-sized Ni powder. This should be the result of higher surface tension in nano-sized Ni powder, which causes higher attractive force among the particles. In samples with high weight fractions of Ni, the possible maximum distance among nano particles is small and hence aggregates can form easily. Consequently, the chains are wider. However, at very low Ni content (1 wt.%), the nano-sized particles have higher mobility and are separated far apart after thorough stirring, which results in thin chains.

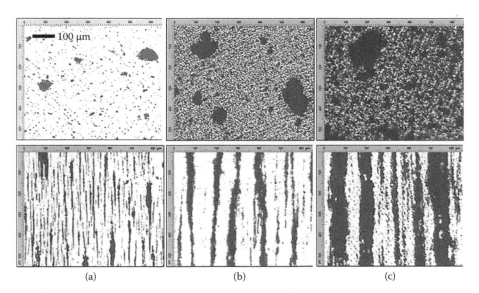

(a)   (b)   (c)

FIGURE 6.11   Magnetic chains formed using different weight ratios of nano-sized Ni powder: (a) 1 wt.% of Ni; (b) 5 wt.% of Ni; (c) 10 wt.% of Ni. Top row: non-chained samples for comparison.

As reported by Bansal et al. (2005), Kropka et al. (2008), and Oh and Green (2009), the distribution of particles, in particular nano-sized ones, can have significant influence on the T_g values of polymer composites. Figure 6.12 summarizes T_g levels of all types of samples obtained from dynamic mechanical analysis (DMA). Unlike DSC, DMA testing allowed us to determine the T_g of a sample along a

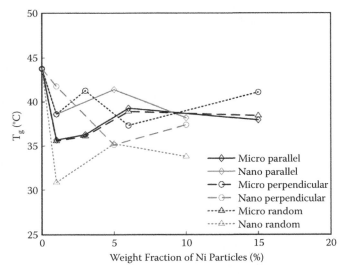

FIGURE 6.12   Weight fraction of Ni particles versus T_g relationships for micro- and nano-sized Ni powders.

particular direction, e.g., parallel or perpendicular to the chain direction. We define T_g as the peak point after differentiating the DMA curve with respect to the temperature. Although we cannot name the exact trend due to limited testing, it is clear that the T_gs of all composites were lower than those of the pure SMPs for both micro and nano Ni powder-filled, chained or non-chained samples. Further systematic study on this issue is necessary.

DMF was used to bond two layers of chained thin films with 20 wt.% micro-sized Ni particles to form a layered composite. First, a thin layer of DMF was spread atop one of the chained thin films. Subsequently, another piece of chained thin film was placed atop it with the chain direction 90 degrees to the direction of the bottom thin film. The pieces were then dried in an oven at 80°C for 60 minutes. Figure 6.13 reveals the resultant double-layered thin film that may exhibit isotropic electro-thermomechanical properties.

6.3.2 VERTICAL CHAINS

In addition to forming in-plane chains, one can produce out-of-plane vertical chains in a similar manner. Figure 6.14 illustrates a typical experimental setup to produce vertical chains. The magnet dimensions are $20 \times 20 \times 40$ mm. The gap between the magnet and the Petri dish is 1 mm. The strength of the applied magnetic field is 0.4 ± 0.05 T within the area of the Petri dish.

We used a few different low-volume fractions of micro-sized Ni powder in the experiments. After drying, vertical chain arrays were formed as shown in Figure 6.15. Figure 6.16 is a top view revealing that the vertical chain is actually a composite formed by Ni particles bonded with SMP. Figure 6.17 summarizes the relationships of the dimensions of chains (diameter and height; separation distance between chains) and volume fractions of Ni powder. We can see that the diameters and heights of the chains and separation distances between chains all increase with the increase of Ni content. The shape memory effect in these vertical chains is demonstrated in Figure 6.18. A single vertical chain was heated to a high temperature. Subsequently, a magnet was placed nearby to generate a magnetic field. Because the SMP was soft at high temperatures, the magnetic force can easily bend the vertical chain. After holding the magnet and cooling the chain back to room temperature, the chain became hard again, but was still curved even after the magnet was removed. The chain returned to its original vertical shape after it was reheated to the high temperatures. This experiment indicates that controlling the temperature and magnetic fields allows control of surface conditions by using these vertical SMP chains.

6.4 ELECTRICALLY CONDUCTIVE SMP

There are two types of electrically conductive polymers. One is intrinsically conductive polymers (ICPs). However, commonly used ICPs have typical conductivity values around only 10^{-10} to 10^5 S/m. Furthermore, they are not stable in terms of mechanical properties and are expensive and difficult to process, in particular

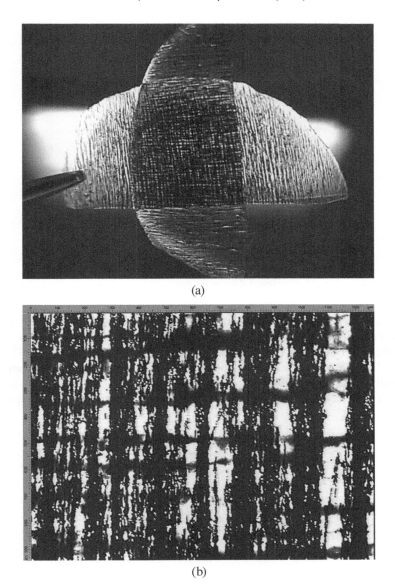

(a)

(b)

FIGURE 6.13   Double-layered thin film. (a) Overall view. (b) Zoom-in view. Area size: 650 × 550 μm.

through polymerization (Cotts and Reyes 1986, Morgan et al. 2001). The second type is a polymer doped or loaded with conductive fillers, such as graphite, carbon black, and metallic particles. Due to low cost and convenience in fabrication, the second approach is more popular in engineering practice (Jäger et al. 2001, Thommerel et al. 2002, Zheng and Wong 2003).

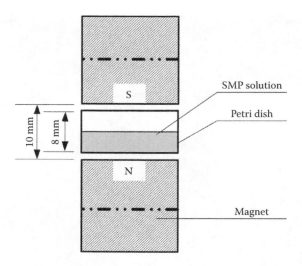

FIGURE 6.14   Typical experimental setup for producing out-of-plane vertical chains.

Electrically conductive SMPs have been developed by blending with various types of conductive fillers such as carbon nanotubes, carbon black, and carbon fibers (Beloshenko et al. 2005, Cho et al. 2005, Li et al. 2000, Sahoo et al. 2007, Singh 2005). To achieve good electrical conductivity, for example, for Joule heating for thermally induced actuation as in shape memory alloys (Huang 2002), the most effective approach is to align the particles into chains to form conductive channels.

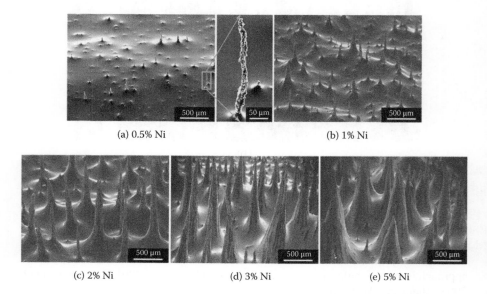

FIGURE 6.15   Typical arrays of vertical protrusive chains at different fractions of Ni content.

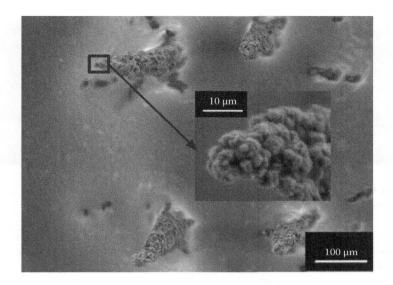

FIGURE 6.16   Top view of vertical chains (2% Ni). Inset: zoom-in view.

6.4.1 Ni Powder

We prepared SMP thin films with different volume fractions of Ni powder (Goodfellow Co.). We used the polyurethane SMP solution (MS5510) consisting of 30 wt.% polyurethane and 70 wt.% DMF. The Ni powder average size was 3 to 7 μm and purity was 99.8%.

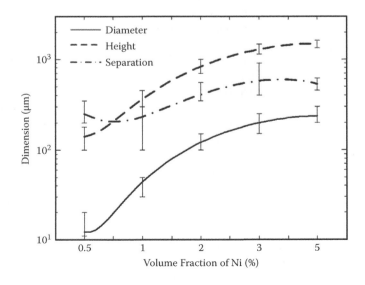

FIGURE 6.17   Dimensions of vertical chains.

FIGURE 6.18   Shape recovery of vertical chain. (Reprinted from Huang WM, Liu N, Lan X et al. *Materials Science Forum,* 614, 243–248, 2009. With permission.)

First, Ni powder was mixed into the solution and stirred. Subsequently, the highly viscous mixture was poured into a Petri dish (40 mm diameter) to a height of ~1.5 to 2.0 mm (Figure 6.19a). As shown in the figure, the Petri dish was placed 12 mm above two magnets (Neoflux® Nd-Fe-B, GSN-35, Alniff Industries, Singapore) set 60 mm apart. The Petri dish and magnets were placed in an air-tight container and kept in an oven at 80°C for 24 hours for curing. To remove moisture, the resultant thin films (~0.5 to 0.7 mm thick) were kept in the oven at 160°C for more than 3 hours. Finally, various sized samples were cut from the films for testing (chained samples). For comparison, another set of thin films were prepared without applying the magnetic field, so that Ni particles were randomly distributed within the SMP matrix (random samples).

Scanning electron microscopy (SEM) was used to study the formation of Ni chains in the chained samples with different Ni contents. As revealed in Figure 6.20, the chains start to form even at only 1 vol.% of Ni. However, with 10 vol.% or more Ni content, it becomes difficult to identify individual chains. At 20 vol.% of Ni, the sample is virtually fully occupied by Ni powder. DSC results from chained samples (Modulated DSC-2920, heating rate of 20°C/minute) are presented in Figure 6.21. With the increase of Ni content, the glass transition shifts slightly toward a lower temperature range, indicating a slight interaction between Ni powder and polymer.

DMA (DMA 2980, TA Instruments) was conducted on both random samples and chained samples (along chained direction) in film mode at a heating rate of 5°C/minute and a constant frequency of 1 Hz. The variation of storage modulus upon heating in all samples is summarized in Figure 6.22. The storage modulus (at 0°C) is plotted against the volume fraction of Ni for both random and chained samples, as shown in Figure 6.23. As expected, at a low Ni content, there is no obvious difference for both types of samples because the amount of Ni is insufficient to produce significant improvement in storage modulus. However, above 5 vol.% of Ni powder, the storage moduli of the chained samples were higher than those of the random samples for the same amount of Ni content; by forming chains, Ni powder more effectively enhances strength along the chain direction.

The glass transition in polymers actually occurs within a temperature range. Among the many ways to define T_g, here we define it as the peak point after

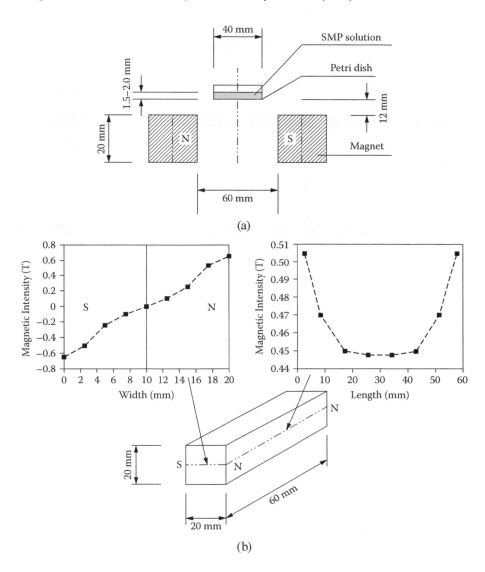

FIGURE 6.19 (a) Sample preparation. (b) Magnetic flux intensity measured by Gauss meter. (Reprinted from Lan X, Huang WM, Leng JS et al. *Rubber Fibers Plastics*, 4, 84–88, 2009. With permission.)

differentiating the DMA curve with respect to temperature. As shown in Figure 6.24, the T_gs of both random and chained samples decreased in a similar linear fashion with an increase in Ni content. At 20 vol.% of Ni—extremely high—the T_g drops by about 15°C to 35°C only, which further confirms the slight chemical interaction between SMP and Ni powder. The electrical resistivity of the random and chained

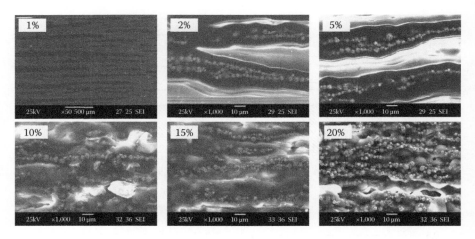

FIGURE 6.20 SEM images showing morphologies of samples with different volume fractions of Ni.

samples was determined. The inset of Figure 6.25 illustrates the experimental setup. The volumetric electrical resistivity, ρ, is calculated by:

$$\rho = \frac{RA}{L} \tag{6.1}$$

where R is the measured resistance, A is the cross-sectional area of sample, and L is the distance between two aluminum electrodes. Note that in chained samples, we measured the resistance in both chained and transverse directions. As

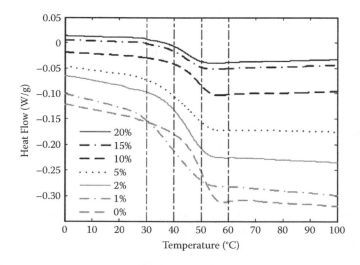

FIGURE 6.21 DSC results for chained samples.

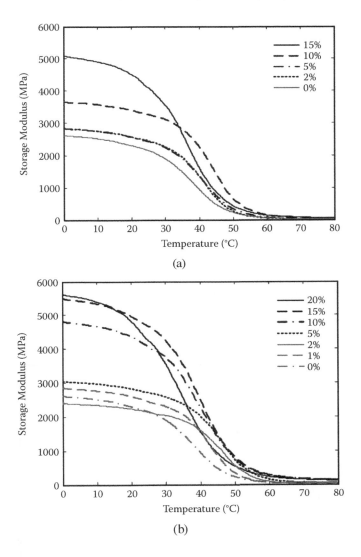

FIGURE 6.22 Storage modulus versus temperature relationships of (a) random and (b) chained samples. (Reprinted from Lan X, Huang WM, Leng JS et al. *Rubber Fibers Plastics*, 4, 84–88, 2009. With permission.)

seen in Figure 6.25, the resistivity of the chained sample in the chain direction is lower than that of random sample for the same volume fraction of Ni content. As expected, the resistivity of the chained samples in the transverse direction is the highest.

With 10 vol.% of Ni, the chained sample (bottom left inset shows sample dimensions and temperature distribution) can be Joule heated from room temperature (about 20°C) to about 55°C under a constant voltage of 6 V—sufficient to trigger

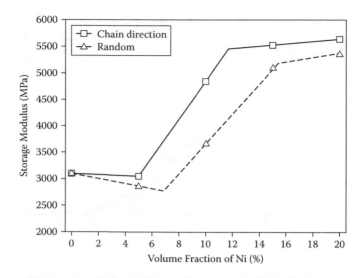

FIGURE 6.23 Storage modulus versus volume fraction of Ni relationship at 0°C. (Reprinted from Leng JS, Lan X, Liu YJ et al. *Applied Physics Letters*, 92, 014104, 2008. With permission.)

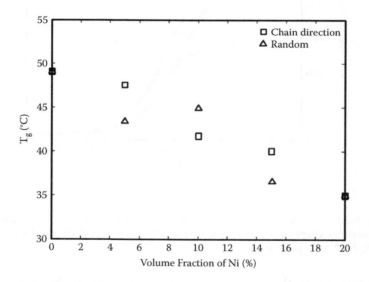

FIGURE 6.24 T_g values of chained and random samples. (Reprinted from Leng JS, Lan X, Liu YJ et al. *Applied Physics Letters*, 92, 014104, 2008. With permission.)

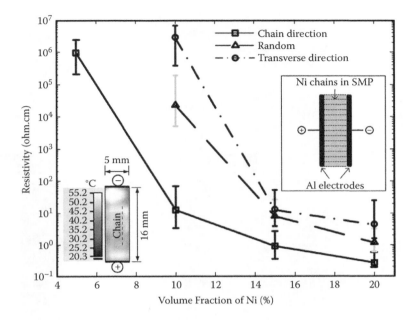

FIGURE 6.25 Electrical resistivity versus volume fraction of Ni powder relationships. Right inset: setup for resistance measurement along chained direction. Bottom left inset: infrared image of temperature distribution in chained sample (10 vol.% of Ni, 6 V). (Reprinted from Leng JS, Lan X, Liu YJ et al. *Applied Physics Letters,* 92, 014104, 2008. With permission.)

shape recovery. Figure 6.26 shows a piece of pre-bent sample that recovered its original straight shape in 90 seconds when 20 V current was applied. However, for the same setup and configuration, the random sample is unable to generate heat sufficient for shape recovery.

6.4.2 POLYURETHANE–CARBON BLACK WITH ADDITIONAL NICKEL POWDER

Nickel powder is not an ideal filler for achieving excellent electrical conductivity because of its high density and high cost. On the other hand, carbon black (CB) has been widely

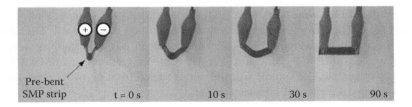

FIGURE 6.26 Shape recovery sequence in chained sample. Sample size 30 × 7 × 1 mm, 10 vol.% Ni, 20 V. (Reprinted from Leng JS, Lan X, Liu YJ et al. *Applied Physics Letters,* 92, 014104, 2008. With permission.)

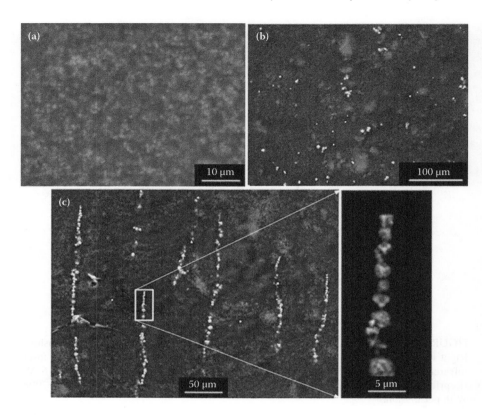

FIGURE 6.27 Dispersion of CB and Ni powder in polyurethane SMP (SEM image, 8 vol.% CB, 0.5 vol.% Ni, if applicable). (a) Without Ni. (b) With randomly distributed Ni. (c) With chained Ni. Grey dots = CB. White dots = Ni. ((a) and (b) reprinted from Huang WM, Liu N, Lan X et al. *Materials Science Forum,* 614, 243–248, 2009. With permission. (c) reprinted from Leng JS, Huang WM, Lan X, et al. *Applied Physics Letters*, 92, 204101, 2008. With permission.)

used in conductive polymers including SMPs (Chapters 4 and 5) due to low cost. Instead of using only Ni powder, we propose to use only a small amount of Ni powder to form chains inside the polymers with randomly distributed CB to significantly reduce the electrical resistivity, while the polymer composite is kept mechanically isotropic.

We utilized the same Ni powder and polyurethane SMP solution examined in Section 6.4.1. The CB was the same as the material used in Chapter 4. The amount of Ni, if applicable, was only 0.5 vol.%. Three types of thin film samples were fabricated, namely SMP–CB–Ni (chained), SMP–CB–Ni (random), and SMP–CB. For each type of sample, the volume fraction of CB varied from 4 to 10%. The detailed steps for preparing chained samples are:

1. CB powder and Ni powder are mixed with SMP–DMF solution. The mixture is stirred for uniform dispersion and then poured into a 50 mm diameter Petri dish.

2. The Petri dish is placed above two magnets (see Figure 6.19 for experimental setup). The magnetic flux intensity at the bottom of the Petri dish is 0.03 ± 0.003T (measured by a Hirst GM05 Gauss meter).
3. The whole setup is then placed into an air-tight container and kept in an oven at 80°C for 24 hours for curing.

The resultant SMP samples were ~0.6 to 1 mm thick with Ni chains inside. The SMP–CB–Ni (random) samples were prepared in the same way but without application of the magnetic field. The SMP–CB samples contained no Ni powder. SEM images of samples containing 8 vol.% of CB are presented in Figure 6.27. The dispersion of CB in the SMP–CB is seemingly uniform (a), but a closer-look at the figure reveals that the CB is not well separated; it aggregates because the size of the powder was about 30 nm. Ni powder was also distributed more or less randomly within the SMP–CB (b). Short, parallel Ni particle chains ~100 to 200 μm long, with typical distances between adjacent parallel chains ~50 to 100 μm, were observed in SMP–CB–Ni (chained; Figure 6.27c). A zoom-in view reveals that the chains formed by packing Ni particles one after another. The sizes of the Ni particles were ~3 to 7 μm and the aspect ratios were ~10 to 70, comparable to those of short microfibers.

The electrical resistivities of samples containing different amounts of CB with and without 0.5 vol.% of Ni were plotted as shown in Figure 6.28. It is apparent that the additional 0.5 vol.% of Ni, if distributed randomly, had limited effect in reducing resistivity. However, using the same quantity of Ni particles aligned into chains

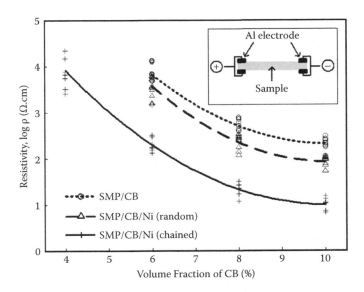

FIGURE 6.28 Resistivity versus volume fraction of CB with and without 0.5 vol.% of Ni powder. The inset shows how resistance was measured. (Reprinted from Leng JS, Huang WM, Lan X, et al. *Applied Physics Letters*, 92, 204101, 2008. With permission.)

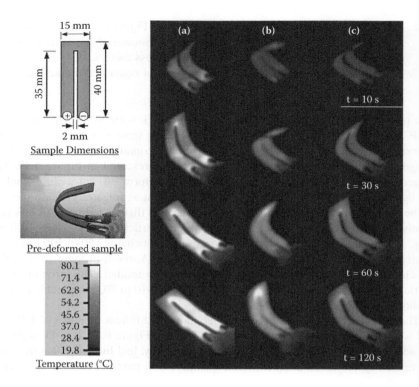

FIGURE 6.29 Sequence of shape recovery and temperature distribution. Top left inset: dimensions of sample. Middle left inset: pre-bent shape. Bottom left inset: temperature bar (°C). Sample (a) 10 vol.% of CB, 0.5 vol.% of chained Ni. Sample (b) 10 vol.% of CB, 0.5 vol.% of randomly distributed Ni. Sample (c) 10 vol.% of CB only. The tests were repeated more than five times on each sample. (Reprinted from Leng JS, Huang WM, Lan X, et al. *Applied Physics Letters*, 92, 204101, 2008. With permission.)

significantly reduced electrical resistivity by more than 10 times. Obviously, the remarkable reduction of electrical resistivity is a result of the conductive Ni chains that serve as conductive channels to bridge the small, isolated CB aggregations. Such bridging effect is more significant if the amount of CB is small because the CB aggregates are relatively small and more isolated. The resistance measured 1 month later for the same samples was unchanged; thus their resistivity was stable.

To demonstrate shape recovery by Joule heating, three samples [SMP–CB–Ni (chained), SMP–CB–Ni (random), and SMP–CB], all with 10 vol.% of CB, 1 mm thick, and shaped as illustrated in Figure 6.29 (left top inset) were bent by about 150 degrees at 80°C and then cooled to room temperature (22°C) as shown in the left middle inset of the figure. Subsequently, 30 V of power was applied (left top inset). An infrared video camera (AGEMA, Thermo-vision 900) was used to monitor the temperature distribution and shape recovery simultaneously.

Figure 6.29 (right) presents four snapshots of each sample. We can see clearly that Sample (a) [SMP–CB–Ni (chained)] reaches the highest temperature (about

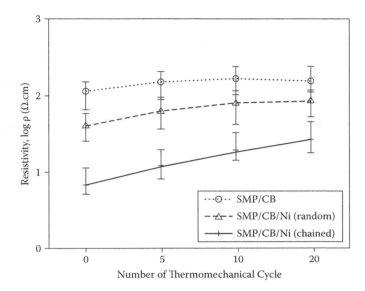

FIGURE 6.30   Evolution of resistivity upon shape-memory cycling. (Reprinted from Leng JS, Huang WM, Lan X, et al. *Applied Physics Letters*, 92, 204101, 2008. With permission.)

80°C throughout—much higher than the T_g of the SMP, so that almost full recovery was observed within 120 seconds). The temperature of Sample (c) [SMP–CB] was the lowest (only ~45°C, slightly higher than the T_g of the SMP; hence, the shape recovery was not obvious). Sample (b) [SMP–CB–Ni (random)] reached ~65°C and the shape recovery was not complete after 120 seconds. Power consumption was ~1.2 W for Sample (a). We also investigated the evolution of resistivity up to 20 shape recovery cycles (20% pre-strained). Figure 6.30 reveals that the degradation of electrical conductivity in Sample (c) is relatively more significant than the degradation of the others.

6.5   SUMMARY

This chapter presented a systematic study of the influences of magnetic particles (Fe_3O_4 and Ni) on the thermomechanical properties of SMP–magnetic particle composites. We explored some unconventional applications of magnetic particles, for example, to alter surface roughness and morphology and/or enhance the electrical conductivity of SMPs.

ACKNOWLEDGMENTS

We thank C.S. Chan, X. Lan, S.Y. Siew, L.H. Tan, Q. Yuan, and Y. Zhao for conducting some of the experiments and Z. Ding for his help in compiling and editing this chapter.

REFERENCES

Aphesteguy JC, Damiani A, DiGiovanni D et al. (2009). Microwave-absorbing characteristics of epoxy resin composites containing nanoparticles of Ni-Zn and Ni-Cu-Zn ferrites. *Physica B*, 404, 2713–2716.

Bansal A, Yang HC, Li CZ et al. (2005). Quantitative equivalence between polymer nanocomposites and thin polymer films. *Nature Materials*, 4, 693–698.

Beloshenko VA, Varyukhin VN, and Voznyak YV (2005). Electrical properties of carbon-containing epoxy compositions under shape memory effect realization. *Composites A*, 36, 65–70.

Buckley PR, McKinley GH, Wilson TS et al. (2006). Inductively heated shape memory polymer for the magnetic actuation of medical devices. *IEEE Transactions in Biomedical Engineering*, 53, 2075–2083.

Cho JW, Kim JW, Jung YC et al. (2005). Electroactive shape-memory polyurethane composites incorporating carbon nanotubes. *Macromolecular Rapid Communications*, 26, 412–416.

Cotts DB and Reyes Z (1986). *Electrically Conductive Organic Polymers for Advanced Applications*. Noyes Data Corporation, Park Ridge, NJ.

Huang W (2002). On the selection of shape memory alloys for actuators. *Materials and Design*, 23, 11–19.

Huang WM, Liu N, Lan X et al. (2009). Formation of protrusive micro/nano patterns atop shape memory polymers. *Materials Science Forum*, 614, 243–248.

Jäger KM, McQueen DH, Tchmutin IA et al. (2001). Electron transport and ac electrical properties of carbon black polymer composites. *Journal of Physics D*, 34, 2699–2707.

Keblinski P, Cahill DG, Bodapati A et al. (2006). Limits of localized heating by electromagnetically excited nanoparticles. *Journal of Applied Physics*, 100, 054305.

Kropka JM, Pryamitsyn V, and Ganesan V (2008). Relation between glass transition temperatures in polymer nanocomposites and polymer thin films. *Physical Review Letters*, 101, 075702.

Lan X, Huang WM, Leng JS et al. (2009). Electric conductive shape-memory polymer with anisotropic electro-thermomechanical properties. *Rubber Fibers Plastics*, 4, 84–88.

Leng JS, Huang WM, Lan X et al. (2008a). Significantly reducing electrical resistivity by forming conductive Ni chains in a polyurethane shape-memory polymer/carbon-black composite. *Applied Physics Letters*, 92, 204101.

Leng JS, Lan X, Liu YJ et al. (2008b). Electrical conductivity of thermoresponsive shape-memory polymer with embedded micron sized Ni powder chains. *Applied Physics Letters*, 92, 014104.

Li FK, Qi LY, Yang JP et al. (2000). Polyurethane/conducting carbon black composites: structure, electric conductivity, strain recovery behavior; and their relationships. *Journal of Applied Polymer Science*, 75, 68–77.

Majewski P and Thierry B (2007). Functionalized magnetite nanoparticles: synthesis, properties, and bio-applications. *Critical Reviews in Solid State and Materials Sciences*, 32, 203–215.

Mohr R, Kratz K, Weigel T et al. (2006). Initiation of shape-memory effect by inductive heating of magnetic nanoparticles in thermoplastic polymers. *Proceedings of the National Academy of Sciences of the United States of America*, 103, 3540–3545.

Morgan H, Foot PJS, and Brooks NW (2001). The effects of composition and processing variables on the properties of thermoplastic polyaniline blends and composites. *Journal of Materials Science*, 36, 5369–5377.

Oh H and Green PF (2009). Polymer chain dynamics and glass transition in athermal polymer/nanoparticle mixtures. *Nature Materials*, 8, 139–143.

Razzaq MY, Anhalt M, Frormann L et al. (2007). Thermal, electrical and magnetic studies of magnetite filled polyurethane shape memory polymers. *Materials Science and Engineering A*, 444, 227–235.

Sahoo NG, Jung YC, Yoo HJ et al. (2007). Influence of carbon nanotubes and polypyrrole on the thermal, mechanical and electroactive shape-memory properties of polyurethane nanocomposites. *Composites Science and Technology*, 67, 1920–1929.

Shimizu T and Matsui M (2003). New magnetic implant material for interstitial hyperthermia. *Science and Technology of Advanced Materials*, 4, 469–473.

Singh M (2005). Thermal characterization of nanocomposite shape memory polymer for their mechanical and thermal properties. MES thesis, Lamar University, Beaumont, TX.

Thommerel E, Valmalette JC, Musso J et al. (2002). Relations between microstructure, electrical percolation and corrosion in metal-insulator composites. *Materials Science and Engineering A,* 328, 67–79.

Zheng W and Wong SC (2003). Electrical conductivity and dielectric properties of PMMA/expanded graphite composites. *Composites Science and Technology*, 63, 225–235.

7 Shape Memory Polymer Nanocomposites

7.1 INTRODUCTION

Shape memory polymers (SMPs) have the capability of recovering their original shapes upon application of external stimuli such as heat, light, and chemical action (Lendlein and Langer 2002, Lendlein and Kelch 2002, 2007, Baer et al. 2006, Lendlein et al. 2005). The advantages of SMPs over other shape memory materials (in particular shape memory alloys or SMAs) include lower cost, lower density, higher shape recoverability (up to 400%), and easier processability (Behl and Lendlein 2007, Dietsch and Tong 2007, Ratna and Karger-Kocsis 2008, Gall et al. 2000, 2004a, 2004b, Liu et al. 2007).

The disadvantages of SMPs are low stiffness and strength as compared to SMAs. Therefore, many studies utilize micro- and nano-sized fillers to enhance the mechanical and recovery properties of SMPs (Gunes et al. 2008a, 2008b, Gunes and Jana 2008, Gall et al. 2005). Polymer composites, in particular those containing nano-sized fillers (nanocomposites), are advanced materials combining polymer matrices [including poly(vinyl) alcohol, styrene–butadiene rubber, epoxy resin, polyethylene, polyurethane, polyamide, polyimide, and others] etc. with nano-sized additives (Chattopadhyay and Raju, 2007).

Recently, organic or inorganic nanocomposites have attracted significant attention due to enhanced performance resulting from their mechanical, optical, electrical, magnetic, and other functional properties (Schmidt et al. 2002, Schmidt and Malwitz, 2003, Schmidt 2006, Sanchez et al. 2003, Morgan 2006, Utracki 2008, Carrado 2000, Hussain et al. 2006, Carastan and Demarquette 2007). Nanocomposites exhibit outstanding mechanical properties such as high elastic stiffness and high strength due to large surface area-to-volume ratios of nano additives when compared to micro and macro additives (Hussain et al. 2006). The advantages of using nano scale rather than micro scale reinforcements include significant surface area improvement, larger polymer chain confinement, higher stiffness and tensile strength, and better tensile modulus and flexural strength. Some nanofillers such as carbon nanotubes (CNTs) and clays provide nanoscale zigzag diffusion paths and large interfacial areas, creating enhanced barriers to thermal, gas, or moisture penetration (Khudyakov et al. 2009). Generally, building blocks of nanocomposites include zero-dimensional (spherical nanoparticles), one-dimensional (clay rods, nanotubes, fibers), two-dimensional (layered clay minerals), or three-dimensional (3-D inorganic networking frames) nanofillers (Sanchez et al. 2001, Amalvy et al. 2001).

Zero-dimensional nanofillers (Figure 7.1a shows an example) such as nanoparticles, nanodots, nanospheres, quantum dots of metals, and oxides or nitrides range

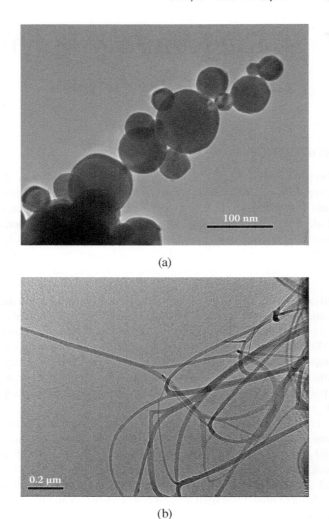

(a)

(b)

FIGURE 7.1 Typical TEM images of (a) nanoparticle of SiO$_2$; (b) CNT; (c) nanoclay rod; (d) clay layers. (Reprinted from Xu B, Fu YQ, Ahmad M et al. *Journal of Materials Chemistry*, 20, 3442–3448, 2010. With permission.)

in diameter from a few nanometers to several hundred nanometers. Common problems associated with nanoparticles are oxidation, agglomeration, and rapid growth. Normally, an organic substance is incorporated as a surface treatment agent to modify a nanoparticle surface to improve compatibility, reduce surface energy, and prevent agglomeration.

Among the one-dimensional nanomaterials, CNTs constitute the hottest research topic because of their unique nanostructures that display remarkable electronic and mechanical properties. An ideal CNT can be visualized as a hexagonal network of carbon atoms rolled to shape a hollow cylinder (Figure 7.1b). The density of a

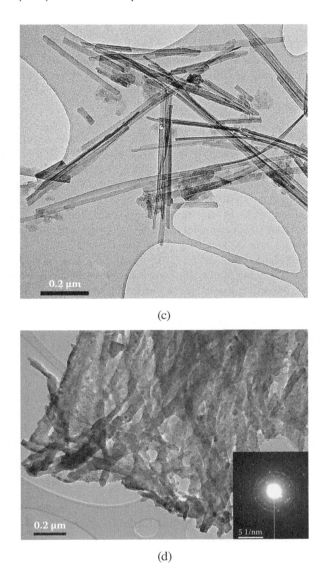

(c)

(d)

FIGURE 7.1 (Continued)

single-wall CNT (SWCNT) is ~1.33 to 1.40 g/cm^3—one-half the density of aluminum (Thostenson et al. 2001). The elastic modulus of an SWCNT is comparable
to that of a diamond. The tensile strength of an SWCNT is much higher than that
of high-strength steel, and its significant resilience under bending to large angles
and then restraightening without damage is dramatically different from the plastic deformations of common metals and brittle fractures of carbon fibers (Curtin
and Sheldon 2004). The electric-current-carrying capability of an SWCNT is estimated to be 1×10^9 Amp/cm^2, whereas copper wires burn out ~1×10^6 Amp/cm^2.

Thermal conductivity is predicted to be 6000 W/m K at room temperature, almost double the 3320 W/m K thermal conductivity of diamond (Thostenson et al. 2001). SWCNTs are stable up to 2800°C in vacuum and 750°C in air. The outstanding thermal and electrical properties along with high specific stiffness and strength and very large aspect ratios stimulated the development of nanotube-reinforced composites for structural and functional applications (Curtin and Sheldon 2004).

Most research of nanocomposites with two-dimensional materials is based on polymer–layered silicate composites (Pavlidou and Papaspyrides 2008). Attapulgite is a natural hydrated magnesium–aluminum silicate clay $[(Mg,Al)_2Si_4O_{10}(OH)\cdot4(H_2O)]$ consisting of a network of densely packed layers or rods (Alexandre and Dubois 2000) that normally have diameters smaller than 100 nm and length ranging from hundreds of nanometers to several micrometers (see Figure 7.1c). The length of each fiber varies from submicrometers to a few micrometers; diameters are about tens of nanometers.

Pan et al. (2008) and Xu et al. (2009) systematically studied the properties and stimulus-response behaviors of polyurethane–attapulgite nanocomposites under different conditions. Montmorillonite (MMT) has a layered clay structure (Carastan and Demarquette 2007) and received a great deal of attention recently because of its potentially high aspect ratio (up to 1000) and its unique intercalation and exfoliation characteristics (one example is shown in Figure 7.1d). Cao and Jana (2007) investigated the shape memory properties of nanoclay-tethered polyurethane nanocomposites. The shape recovery stress was found to increase significantly as a function of exfoliated nanoclay content.

Typical examples of three-dimensional nanomaterials in hybrids are Y zeolite and Si molecular sieves. Spange et al. (2001) introduced a flexible polymer directly into an ordered mesoporous host material (Y zeolite, MCM-41), after dissolution of the hybrid material in an aqueous potassium hydroxide solution. The formation of conjugated sequences of poly(vinyl ether) segments inside the Y zeolites arose from successive ether cleavage along the encapsulated polymer chains (Spange et al. 2001).

Significant progress has been made in recent decades in designing, synthesizing, characterizing, and applying SMP nanocomposites by adding various types of nanofillers. Consequently, in thermo-responsive SMPs, the shape memory effect (SME) can be triggered by direct application of heat and also by Joule (electrical) heating and inductive (magnetic or electromagnetic) heating. This chapter summarizes the synthesis techniques, designs, and characterizations that focus on polyurethane SMP-based nanocomposites.

7.2 SYNTHESIS TECHNIQUES OF SMP NANOCOMPOSITES

7.2.1 SOLUTION MIXING

In the solution mixing method, nanoparticles and polymer are mixed in a suitable solvent, followed by precipitation or evaporation of the solvent. The advantages of using this method include the following (Kirk et al. 2003): (1) nanometric materials are obtained first; (2) the influences of reaction conditions on the formation of inorganic structures (a serious problem via traditional methods) can be avoided; and (3) the fixed inorganic structure benefits the characterization of the final composites. Direct mixing usually does not provide uniform solutions because of the aggregations of nanoparticles or CNTs. Agitation with ultrasonic, magnetic, and shear mixing methods can be used to enhance dispersion. However, nanoparticles and CNTs are not well dispersed using simple solvent mixing methods (Watts and Hsu 2003).

A high-power ultrasonication process is more effective in forming a good dispersion of nanoparticles or CNTs. Ultrasonic irradiation has been extensively used in dispersing, emulsifying, crushing, and activating the particles (Sahoo et al. 2006). For example, by taking advantage of the multiple effects of ultrasound, the aggregates and entanglements of CNTs can be effectively broken down. Cho and Lee (2004) and Cho et al. (2005) successfully prepared polyurethane–multiwall carbon nanotube (MWCNT) composites with better dispersion of CNTs up to 20 wt.% in polyurethane (Sahoo et al. 2006). Many polymer composites, including polyurethane–CNTs (Chen and Tao 2005), polystyrene–CNTs (Kota et al. 2007), epoxy–CNTs (Song and Youn 2005), poly(vinyl alcohol)–CNTs (Shaffer and Windle 1999), and polyethylene–CNTs (Bin et al. 2003) have been fabricated by this method.

In solution mixing, the sol–gel process is convenient and versatile for preparing nanocomposites at ambient conditions and it allows *in situ* entrapment of inorganic nanoparticles or organic, organometallic, and biological molecules within microporous networks of sol–gel-derived matrices (Wen and Wilkes 1996, Chaudhury and Gupta 2007). In general, the sol–gel process involves the transition of a system from a liquid sol (mostly colloidal) into a solid gel phase. Sol–gel materials have controllable surface areas of nanoparticles and average particle or pore sizes. They can be miniaturized to micron or nano sizes and allow control of conductivity via a choice of metal or metal oxide (Díaz-García and Badía-Lamo 2005).

7.2.2 Melting Mixing

Many polymers are non-soluble in liquids, and melting mixing may be one option. Melt mixing is a common and simple method that is particularly useful for thermoplastic polymers. It involves the melting of polymers to form a viscous liquid, then additives are included for shear mixing, followed by molding or extrusion to form samples (Jin et al. 2001). Special care should be taken as thermal processing may be detrimental to nanofillers such as CNTs or cause polymer degradation (Andrews et al. 2002, Andrews and Weisenberger 2004). This approach is simple and compatible with current industrial practices. The large shear forces can be applied to break nanofiller aggregates or prevent their formation.

Zhang et al. (2004) prepared nylon-6–MWCNT composites containing 1 wt.% of MWCNT via a melt compounding method using a twin-screw mixer. The disadvantage was that the dispersion of CNTs in the polymer matrix was poor compared to the dispersion achieved by solution mixing, along with other potential problems such as changes of nanofiller properties during melting and high viscosities of composites at higher loading of nanofillers such as CNTs.

7.2.3 IN SITU OR INTERACTIVE POLYMERIZATION

This method is crucial for preparing composites with polymers that cannot be processed by solution or melt mixing, e.g., insoluble and thermally unstable polymers. *In situ* polymerization can be applied for most nanocomposites. This section focuses on CNT-based nanocomposites. Modification may involve non-covalent or covalent bonding between CNT and polymer. Non-covalent CNT modification applies physical adsorption and/or wrapping of polymer molecules to the surfaces of the CNTs (Velasco-Santos et al. 2003, Cadek et al. 2004) through van der Waals and π–π interactions. The graphitic sidewalls of CNTs provide possibilities for π-stacking interactions with conjugated polymers and organic polymers containing heteroatoms with free electron pairs (Sahoo et al. 2007a, 2007b). The advantage is that non-covalent functionalization does not destroy the conjugated system of CNT sidewalls and therefore does not affect the final structural properties of the nanocomposites.

Conductive polymers may be attached to CNT surfaces by *in situ* polymerization to improve processability and also electrical, magnetic, and optical properties (An et al. 2004, Yoo et al. 2006, Sahoo et al. 2007a, 2007b). For example, MWCNTs are first dispersed in poly(ε-caprolactone)diol (PCL), followed by adding methylene bis(phenylisocyanate) (MDI) to the mixture. A butanediol (BD) chain extender is added to this prepolymer, and finally a polyurethane–MWCNT composite is synthesized. Xia and Song (2005, 2006) found that SWCNTs are not dispersed well by this method compared with MWCNTs. Therefore, they synthesized polyurethane–SWCNT nanocomposites using polyurethane-grafted SWCNTs. This improved the dispersion of SWCNTs in the polyurethane matrix and strengthened the interfacial interaction between polyurethane and SWCNT (Xia and Song 2005).

Guo et al. (2006, 2009) introduced iron oxide–polyurethane nanocomposites based on surface initiated polymerization that may improve the structural integrity of a nanocomposite through better chemical bonding between nanoparticles and polymer matrix with a uniform particle distribution. Jung et al. (2006) synthesized polyurethane–MWCNT nanocomposites following a two-step process. In the first step, a pre-polymer was prepared from a reaction of MDI and PCL at 80°C for 90 minutes. In the second stage, carboxylated MWCNTs were added to the prepolymer at 110°C, and then reacted for 150 minutes to obtain the final cross-linked polyurethane–MWCNT nanocomposites.

In brief, incorporating a functional group is an effective way to reduce the surface energies of nanofillers and improve the compatibility of the organic and inorganic interfaces (Cui et al. 2004). End grafting of polymers onto a solid surface is an important technique in many areas of science and technology, e.g., colloidal stabilization, adhesion, lubrication, tribology, and rheology. Recently, new techniques including anionic polymerization (Zhou et al. 2002), cationic polymerization (Rusa et al. 2004), and atom transfer radical polymerization (ATRP) (Kotal et al. 2005, Marutani et al. 2004, Werne and Patten 2001) were proposed and successfully applied to surface-initiated graft polymerization to synthesize nanocomposites.

Generally, the two main strategies for the covalent grafting of polymers to nanoparticles are "grafting to" and "grafting from" (Coleman et al. 2006). The "grafting to" technique is based on the attachment of preformed end functionalized polymer molecules to functional groups on CNT surfaces by chemical reactions (Rozenberg and Tenne 2008). In the "grafting to" method, the main approaches are radical or carbanion additions and cycloaddition reactions to the CNT double bonds.

The method of "grafting to" CNT defect sites indicates that polymers with reactive end groups can react with the functional groups on nanotube surfaces. An advantage of the "grafting to" method is that pre-formed commercial polymers of controlled molecular weights may be used. The main limitation is that the initial binding of polymer chains hinders the diffusion of additional macromolecules to the CNT surfaces, leading to a low grafting density (Cui et al. 2004). Also, only polymers containing reactive functional groups can be used. The "grafting from" technique involves the polymerization of monomers from surface-derived initiators on the MWCNTs or SWCNTs. These initiators are covalently attached via various functionalization reactions developed for small molecules including acid-defect group chemistry and sidewall functionalization of CNTs (Xie et al. 2005). The advantage of "grafting from" is that the polymer growth is not limited by steric hindrance, allowing efficient grafting of high-molecular-weight polymers. However, this method requires strict control of the amounts of initiator and substrate along with accurate control of conditions required for the polymerization reaction (Rozenberg and Tenne 2008).

7.2.4 ELECTROSPINNING

Electrospinning has been recognized as an efficient technique for fabricating polymer nanofibers, and various polymers have been successfully electrospun into ultrafine fibers for use as reinforcements in nanocomposite development (Huang et al. 2003). Many research groups tried to incorporate CNTs into polymer nanofibers by electrospinning (Kim and Reneker 1999, Hou et al. 2005). The spinning process is expected to align CNTs or their bundles along the fiber direction by high shear forces induced by the spinning and dielectrophoretic forces caused by a dielectric or conductivity mismatch between CNTs and the polymer solution. Polymer composites reinforced with electrospun nanofibers have been developed mainly for good physical (optical and electrical) and chemical properties while maintaining their appropriate mechanical performance (Lau and Hui 2002).

7.2.5 TECHNIQUES TO ENHANCE DISPERSION OF CNTS

Several techniques such as physical blending, *in situ* polymerization, and chemical functionalization improve the dispersion of CNTs in polymer matrices (Jordan et al. 2005, Cadek et al. 2004). For polymer–CNT composites, high-power dispersion methods such as ultrasound and high-speed shearing are the simplest and most convenient ways to improve the dispersion of CNTs (Qian et al. 2000, Sandler et al. 1999, 2003). Surfactants are often used as dispersing agents to improve CNT dispersion in processing polymer–CNT composites (Gong et al. 2000). The surfaces of CNTs can be chemically functionalized (including grafting copolymerization) to achieve good dispersion in composites and strong interface adhesion between surrounding polymer chains (Song and Xia 2006).

CNTs are assembled as ropes or bundles containing many defects such as catalyst residuals, bucky onions, spheroidal fullerenes, amorphous carbon, polyhedron graphite nanoparticles, and other forms of impurities (Lin et al. 2007). Therefore, purification, cutting, or disentanglement and activation are needed before chemical functionalization. The popular method of purifying MWCNTs is based on oxidation. The MWCNTs are purified by burning away the tube ends, defects, and amorphous carbon at a temperature above 700°C in air or oxygen (Lin et al. 2007). However, the yield is extremely low (<5%). Hiura et al. (1995) purified CNTs by a mixture of concentrated sulfuric acid and potassium permanganate, but the method may not be suitable for large-scale production.

Tohji et al. (1997) used a purification method that included hydrothermal treatment, thermal oxidation, and hydrochloric acid etching. To prevent the destruction of CNTs during purification, Bonard et al. (1997) dispersed the CNTs in polar solvents assisted by surfactants such as sodium dodecyl sulfate, followed by micro filtration and size exclusion chromatography. Yamamoto et al. (1998) used AC electrophoresis to treat the CNTs dispersed in isopropyl alcohol and found that the separation from impurity particles depended on the frequency of the applied field.

7.3 SHAPE MEMORY POLYMER NANOCOMPOSITES

7.3.1 Nanocomposites for Mechanical Enhancement

Most nanofillers are beneficial for enhancing the mechanical properties of SMPs. The effects of the nanoparticles depend on many variables, especially on the crystalline or amorphous nature of the polymer matrix and the interaction of filler and matrix (Huang et al. 2010, Jordan et al. 2005). Good dispersion of nanoparticles can also improve the fracture, fatigue, and creep resistance of the nanocomposites because they can restrict the slippage, reorientation, and motion of the polymer chains.

7.3.1.1 Nanoparticle-Based SMP Nanocomposites

Nanoparticles as filler materials can improve the elastic moduli of SMPs (Liu et al. 2004). SMP nanocomposites with SiC and Al_2O_3 additions exhibit higher elastic moduli and are capable of generating higher recovery forces in comparison with pure SMPs. Xu et al. (2010a, 2010b) confirmed that the addition of nanofillers of alumina and silica can significantly enhance the hardness of polystyrene. Alumina nanoparticles exhibited better enhancing effects than SiC due to their high intrinsic hardness (see Figure 7.2).

Gall et al. (2000, 2004a, 2004b) found that nanoparticulate reinforcement can increase the stiffness and recoverable stress levels for SMP materials, but deteriorates the SME of SMPs even at very low loading. Gunes et al. (2008a, 2008b, 2009) also observed a negative effect of SiC on the SMEs of polyurethane SMPs. They ascribed the effect to the dramatic decrease of soft segment crystallinity by the addition of SiC. Cho and Lee (2004) used a sol–gel process to incorporate silica from tetraethoxysilane (TEOS) into a shape memory polyurethane matrix, and the maximum fracture stress and modulus were obtained at 10 wt.% of TEOS content.

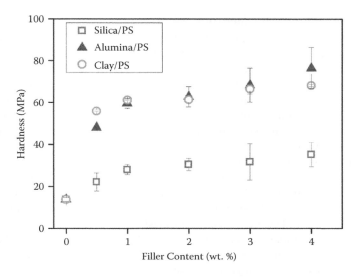

FIGURE 7.2 Comparison of mechanical properties of SMP nanocomposites: Vickers hardness results with 25 g load and 20-second load time. (Reprinted from Xu B, Fu YQ, Ahmad M et al. *Journal of Materials Chemistry*, 20, 3442–3448, 2010. With permission.)

Silica nanoparticles interact relatively strongly with polyurethane matrices. Petrovic et al. (2000) found that the tensile strength varied a little between the composites made of micro- and nano-sized particles, up to a filler composition of 20% weight fraction. Above 20%, the tensile strength was higher for the nanocomposites than for the pure polyurethane polymer. In the nanocomposites, the maximum strain-to-failure increased by 500% over the pure system. For composites with micro-sized inclusions, the increase was only 100%. The size of the nanosilica particles may have a significant influence on the mechanical property enhancement. Maximum effects on glass transition temperature, tensile properties, and wear resistance were obtained at a particles size of 28 nm (Chen et al. 2005, 2006). Solution modification may be used for silica-based nanocomposites (Jang et al. 2009), in which the SMP is solubilized in an organic solvent and then end-capped with glycidol. Silica is exfoliated in the solvent and functionalized with amino groups. Upon mixing of the silica with the end-capped polyurethane, the amine groups on the silica react with the epoxy end groups in the polymer. Because each amine-functionalized silica particle contains more than two amine groups, the silica nanoparticles effectively act as multifunctional cross-linkers (Jang et al. 2009). A detailed review of the preparation, characterization, properties, and applications of the polymer–silica nanocomposites appears in Zou et al. (2008).

Celite (World Minerals) is a mined product composed of silica and alumina. It has surface hydroxyl groups that may be coupled with shape memory polyurethane chains. Park et al. (2008) fabricated Celite–shape memory polyurethane composites with Celite as a cross-linker added in the middle of polymerization. The Celite improved the shape memory and mechanical properties of shape memory polyurethanes. The best mechanical properties and good SME were obtained at 0.2 wt.% of Celite.

7.3.1.2 Clay-Based SMP Nanocomposites

Montmorillonite, saponite, and synthetic mica are commonly used clay materials. Silicates have a characteristic distance between galleries of 1 nm and the basal spacing of a gallery is also about 1 nm (Pavlidou and Papaspyrides 2008). Inorganic ions like Na between galleries hold negatively charged galleries together. The replacement of the inorganic ions in the galleries of the native clay by alkylammonium (onium) salts or quaternary amines with long alkyl substituents (surfactants) leads to a better compatibility between the inorganic clay and hydrophobic polymer matrix (Pavlidou and Papaspyrides 2008). The replacement leads to an increase of the space between galleries, facilitating intercalation of polymer molecules into the clay.

The advantages of polymer-based clay nanocomposites include improved stiffness, strength, toughness, and thermal stability along with reduced gas permeability and coefficient of thermal expansion (Alexandre and Dubois 2000, Carastan and Demarquette 2007). The improvement in elasticity is attributed to the plasticizing effects of onium ions that contribute to dangling chain formation in the matrix and conformational effects on the polymer at the clay–matrix interface. Loading 5 to 10 wt.% of clay into polyurethane results in a twofold to threefold improvement in tensile strength (Wang and Pinnavaia 1998). The toughness of the polymer also improves because the physical linking between polymer and clay nanofillers increases the fracture strength of the polymer and triggers a toughening mechanism, via multiple crazing and shear yielding (Xu et al. 2008).

For clay nanocomposites, the specific choice of processing steps depends on the final morphology required in the composite, i.e., exfoliated or intercalated form (see Figure 7.3). For nanoclay, realization of full exfoliation is a problem. Many methods increase the degree of the exfoliation ratio, for example, *in situ* polymerization, sonication, and high shear or melting intercalation (Choi et al. 2001). In the intercalated form, matrix polymer molecules are introduced between the ordered layers of clay, resulting in an increase in interlayer spacing, but the clay still maintains the layer order. In the exfoliated form, clay layers are separated and distributed within the matrix (see Figure 7.3). Intercalated nanocomposites are generally formed by melt mixing or by *in situ* polymerization (Khudyakov et al. 2009). Exfoliation level depends on the nature of the clay, blending process, and agents used for curing. The final structure of a clay composite has a wide range of variations, depending on the degrees of intercalation and exfoliation (Alexandre and Dubois 2000, Carastan and Demarquette 2007).

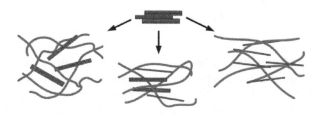

FIGURE 7.3 Top: three types of polymer composites with clay. Bottom left: conventional composite. Bottom center: intercalated nanocomposite. Bottom right: exfoliated nanocomposite.

Cao and Jana (2007) systematically studied the properties of nanoclay–polyurethane nanocomposites using MDI, BDO, and PCL 4000 as soft segments, focusing on the shape memory property. They fabricated nanocomposites with nano clay contents of 1, 3, and 5 wt.% by bulk polymerization. The results showed that clay particles exfoliated well in the matrix, decreased the crystallinity of the soft segment phase, and promoted phase mixing between the hard and soft segments (Cao et al. 2007). A 20% increase in the magnitude of shape recovery stress was obtained with the addition of 1 wt.% nanoclay, and the tensile modulus increased with clay content at temperatures above the melting point of the soft segment crystals.

Mechanical properties of the attapulgite clay-reinforced polyurethane shape memory nanocomposites depend strongly on pre-treatment of the nano clay powder (Xu et al. 2009). After the clay powders were heat treated at 800°C for 2 hours, the loss of moisture and most surface hydroxyl groups resulted in a crystallized and bundled structure (Xu et al. 2009, Pan et al. 2008). Improved interfacial bonding between the polymer and filler enhanced the mechanical properties of the nanocomposites. In comparison, untreated polyurethane–clay nanocomposites showed decreases in both glass transition temperature (T_g) and strength (Xu et al. 2009).

Thermogravimetric analysis (TGA) results for the original and treated clay powders are shown in Figure 7.4. Note the three-stage weight loss of the original clay during heating (Xu et al. 2009). The first stage at a temperature of about 100°C corresponds to the loss of moisture that may exist in attapulgite powder as free water. The second stage occurs at ~200°C when the zeolitic tube is destroyed, coinciding with the loss of hygroscopic water and zeolitic water (Frost and Ding 2003). The third stage beyond 450°C is the point at which the hydroxyl group is gradually reduced. The total weight loss approached 15.84%. Compared with the original clay, the heat-treated one showed no significant drop in weight; re-adsorption of

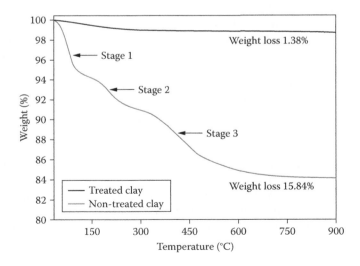

FIGURE 7.4 TGA results of original and heat-treated attapulgite. (Reprinted from Xu B, Huang WM, Pei YT et al. *European Polymer Journal*, 45, 1904–1911, 2009. With permission.)

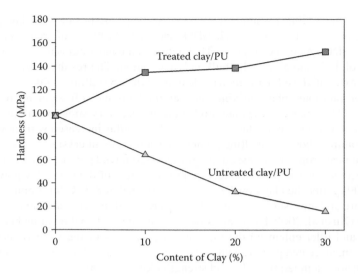

FIGURE 7.5 Microhardness results for thermally treated and untreated polyurethane–clay samples. (Reprinted from Xu B, Huang WM, Pei YT et al. *European Polymer Journal*, 45, 1904–1911, 2009. With permission.)

water molecules after heat treatment was negligible. The results confirm that thermal treatment removes water molecules and most hydroxyl groups in the natural clay powder.

Figure 7.5 shows microhardness data for the pure polyurethane and treated and untreated clay reinforced composites as a function of applied normal load (Xu et al. 2009). The pure polyurethane sample showed the most significant decrease in hardness as a function of normal load (Figure 7.5). This may be explained by the apparent elastic recovery of the polyurethane that may "artificially" enhance the microhardness value at a low load due to shrinkage of the indentations. At a high load, this artificial enhancing is not significant as the indention is quite large. Figure 7.5 clearly shows that with the addition of heat-treated clay powder, the hardness of the composites significantly increases with clay powder content. Conversely, adding untreated clay powder led to a tremendous decrease in the hardness of the nanocomposites—up to nearly 85% for a 30 wt.% content of untreated clay powder.

Figure 7.6a shows the storage moduli of the composites with different clay contents based on dynamic mechanical thermal analysis (DMTA; Xu et al. 2010b). The storage modulus has a maximum value for the SMP nanocomposite with 30 wt.% of clay, indicating that the stiffness of the 30 wt.% polyurethane–clay nanocomposite was the highest among all the tested samples. A sharp drop in modulus was observed above T_g within a narrow temperature range due to the softening effects of the polymer nanocomposites. The tan δ curves shown in Figure 7.6b reveal that the transition temperatures of the nanocomposites increase with clay content. The nanocomposite with 30 wt.% of clay also shows the highest tan δ value, revealing the best energy absorption capacity among these samples.

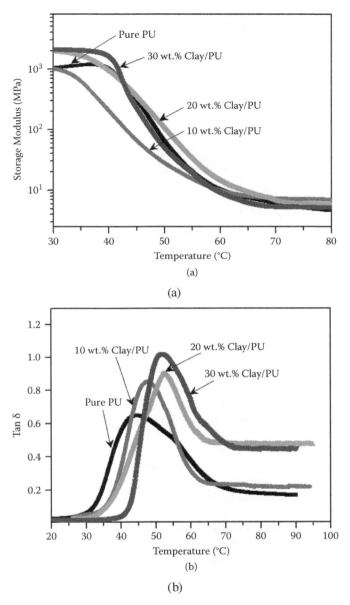

(a)

(b)

FIGURE 7.6 DMTA results of nanoclay-filled SMP composites. (a) Storage modulus versus temperature curves. (b) Tan δ versus temperature curves. (Reprinted from Xu B, Fu YQ, Huang WM et al. *Polymers*, 2, 31–39, 2010. With permission.)

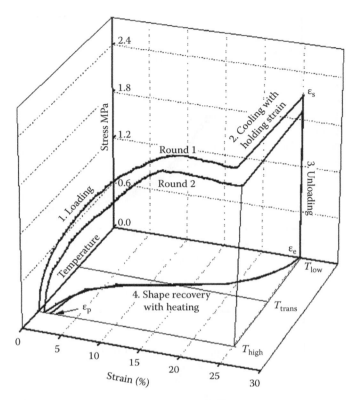

FIGURE 7.7 Typical thermo-uniaxial tensile cyclic results of 30 wt.% clay–polyurethane nanocomposite. (Reprinted from Xu B, Fu YQ, Huang WM et al. *Polymers*, 2, 31–39, 2010. With permission.)

Figure 7.7 shows the thermal cyclic test results for the 30 wt.% polyurethane–clay composite (Xu et al. 2010b). Good shape recovery is observed. The recovery rate was 99.2% in the first tensile cycle and 97% in the second. The maximum stress decreased by 8%, probably because some defects were generated and creep occurred during continuous loading at an air temperature of 60°C.

Thermal annealing (heat treatment) of attapulgite clay may have significantly influenced the transformation temperatures of nanocomposites as seen from DSC results in Pan et al. (2008). Figure 7.8a shows the DSC results for dry polyurethane, a polyurethane–non-treated clay (NTC) composite, and a polyurethane–treated clay (TC) composite upon heating in the first two thermal cycles. For the polyurethane sample, T_g increased by ~17.5°C from 22.93°C in the first cycle to 40.61°C in the second. The T_g of NTC10 was 14.47°C in the first cycle; that of the second was 29.28°C (a 15°C increase). However, for NTC20 (with 20% NTC) and NTC30 (30% NTC), the T_g values in the first cycle were below 0°C (–8.92°C and –13.90°C, respectively). After one cycle, the T_g of NTC20 was 14.47°C (increase of 23.5°C); that of NTC30 was 8.82°C (increment ~22.5°C). The results indicate that the samples were not completely dry. The moisture in the dry samples should have been absorbed from the air

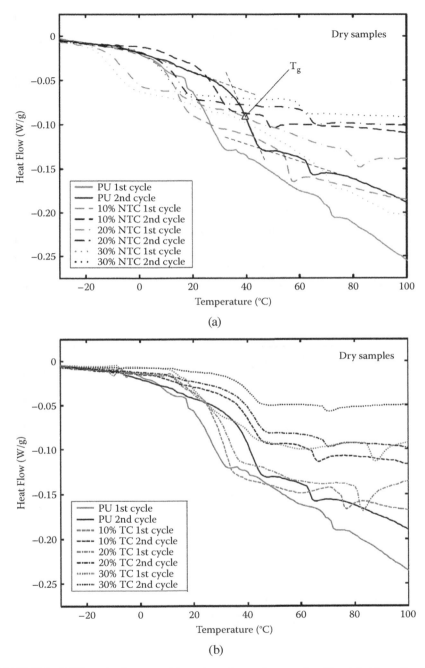

FIGURE 7.8 DSC results for wet and dry polyurethane–clay samples. (Reprinted from Pan GH, Huang WM, Ng ZC et al. *Smart Materials and Structures*, 17, 045007, 2008. With permission.)

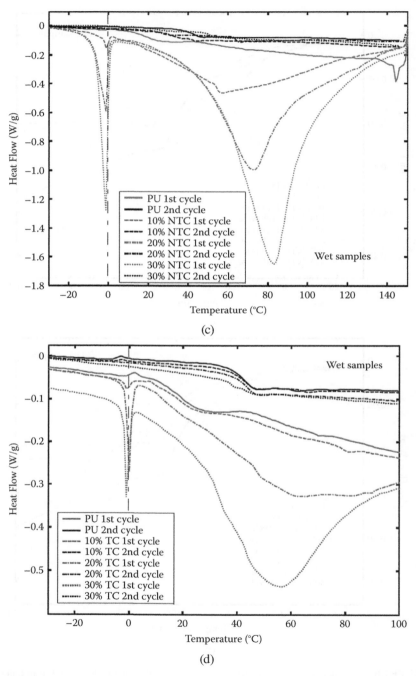

FIGURE 7.8   (Continued)

during sample preparation and storage in containers. NTC content greatly influenced the T_g of the polyurethane–NTC composite. The more NTC in the composite, the lower the T_g.

Figure 7.8b shows the DSC results of dry polyurethane–TC upon heating in the first two cycles. The results of polyurethane are also plotted for comparison. The T_gs of all composites in the first cycle are low. After one cycle, they are about the same as that of polyurethane, indicating the negligible influence of treated clay on the T_g of polyurethane–TC (Pan et al. 2008). A lower T_g in the first cycle should be due largely to moisture.

Figures 7.8c and d present DSC results of wet polyurethane–NTC and polyurethane–TC. We can see noticeable differences in all polyurethane–NTC composites in the first cycle—two apparent troughs. The first trough at ~0°C should be the result of the phase transformation from ice to water. The bottom of the second trough is well above the T_g of even dry polyurethane and spans a very wide temperature range. On the other hand, the results of the second cycle return to normal—only one trough upon heating. The T_g is much higher than that of dry polyurethane–NTC in the second cycle, in particular NTC10 (10% NTC) exhibits about the same T_g as polyurethane.

In brief, NTC exhibited much greater water absorption ability than TC in polyurethane–TC composites. However, the water absorption ability of wet polyurethane–NTC composites is still much lower than that of pure NTC and is significantly higher than that in dry polyurethane–NTC. It appears reasonable to say that high internal stress in a composite can result from expansion of NTCs upon wetting (Pan et al. 2008). On the other hand, this high internal stress can prevent further water absorption in NTCs. If it exceeds a critical point, it may cause permanent deformation of the polyurethane, in particular upon cooling to sub-ambient temperature (due to further volume expansion during the transition from water to ice, while polyurethane is in the hard glass state at low temperature). Subsequently, the significant shrinkage of NTC upon drying may cause its partial detachment from the polyurethane (Pan et al. 2008). The contact and thus the interaction of polyurethane and NTC then reduce dramatically. Consequently, the T_g of the polyurethane–NTC composite is largely determined by that of polyurethane, indicating a much higher T_g than the dry polyurethane–NTC. The highest value may match that of polyurethane.

7.3.1.3 CNT-Based Nanocomposites

The excellent electrical and mechanical properties of carbon nanotubes have aroused great interest in the fabrication of advanced nanocomposites (Geng et al. 2002, Gunes and Jana 2008). CNT-filled polymer composites are expected to improve the mechanical properties of the matrix polymer. SMP–CNT nanocomposites show increased storage elastic moduli, improved shape memory effects, good toughening, and better fatigue strength than pure SMPs (Esawi and Farag 2007). For example, the recovery forces of SMP–CNT nanocomposites filled with 3.3% weight fractions of CNTs improves to almost twice the level of the SMP alone (Gunes and Jana 2008).

For the best performance enhancement, a large aspect ratio, sufficient dispersion, good alignment, and interfacial stress transfer are critical (Coleman et al. 2006). Alignment of CNTs in the loading direction may enhance the modulus by a factor of five or more. The interfacial strength of the CNT–polymer is a critical issue. CNTs in nanocomposites may withstand large forces, but good interfacial strength between CNT and polymer matrix is a pre-requisite, so that the force can be transferred to the CNTs. With an increase of external stress, the matrix in the interface or the matrix–nanotube bond would be broken, resulting in debonding; the strength would decrease substantially. Koerner et al. (2004, 2005) used CNTs to reinforce polyurethane that showed both a significant increase in shape fixity and almost 100% shape recovery. For comparison, the polyurethanes reinforced with carbon black exhibited only limited shape recovery (30%). The difference in the enhancing effect arises from the interactions between the anisotropic CNTs and the crystallizing polyurethane switching segments (Koerner et al. 2004).

The addition of carboxylated MWCNTs to a polyurethane matrix by solvent mixing improves the tensile strength and modulus of the matrix (Sahoo et al. 2006, 2007b). The tensile strength of a composite containing 10 wt.% of carboxylated MWCNT was enhanced by 108% as compared to pure polyurethane. An increase of 68% was achieved by incorporating the same amount of raw MWCNT into the polyurethane matrix. Tensile strengths and moduli of nanocomposites increased from 7.6 MPa in pure polyurethane to 21.3 MPa (increase of 180%) and 50 to 420 MPa (increase of 740%) in the composites, respectively (Sahoo et al. 2005). The hydrophilic functional groups on the MWCNTs help improve interactions with –CONH– groups in the polyurethanes. Therefore, the strong interaction between the functionalized MWCNTs and the polyurethane matrix greatly enhances both dispersion and interfacial adhesion, thus improving the mechanical performance of the composite (Sahoo et al. 2005).

The mechanical properties of nanocomposites also depend on the CNT acid treatment temperature (Sahoo et al. 2010). Kuan et al. (2005) incorporated amino functionalized MWCNTs into waterborne polyurethane. They found increases in modulus from 77 to 131 MPa for a 4% MWCNT composite (increase of 70%) and in tensile strength from 5.1 to 18.9 MPa (increase of 270%) at the same loading level.

Covalent bond formation between amino functionalized MWCNTs and polyurethane promote increases in interfacial strength and tensile strength. MWCNTs are more effective for improving modulus and SWCNTs are better for elongation and tensile strength (Sahoo et al. 2010). The different reinforcing effects of MWCNTs and SWCNTs on polyurethane are attributed to the shear thinning exponent and the shape factor of a CNT in a polyol dispersion (Sahoo et al. 2010). Wang and Tseng (2007) found that adding 1 to 10 wt.% polyurethane-functionalized MWCNT to polyurethane increased the tensile strength by 63 to 210%. The storage modulus and soft segment Tg (from tan δ) increased with increasing polyurethane-functionalized MWCNT in the polyurethane (Kuan et al. 2005, Kwon and Kim 2005).

Polymer grafting is effective for increasing dispersion and mechanical properties of composites due to its strong chemical bonding of polymer and CNT (Sahoo et al. 2010, Lin et al. 2003). McClory et al. (2007) reported polyurethane–MWCNT nanocomposites via a polymerization reaction. The Young's modulus increased by 97 and 561% on the addition of 0.1 and 1 wt.% MWCNTs in the polyurethane, respectively, whereas ultimate tensile strength increased by 397% when 0.1 or 1 wt.% of MWCNTs was added to polyurethane. Xia et al. (2006) studied polycaprolactone-based polyurethane-grafted SWCNTs (SWCNT-g-polyurethane) and poly(propylene glycol)-grafted MWCNTs into polyurethane by *in situ* polymerization. The incorporation of 0.7 wt.% SWCNT-g-polyurethane into polyurethane improved the Young's modulus by 278 and 188% compared to the pure polyurethane and ungrafted pristine polyurethane–SWCNT composites, respectively. The reasons are the better dispersion of SWCNT-grafted polyurethane and MWCNT-grafted polyurethane and the stronger interfacial interaction between the CNTs and polyurethane (Xia et al. 2006).

Improvement of the mechanical properties was reported for melt-processed CNT–polyurethane composite fibers. Sen et al. (2004) studied the fabrication of membranes of SWCNT-filled polyurethane by the electrospinning technique. The tensile strength of ester-functionalized polyurethane–SWCNT membranes was enhanced by 104%, and the tangent modulus improved by 250% compared to polyurethane membranes. The enhancements in mechanical properties may be attributed to the high dispersion of CNTs through the polymer matrix and good interfacial interactions of CNTs and polyurethane.

7.3.1.4 Carbon Nanofibers

The dimensions of carbon nanofibers vary from 5 to 100 microns in length and from 5 to 100 nm in diameter. Conventional fibers have Young's moduli equal to 0.9 TPa, depending on the fabrication processes and structure designs. Due to the exceptional mechanical strength with high elastic modulus and high aspect ratio, carbon nanofibers (CNFs) are effective for improving the mechanical strength and shape recovery stress of SMPs (Meng et al. 2007, 2008, Meng and Hu 2009). Some benefits from CNF nanocomposites include the following (Lan et al. 2009):

- Lower fiber loadings
- Easier fabrication and processing
- More uniform thermal and electrical conductive properties
- Higher ratio of length to diameter
- Shorter cycle and dispersion times
- Lower specific gravity and high strength
- Lower cost

Gunes et al. (2008a, 2008b) fabricated shape memory polyurethane–CNF composites by melt mixing after the chain extension of a T_m-type shape memory polyurethane (melting transition temperature as the switch temperature). The CNFs diminished the shape memory function of the SMPs and the change was ascribed to the interference of CNFs on the crystallization of the soft segment.

Polyurethanes filled with surface-oxidized CNF showed better dispersion, crystallinity, tensile properties, and shape recovery force than their counterparts containing untreated CNFs (Meng and Hu 2009). Koerner et al. (2004) fabricated shape memory polyurethane–CNF composites by solution mixing in a polar solvent, followed by slow evaporation of the solvent. The CNFs had an average diameter of 100 nm and length above 10 µm. Anisotropically distributed CNFs were obtained, thus increasing the rubber modulus by a factor of 2 to 5 (see Figure 7.9 for an example

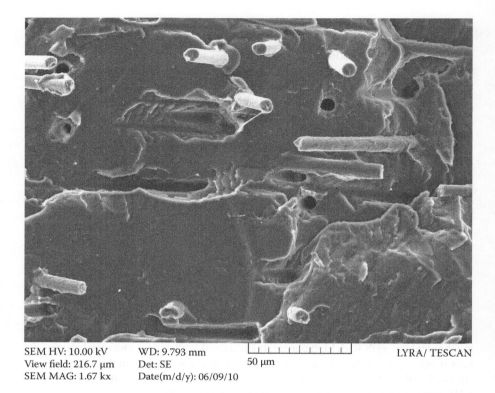

SEM HV: 10.00 kV   WD: 9.793 mm   [scale bar] 50 μm   LYRA/ TESCAN
View field: 216.7 μm   Det: SE
SEM MAG: 1.67 kx   Date(m/d/y): 06/09/10

FIGURE 7.9   Pull-out for SMP–CNF nanocomposite showing mechanical enhancing effect.

of enhancing effect). Shape fixity was improved via the enhanced strain-induced crystallization. In comparison with pure SMPs, shape memory composites with uniform dispersions of 1 to 5 vol.% of CNFs produced up to 50% more recovery stress. Ni et al. (2007) prepared shape memory polyurethane–CNF composites by solution mixing with ultrasonic distribution. The CNFs had a diameter of ~150 nm and lengths of 10 to 20 μm. The recovery stress of the composites at 3.3 wt.% of CNF loading increased almost twofold over pure SMPs.

Leng et al. (2007, 2008a, 2008b, 2008c) fabricated SMP–CNF nanocomposites with volume fractions of 5, 10, 15, and 20%, and studied the effects of different concentrations of nanofibers and chopped carbon fibers on the SME. As the nanofiber concentrations in the SMP matrix increased, the samples showed increasing strength and decreasing damping effect. When the samples were filled with the same amounts of nanofibers and conventional chopped carbon fibers for comparison, the nanofiber-filled sample exhibited relatively higher strength than the sample filled with chopped carbon fiber.

Carbon-fiber-reinforced composites have limits on achievable properties due to the low modulus and strength of the matrix phase compared with CNT-based nanocomposites (Gall et al. 2000). Modification of the matrix phase with carbon nanotubes on a lower scale of dimensions and carbon nanofibers on a higher dimensional scale would allow significant increases in the modulus and strength contributions of the matrix to the overall composite properties. Leng et al. (2008a, 2008b, 2008c) prepared carbon black and short carbon fiber SMP nanocomposites. The synergic effects of CB and CNF networks resulted in significant improvement in both mechanical and electrical properties. The fibrous filler could combine well with particulate filler and enhance the formation of a conductive network.

Lu et al. (2010) studied the synergistic effects of carbon nanofibers and carbon nanopaper on the shape recovery of SMP nanocomposites. The SMP surface was coated with carbon nanopaper to achieve good electrical resistive heating, and CNF was blended into the SMP to improve mechanical and thermal conductivity. Esawi and Farag (2007) also proposed the addition of carbon nanotubes in small quantities to carbon-nanofiber-reinforced nanocomposites to produce components with much enhanced performance at reduced cost.

Another good method to reinforce SMPs is using high modulus inorganic or organic fillers. Chopped and continuous microfibers and fabrics are superior to micro- and nano-sized particles to improve SMP mechanical strength. The studies of the reinforcement effect of chopped glass fibers, unidirectional Kevlar® fibers, and woven fiberglass on thermoplastic SMPs indicated that fibers and fabrics may significantly increase the strength and stiffness levels of SMPs (Meng and Hu 2009, Ohki et al. 2004, Liang et al. 1997).

For the thermo-responsive SMP components of biomedical and healthcare devices, the requirement of external heat for actuation may severely restrict their applications because high T_g levels of SMPs in the range of 50 to 90°C are harmful to the human body. To solve this problem, attempts were made to develop novel SMP composites that are actuated by other external stimuli such as light, alternative magnetic field (AMF), change in pH, and chemical reactions (Jang et al. 2009, Schmidt 2006). Synthesis of such SMP composites typically involves the inclusion of molecules or nanoparticles that are sensitive to heat, light, and AMF into the polymer matrices (Meng and Hu 2009). The next sections focus on these functional applications of shape memory nanocomposites.

7.3.2 SMP Nanocomposites for Electrical Actuation

A polymer is normally considered a non-conductive material because of its extremely low electrical conductivity (10^{-10} to 10^{-15} S/m). Dispersing conductive particles into a non-conductive matrix can form conductive composites. The electrical conductivity of a composite depends strongly on the volume fraction of the conductive phase (Liu et al. 2009). At low volume fractions, the conductivity remains very close to the conductivity of the pure matrix.

With the increased use of conductive fillers, the conductivity of composites drastically increased by many orders of magnitude. Conductive SMP nanocomposites are normally mixed with fillers such as carbon nanotubes (Cho et al. 2005, Paik et al. 2006), carbon nanoparticles (Yang et al. 2005a, 2005b), conductive fibers (Leng et al. 2007), and metals such as Au, Ni, and nickel–zinc–ferrite ferromagnetic particles (Schmidt 2006, Leng et al. 2008a, 2008b, 2008c). For example, the electrical conductivity of a shape memory polyurethane filled with 30 wt.% of carbon black is

~1 to 10^{-1} S/cm (Liu et al. 2007). Another example is a composite SMP incorporating 5 wt.% of surface-modified MWCNTs exhibiting an electrical conductivity as high as 10^{-3} S/cm (Cho et al. 2005). Paik et al. (2006) prepared polyurethane with MWCNTs, and the electrical conductivity was about 2.5×10^{-3} S/cm.

7.3.2.1 Carbon Nanoparticle-Based Nanocomposites

Nano-sized carbon black (CB) is much cheaper than CNTs or CNFs and thus interesting to use for electrically conductive nanocomposites (Meng and Hu 2009, Li et al. 2000 and 2008) triggered by Joule heating (Koerner et al. 2004). However, Paik et al. (2006) found that CB is not so effective in improving the mechanical strength and shape recovery stress of SMPs (Yang et al. 2005a, 2005b, 2006). Gunes et al. (2008a, 2008b) prepared polyurethane–CB composites by melt mixing and found that the soft-segment crystallinity decreased due to the constraining effects of CBs on the mobility of soft segments during crystallization.

Xu et al. (2011) studied polystyrene (PS)-based carbon nanoparticle nanocomposites (Figure 7.10a). Figure 7.10b shows the measured conductivity versus CB

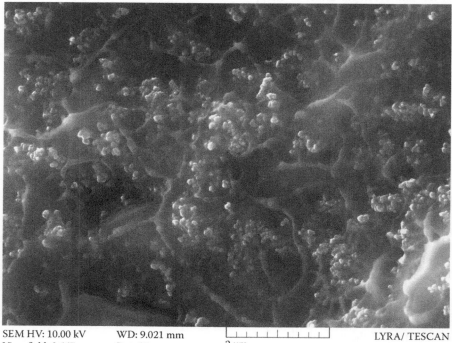

SEM HV: 10.00 kV	WD: 9.021 mm	LYRA/ TESCAN
View field: 8.667 μm	Det: SE	2 μm
SEM MAG: 41.67 kx	Date(m/d/y): 06/09/10	

(a)

FIGURE 7.10 Surface morphology of PS–CB nanocomposite (a) and conductivity of PS–CB nanocomposites versus CB content (b). (From Xu B, Zhang L, Tao SW et al. (2011). Dielectric property of thermo electroactive shape memory polymer nanocomposites. Submitted.)

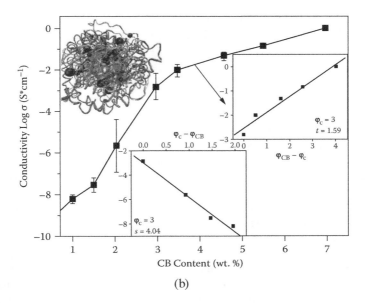

FIGURE 7.10 (Continued)

concentrations at room temperature. With CB content, φ_{CB}, increases from 1 to ~3 wt.%, the measured conductivity dramatically changes from 1×10^{-8} to 10^{-2} S/cm, after which the increase rate drops with a CB concentration above 3 wt.%. At low CB concentrations, the conduction is dominated by hopping among the fillers with electrical conductivity closer to that of an insulator. A material becomes more conductive as the filler concentration increases to a critical value, the percolation threshold (φ_c), which is affected by the filler properties and dispersion in the matrix. Figure 7.10b clearly shows two segments with different slopes, dominated by insulative behavior at $\varphi_{CB} < 3$ wt.% and conductive behavior at $\varphi_{CB} > 3$ wt.%. The conductivity and dielectric constants change dramatically as a function of frequency, temperature, and CB concentration. Measurement of dielectric constant and dielectric loss as a function of temperature and frequency indicated stable thermal electrical performance from room temperature to 100°C (Xu et al. 2011).

7.3.2.2 CNF- and CNT-Based Nanocomposites

CNFs are useful for reinforcement applications, as electrically conductive fillers, and as catalyst supports. The temperature and strain significantly influence the electrical resistivity of CNF-filled SMPs. Gunes et al. (2009) studied the relationship between electrical resistivity and sample temperatures of SMPs filled with CNFs, oxidized CNFs, and CBs. The CNF- and CB-filled SMPs showed pronounced positive temperature coefficients of resistivity. Although fibrous conductive fillers can significantly improve electrical conductivity and enhance the stiffness and strength of SMPs in comparison with CBs, they cause lower maximum bending strain in comparison with conductive particles. A balance must be achieved among filler content, filler type, mechanical properties, and shape memory properties for specific applications. According to the percolation theory, the following equation can be used:

$$\sigma_c = A(V - V_c)^\beta \tag{7.1}$$

where σ_c represents the conductivity of the composites, V is the CNF or CNT volume fraction, V_c is the CNF or CNT volume fraction at the percolation threshold, and A and β are fitted constants (Sahoo et al. 2010). Because CNTs have tremendously large aspect ratios (100 to 10,000), many researchers observed exceptionally low electrical percolation thresholds for both MWCNTs and SWCNTs (Sandler et al. 2003, Bryning et al. 2005). Depending on the matrix, processing technique, and nanotube type used, percolation thresholds ranging from 0.001 to more than 10 wt.% have been reported (Li et al. 2008, Sandler et al. 2003). The percolation threshold for the electrical conductivity in polymer–CNT nanocomposites depends significantly on polymer type, composite processing method, disentanglement of CNT agglomerates, uniform spatial distribution of individual CNTs, degree of alignment, dispersion, alignment (Choi et al. 2003), aspect ratio, and degree of surface modification of CNTs (Sahoo et al. 2010).

Chemical functionalization may disrupt the extended π-conjugation of nanotubes and hence reduce the electrical conductivity of functionalized CNTs (Sahoo et al. 2010). Silane-functionalized CNT–epoxy nanocomposites show lower electrical conductivity than untreated CNT composites at the same nanotube content (Ma et al. 2007, Cho et al. 2005). This is attributed to the increased defects in the lattice structures of carbon–carbon bonds on the nanotube surfaces as a result of acid treatment. However, Tamburri et al. (2005) found that the surface functionalization of CNTs could improve the electrical conductivity of composites.

In brief, adding different electrically conductive components into a thermo-responsive SMP matrix can achieve Joule heating to trigger shape recovery. Such composites can be used in electronic communication, microelectromechanical systems (MEMS), and electromagnetic fields. For example, nanotube-filled polymers are useful for transparent conductive coatings, electrostatic dissipation, electrostatic painting, and electromagnetic interference shielding applications (Sahoo et al. 2010).

7.3.2.3 Graphene-Based Nanocomposites

Graphene is composed of sp^2-bonded carbon atoms arranged in a two-dimensional (2-D) honeycomb lattice consisting of two interpenetrated triangular sub-lattices (Geim 2009, Terrones et al. 2010). Graphene sheets have unique electrical properties and are extremely strong (Lee et al. 2008). Graphene is an ideal material for electrochemistry because of its very large 2-D electrical conductivity, large surface area, and low cost; it is also a good material for fabricating polymer-based nanocomposites (Zhang et al. 2010) and has many distinct advantages over CNT-based nanocomposites (Kim et al. 2010, Choi et al. 2010).

Graphene is a biocompatible material with few metallic impurities; in contrast, serious toxicological effects were reported for CNTs mainly because they contain metallic impurities.

Graphene can be fabricated from inexpensive graphite. CNTs are synthesized using nanoparticles (NPs) as templates from carbon-containing gases; graphene is prepared by "peeling off" highly oriented pyrolytic graphite (HOPG) or grown epitaxially on silicon wafers (Cai and Song 2010). Bulk quantities of graphene can be produced by chemical vapor deposition (CVD), intercalation of small molecules in a graphite lattice and exfoliation by thermal shock, ultrasonication or "unzipping" CNTs via hypermanganate chemical oxidation, or plasma etching of multi-walled CNTs (Patole et al. 2010, Stankovich et al. 2006). However, these bulk-scale production methods for graphene produce 99% of the material as multi-layer graphene nano platelets and only about 1% as true monolayer graphene sheets.

The graphene surface is rough and wrinkled due to the very high density of surface defects resulting from the thermal exfoliation process used to manufacture bulk quantities of platelets from graphite (Rafiee et al. 2009). The wrinkled surfaces interlock extremely well with the surrounding polymer material, helping to enhance the interfacial load transfer between graphene and polymer. Graphene in planar sheets offers considerably more contact with the polymer than tube-shaped CNTs.

When microcracks in a composite structure encounter a 2-D graphene sheet, they are deflected or forced to tilt and twist around the sheet. This process helps absorb the energy responsible for propagating cracks (Rafiee et al. 2009, 2010a, 2010b). Crack deflection processes are far more effective for 2-D sheets with high aspect ratios (e.g., grapheme) than for 1-D nanotubes. Therefore, functionalized graphene sheets are remarkably effective for enhancing fracture toughness, fracture energy, stiffness, strength, and fatigue resistance of epoxy polymers at significantly lower nanofiller loading fractions in comparison to CNTs, nanoparticles, and nanoclay additives. This can be attributed to their enhanced specific surface area, 2-D geometry, and strong nanofiller–matrix adhesion (Liang et al. 2009, Gong et al. 2010).

Graphene fillers have been successfully dispersed in poly(styrene), poly-(acrylonitrile), and poly(methyl methacrylate) matrices and the responses of their Young's moduli, ultimate tensile strength, and glass transition temperatures were characterized by Stankovich et al. (2006). They synthesized graphene–polymer composites via complete exfoliation of graphite and molecular-level dispersion of individual, chemically modified graphene sheets within polymer hosts. The resulting composite exhibited a percolation threshold of 0.1 vol.% for room-temperature electrical conductivity. PET–graphene nanocomposites prepared by melt compounding exhibited superior electrical conductivity and a low percolation threshold of 0.47 vol.% (Zhang et al. 2010). A high electrical conductivity of 2.11 S/m of PET nanocomposite was achieved with only 3.0 vol.% of graphene. The low percolation threshold and superior electrical conductivity are attributed to the high aspect ratio, large specific surface area, and uniform dispersion of the graphene nanosheets in a PET matrix (Zhang et al. 2010).

Ramanathan et al. (2008) prepared graphene–PS nanocomposites revealing a modulus increase of 33% and a percolation threshold for conductivity as low as

0.1 vol.% of graphene. A significant increase (up to 52%) in critical buckling load was also observed in epoxy–graphene nanocomposites with the addition of only 0.1% weight fraction of graphene platelets into the epoxy matrix (Rafiee et al. 2010a, 2010b). The significant increase in buckling load suggests a significant enhancement in load transfer effectiveness between the matrix and graphene platelets under compressive load. Xiao et al. (2010) prepared SMP nanocomposites using free-standing nanolayer graphene by a microwave CVD method. The graphene could be directly used as nanofiller without pre-treatment. The fabricated nanocomposites displayed significant enhancement in modulus, scratch resistance, and thermal healing capability, even at ultra-low filler concentrations of 0.0025 and 0.0125 vol.%. Xiao et al. (2010) attributed the improvement to the good separation and dispersion of nano-graphene and sufficient interfacial adhesion between filler and polymer matrix, thus improving load transfer. The good in-plane fracture strength of individual graphene sheets was one possible reason for the enhancement in properties.

7.3.3 SMP NANOCOMPOSITES FOR MAGNETIC FIELD ACTUATION

Ferromagnetic and ferrite nanoparticles such as FeO_x, $MgZnO_x$, and $NiZnFeO_x$ can be inductively heated up to ~100°C by a magnetic field (Kainuma et al. 2006). The electromagnetic energy from the external high frequency field is transformed to heat via a relaxation process (Spinks et al. 2006, Ahir and Terentjev 2005, Ahir et al. 2006). The incorporation of magnetic nanoparticles such as Fe_3O_4 with a shell of oligo(e-caprolactone) into SMP material may result in the heating of SMP nanocomposites in the presence of an alternating magnetic field (Mohr et al. 2006).

Schmidt (2006) and Yakacki et al. (2009) used a magnetic field to remotely actuate SMP composites by incorporating nano iron (II, III) oxide nanoparticles into thermoplastics and thermoset SMPs. To improve particle dispersion in the matrix, the iron (III) oxide nanoparticles were coated with silica (Weigel et al. 2009). By selecting a ferromagnetic particle material with a Curie temperature within safe medical limits, overheating can be avoided because the material heating is induced only by a magnetic hysteresis loss mechanism instead of an eddy current mechanism. Complex shapes are also applicable because this method produces uniform heating for any type of device geometry. Selective heating of specific areas is also possible. Remote actuation allows the actuation of embedded devices by an externally applied magnetic field (Meng and Hu 2009, Zhang et al. 2007), thus eliminating connection wires, power transmission lines, and fiber optics (Buckley et al. 2006). Medical devices such as expandable stents and intravascular microactuators of SMP nanocomposites have utilized magnetic actuation (Buckley et al. 2006).

7.3.4 SMP NANOCOMPOSITES FOR OPTICAL AND PHOTOVOLTAIC ACTUATION

Both SWCNTs and MWCNTs have been used for photonic devices to control light frequency and intensity in a predictable manner (Yu and Ikeda 2005). Goh et al. (2003) found that an aqueous MWCNT suspension showed only a weak optical limit toward a laser at 532 nm operating at 20 Hz, but its mixture with double-C60 end-capped poly(ethylene oxide) (FPEOF) solution displayed enhanced optical limiting responses at 532 and 1064 nm. Polymer–CNT composites may be useful for eye protection, optical elements, optical sensors, and optical switching.

CNTs are also widely used in organic photovoltaic devices (Xie et al. 2005). Doping with 6 wt.% of chemically functionalized MWCNTs by grafting dodecylamine chains increased the photosensitivity of oxotitanium phthalocyanine (TiOPc) five times that of undoped TiOPc when exposed to 570 nm wavelength (Cao et al. 2003). It is beneficial to design photoconductive devices with highly efficient charge carrier generation, and polymer–CNT nanocomposites represent an alternative class of organic semi-conducting materials. They are promising for organic photovoltaic cells and devices requiring improved performance (Kymakis and Amaratunga 2002, Kymakis et al. 2003).

Noble metal (Au and Ag) nanoparticles have strong surface plasmon resonance (SPR) with distinctive absorption wavelengths. This characteristic is widely utilized in biotechnology for biomarker monitoring, controlled drug release, and hyperthermia therapy (Pitsillides et al. 2003). Impregnation of such nanoparticles in a medium increases its ability to absorb light and convert most of the light into localized "heat" with a small portion converted into photoluminescence. The absorption efficiency and SPR frequency strongly depend on the sizes, shapes, and concentrations of the nanoparticles. The resonant frequency may shift from deep UV to near IR light; smaller nanoparticles absorb short-wavelength light and larger ones absorb longer wavelengths. Au-Au_2S nanoparticles of ~20 nm, consisting of an Au_2S dielectric core encapsulated by a thin gold shell, were reported to reach absorption peak at near-IR light and demonstrated controlled drug release (Ren and Chow 2003).

7.3.5 THERMAL PROPERTIES OF SMP NANOCOMPOSITES

The organic nature of SMPs allows them to function as thermal insulators with thermal conductivities usually below 0.30 W/m K. Rapid heating to release and rapid cooling to fix a deformed SMP is a challenge. Fillers, such as alumina, fused silica, SiC, boron nitride (BN), and glass fibers, can be used to increase the thermal conductivities of SMPs.

Razzaq and Frormann (2007) used aluminum nitride fillers to increase the thermal conductivities of SMPs. An addition of 40 wt.% of AlN particles increased the thermal conductivity by ~50% at room temperature. The addition of CNTs may increase the glass transition, melting, and thermal decomposition temperatures of the polymer matrix due to their constraint effects on the polymer segments and chains. For example, adding 1 wt.% of CNTs to epoxy increased the glass transition temperature from 63 to 88°C (Haddad and Lichtenhan 1996). The incorporation of CNTs may improve the thermal transport properties of polymer composites due to their excellent thermal conductivity. Polymer–CNT composites can be used in printed circuit boards, connectors, thermal interface materials, heat sinks, lids and housings, and high-performance thermal management devices, from complex satellite structures to simple electronic device packaging (Meng and Hu 2009).

Polymer-based clay nanocomposites exhibit good thermal stability and reduced gas permeability and coefficients of thermal expansion (Carastan and Demarquette

2007). The thermal stability, fire resistance, and gas barrier properties of polymer–clay nanocomposites may be enhanced through the addition of nanometer-scale clay reinforcement (Cao and Jana 2007). Yao et al. 2002 found that the thermal conductivity of PU–layered clay nanocomposites decreased slightly with an increase in clay loading. Clay-based PU nanocomposites also showed a dramatically improved permeability for water vapor when compared to the PU polymer (Xu et al. 2003). Adding ~3 to 5 wt.% of nanodispersed clay may reduce heat release rate up to 70%, significantly improving the fire resistance of the polymer (Zhu et al. 2001).

7.4 SUMMARY

Significant progress has been made recently in the design, synthesis, characterization, and application of SMP nanocomposites to enhance the mechanical, electrical, magnetic, optical, and thermal properties of polyurethanes. Zero-dimensional (spherical nanoparticles), one-dimensional (clay rods, nanotubes, fibers), two-dimensional (layered clay minerals), and three-dimensional (inorganic networking frames) types of nanofillers have been used. Shape recovery can be triggered thermally, electrically, magnetically, and even electromagnetically. Some critical issues in shape memory nanocomposites (Tonge and Tighe 2001, Spitalsky et al. 2010, Tamburi et al. 2005) are

- Uncertainty in theoretical modeling and experimental characterization of nanocomposites, particularly nanotube-based composites.
- Lack of understanding of the interfacial bonding between reinforcements and matrix materials. Adding nanofillers may change the microstructure or crystallinity of the polymer, thus producing a negative effect (Gunes et al. 2008a, 2008b).
- Difficulty in dispersion or alignment of nanoparticles and nanotubes against their agglomerations along with exfoliation of clay and graphene layers.
- Lack of cost-effectiveness and efficiency in high-volume production of nanocomposites.
- New functions and new challenges such as the SME in cyclic actuation, multiple SMEs (Bellin et al. 2007, Chung et al. 2008), and multiple stimuli in activation.

ACKNOWLEDGMENT

We wish to thank C. Tang for his help in compiling this chapter.

REFERENCES

Ahir SV and Terentjev EM (2005). Photomechanical actuation in polymer–nanotube composites. *Nature Materials*, 4, 491–495.
Ahir SV, Tajbakhsh AR, and Terentjev EM (2006). Self-assembled shape-memory fibre of triblock liquid–crystal polymers. *Advanced Functional Materials*, 16, 556–560.

Alexandre M and Dubois P (2000). Polymer-layered silicate nanocomposites: preparation, properties and uses of a new class of materials. *Materials Science and Engineering R Reports*, 28, 1–63.

Amalvy JI, Percy MJ, and Armes SP (2001). Synthesis and characterization of novel film-forming vinyl polymer/silica colloidal nanocomposites. *Langmuir*, 16, 4770–4778.

An KH, Jeong SY, Hwang HR et al. (2004). Enhanced sensitivity of a gas sensor incorporating single-walled carbon nanotube-polypyrrole nanocomposites. *Advanced Materials*, 16, 1005–1009.

Andrews R and Weisenberger MC (2004). Carbon nanotube polymer composites. *Current Opinions in Solid State and Materials Science*, 8, 31–37.

Andrews R, Jacques D, Minot M et al. (2002). Fabrication of carbon multiwall nanotube/polymer composites by shear mixing. *Macromolecular Materials and Engineering*, 287, 395–403.

Baer G, Wilson TS, Matthews DL et al. (2006). Shape-memory behaviour of thermally stimulated polyurethane for medical applications. *Journal of Applied Polymer Science*, 103, 3882–3892.

Behl M and Lendlein A (2007). Shape-memory polymers. *Materials Today*, 10, 20–28.

Bellin I, Kelch S, and Lendlein A (2007). Dual-shape properties of triple-shape polymer networks with crystallizable network segments and grafted side chains. *Journal of Materials Chemistry*, 17, 2885–2891.

Bin Y, Kitanaka M, Zhu D et al. (2003). Development of highly oriented polyethylene filled with aligned carbon nanotubes by gelation/crystallization from solutions. *Macromolecules*, 36, 6213–6219.

Bonard JM, Stora T, Salvetat JP et al. (1997). Purification and size-selection of carbon nanotubes. *Advanced Materials*, 9, 827.

Bryning MB, Islam MF, Kikkawa JM et al. (2005). Very low conductivity threshold in bulk isotropic single-walled carbon nanotube–epoxy composites. *Advanced Materials*, 17, 1186–1191.

Buckley PR, McKinley GH, Wilson TS et al. (2006). Inductively heated shape memory polymer for the magnetic actuation of medical devices. *IEEE Transactions in Biomedical Engineering*, 53, 2075–2083.

Cadek M, Coleman J, Ryan K et al. (2004). Reinforcement of polymers with carbon nanotubes: the role of nanotube surface area. *Nano Letters*, 4, 353–356.

Cai D and Song M (2010). Recent advance in functionalized graphene/polymer nanocomposites. *Journal of Materials Chemistry*, 20, 7906–7916.

Cao F and Jana SC (2007). Nanoclay-tethered shape memory polyurethane nanocomposites. *Polymer*, 48, 3790–3800.

Cao J, Sun JZ, Wang M et al. (2003). Effect of different substituents on conducting character of azos and local states of azo/TiOPc composites. *Thin Solid Films*, 429, 152–158.

Carastan DJ and Demarquette NR (2007). Polystyrene/clay nanocomposites. *International Materials Reviews*, 52, 345–380.

Carrado KA (2000). Synthetic organo- and polymer-clays: preparation, characterization, and materials applications. *Applied Clay Science*, 17, 1–23.

Chattopadhyay DK and Raju KVSN (2007). Structural engineering of polyurethane coatings for high performance applications. *Progress in Polymer Science*, 32, 352–418.

Chaudhury NK and Gupta R (2007). Entrapment of biomolecules in sol–gel matrix for applications in biosensors: problems and future prospects. *Biosensors and Bioelectronics*, 22, 2387–2399.

Chen W and Tao XM (2005). Self-organizing alignment of carbon nanotubes in thermoplastic polyurethane. *Macromolecular Rapid Communications*, 26, 1763–1767.

Chen Y, Zhou S, Yang H et al. (2005). Structure and properties of polyurethane/nanosilica composites. *Journal of Applied Polymer Science*, 95, 1032.

Chen J, Ramasubramaniam R, Xue C et al. (2006). A versatile molecular engineering approach to simultaneously enhanced multifunctional CNT–polymer composites. *Advanced Functional Materials*, 16, 114–119.

Cho JW and Lee SH (2004). Influence of silica on shape memory effect and mechanical properties of polyurethane–silica hybrids. *European Polymer Journal*, 40, 1343–1348.

Cho JW, Kim JW, Jung YC et al. (2005). Electroactive shape-memory polyurethane composites incorporating carbon nanotubes. *Macromolecular Rapid Communications*, 26, 412–416.

Choi ES, Brooks JS, Eaton DL et al. (2003). Enhancement of thermal and electrical properties of carbon nanotube polymer composites by magnetic field processing. *Journal of Applied Physics*, 94, 6034–6039.

Choi YS, Choi MH, Wang KH et al. (2001). Synthesis of exfoliated PMMA/Na-MMT nanocomposites via soap-free emulsion polymerization. *Macromolecules*, 34, 8978–8985.

Choi W, Lahiri I, Seelaboyina R et al. (2010). Synthesis of graphene and its applications: a review. *Critical Reviews in Solid State and Materials Sciences*, 35, 52–71.

Chung T, Romo-Uribe A, and Mather PT (2008). Two-way reversible shape memory in a semicrystalline network. *Macromolecules*, 41, 184–192.

Coleman JN, Khan U, Blau J et al. (2006). A review of the mechanical properties of carbon nanotube–polymer composites. *Carbon*, 44, 1624–1632.

Cui J, Wang WP, You YZ et al. (2004). Functionalization of multiwalled carbon nanotubes by reversible addition fragmentation chain-transfer polymerization. *Polymer*, 45, 8717–8721.

Curtin WA and Sheldon BW (2004). CNT-reinforced ceramics and metals. *Materials Today*, 7, 44–49.

Díaz-García ME and Badía-Lamo R (2005). Molecular imprinting in sol-gel materials: recent developments and applications. *Microchimica Acta*, 149, 19–36.

Dietsch B and Tong T (2007). A review—features and benefits of shape memory polymers (SMPs). *Journal of Advanced Materials*, 39, 3–12.

Esawi AMK and Farag MM (2007). Carbon nanotube reinforced composites: potential and current challenges. *Materials and Design*, 28, 2394–2401.

Frost RL and Ding Z (2003). Controlled rate thermal analysis and differential scanning calorimetry of sepiolites and palygorskites. *Thermochimica Acta*, 397, 119–128.

Gall K, Mikulas M, Munshi NA et al. (2000). Carbon fibre reinforced shape memory polymer composites. *Journal of Intelligent Material Systems and Structures*, 11, 877–886.

Gall K, Dunn ML, Liu YP et al. (2004a). Internal stress storage in shape memory polymer nanocomposites. *Applied Physics Letters*, 85, 290–292.

Gall K, Kreiner P, Turner D et al. (2004b). Shape-memory polymers for microelectromechanical systems. *Journal of Microelectromechanical Systems*, 13, 472–483.

Gall K, Dunn ML, Liu Y et al. (2005). Shape memory polymer nanocomposites. *Acta Materialia*, 50, 5115–5126.

Geim AK (2009). Graphene: status and prospects. *Science*, 324, 1530–1534.

Geng HZ, Rosen R, Zheng B et al. (2002). Fabrication and properties of composites of poly(ethylene oxide) and functionalized carbon nanotubes. *Advanced Materials*, 14, 1387–1390.

Goh HW, Goh SH, Xu GQ et al. (2003). Optical limiting properties of double-C-60-end-capped poly(ethylene oxide), double-C-60-end-capped poly(ethylene oxide)/poly(ethylene oxide) blend, and double-C-60-end-capped poly(ethylene oxide)/multiwalled carbon nanotube composite. *Journal of Physical Chemistry B*, 107, 6056–6062.

Gong XY, Liu J, Baskaran S et al. (2000). Surfactant-assisted processing of carbon nanotube/polymer composites. *Chemistry of Materials*, 12, 1049–1052.

Gong L, Kinloch LA, Young RJ et al. (2010). Interfacial stress transfer in a graphene monolayer nanocomposite. *Advanced Materials*, 22, 2694.

Gunes IS and Jana SC (2008). Shape memory polymers and their nanocomposites: a review of science and technology of new multifunctional materials. *Journal of Nanoscience and Nanotechnology*, 8, 1616–1637.

Gunes IS, Cao F, and Jana SC (2008a). Effect of thermal expansion on shape memory behavior of polyurethane and its nanocomposites. *Journal of Polymer Science B*, 46, 1437–1449.

Gunes IS, Cao F, and Jana SC (2008b). Evaluation of nanoparticulate fillers for development of shape memory polyurethane nanocomposites. *Polymer*, 49, 2223–2234.

Gunes IS, Jimenez GA, and Jana SC (2009). Carbonaceous fillers for shape memory actuation of polyurethane composites by resistive heating. *Carbon*, 47, 981–997.

Guo Z, Pereira T, Choi O et al. (2006). Surface functionalized alumina nanoparticle filled polymeric nanocomposites with enhanced mechanical properties. *Journal of Materials Chemistry*, 16, 2800.

Guo Z, Lee SE, Kim H et al. (2009). Fabrication, characterization and microwave properties of polyurethane nanocomposites reinforced with iron oxide and barium titanate nanoparticles. *Acta Materialia*, 57, 267–277.

Haddad TS and Lichtenhan JD (1996). Hybrid orabin-inorganic thermoplastic: styryl-based polyhedral oligomeric silsesquioxane polymers. *Macromolecules*, 29, 7302–7304.

Hiura H, Ebbesen TW, and Tanigaki K (1995). Opening and purification of carbon nanotubes in high yields. *Advanced Materials*, 7, 275.

Hou HQ, Ge JJ, Zeng J et al. (2005). Electrospun polyacrylonitrile nanofibers containing a high concentration of well-aligned multiwall carbon nanotubes. *Chemistry of Materials*, 17, 967–973.

Huang ZM, Zhang YZ, Kotaki M et al. (2003). A review on polymer nanofibers by electrospinning and their applications in nanocomposites. *Composites Science and Technology*, 63, 2223–2253.

Huang WM, Fu YQ, and Zhao Y (2010). High performance PU shape memory polymer and composites. *PU Magazine*, 1, 49–53.

Hussain F, Hojjati M, Okamoto M et al. (2006). Polymer-matrix nanocomposites, processing, manufacturing, and application: an overview. *Journal of Composite Materials*, 40, 1511–1575.

Jang MK, Hartwig A, and Kim BK (2009). Shape memory polyurethanes cross-linked by surface modified silica particles. *Journal of Materials Chemistry*, 19, 1166–1172.

Jin ZX, Pramoda KP, Xu GQ et al. (2001). Dynamic mechanical behavior of melt-processed multi-walled carbon nanotube/poly(methyl methacrylate) composites. *Chemical Physics Letters*, 337, 43–47.

Jordan J, Jacob KI, Tannenbaum R et al. (2005). Experimental trends in polymer nanocomposites—a review. *Materials Science and Engineering A*, 393, 1–11.

Jung YC, Sahoo NG, and Cho JW (2006). Polymeric nanocomposites of polyurethane block copolymers and functionalized multiwalled carbon nanotubes as crosslinkers. *Macromolecular Rapid Communications*, 27, 126–131.

Kainuma R, Imano Y, Ito W et al. (2006). Magnetic-field-induced shape recovery by reverse phase transformation. *Nature*, 439, 957–960.

Khudyakov IV, Zopf RD, and Turro NJ (2009). Polyurethane nanocomposites. *Designed Monomers and Polymers*, 12, 279–290.

Kim JS and Reneker DH (1999). Mechanical properties of composites using ultrafine electrospun fibers. *Polymer Composites*, 20, 124–131.

Kim H, Abdala AA, and Macosko CW (2010). Graphene/polymer nanocomposites. *Macromolecules*, 43, 6515–6530.

Kirk WP, Wouters KL, Basit NA et al. (2003). Electrical and optical effects in molecular nanoscopic-sized building blocks. *Systems and Nanostructures*, 19, 126–132.

Koerner H, Price G, Pearce N et al. (2004). Remotely actuated polymer nanocomposites: stress recovery of carbon nanotube-filled thermoplastic elastomers. *Nature Materials*, 3, 115–120.

Koerner H, Liu W, Alexander M et al. (2005). Deformation–morphology correlations in electrically conductive CNT–thermoplastic polyurethane nanocomposites. *Polymer*, 46, 4405–4420.

Kota AK, Cipriano BH, Duesterberg MK et al. (2007). Electrical and reheological percolation in polystyrene/MWCNT nanocomposites. *Macromolecules*, 40, 7400–7406.

Kotal A, Mandal TK, and Walt DR (2005). Synthesis of gold-poly(methyl methacrylate) core-shell nanoparticles by surface-confined atom transfer radical polymerization at elevated temperature. *Journal of Polymer Science A*, 43, 3631–3642.

Kuan HC, Ma CCM, Chang WP et al. (2005). Synthesis, thermal, mechanical and rheological properties of multiwall carbon nanotube/waterborne polyurethane nanocomposite. *Composites Science and Technology*, 65, 1703–1710.

Kwon J and Kim H (2005). Comparison of the properties of waterborne polyurethane–multiwalled carbon nanotube and acid-treated multiwalled carbon nanotube composites prepared by in situ polymerization. *Journal of Polymer Science A*, 43, 973–985.

Kymakis E and Amaratunga GAJ (2002). Single-wall carbon nanotube–conjugated polymer photovoltaic devices. *Applied Physics Letters*, 81, 112.

Kymakis E, Alexandrou I, and Amaratunga GAJ (2003). Carrier tunneling and device characteristics in polymer light-emitting diodes. *Journal of Applied Physics*, 93, 1764.

Lan X, Liu Y, Lv H et al. (2009). Fibre reinforced shape-memory polymer composite and its application in a deployable hinge. *Smart Materials and Structures*, 18, 024002.

Lau KT and Hui D (2002). The revolutionary creation of new advanced materials-carbon nanotube composites. *Composites B*, 33, 263–277.

Lee C, Wei XD, Kysar JW et al. (2008). Measurement of the elastic properties and intrinsic strength of monolayer graphene. *Science*, 321, 385–388.

Lendlein A and Kelch S (2002). Shape-memory polymers. *Angewandte Chemie-International Edition*, 41, 2034–2057.

Lendlein A and Kelch S (2007). Shape memory polymers. *Materials Today*, 10, 20–29.

Lendlein A and Langer R (2002). Biodegradable, elastic shape-memory polymers for potential biomedical applications. *Science*, 96, 1673–1676.

Lendlein A, Jiang H, Junger O et al. (2005). Light-induced shape-memory polymers. *Nature*, 434, 879–882.

Leng JS, Lv HB, Liu YJ et al. (2007). Electroactivated shape-memory polymer filled with nanocarbon particles and short carbon fibers. *Applied Physics Letters*, 91, 144105.

Leng JS, Huang WM, Lan X et al. (2008a). Significantly reducing electrical resistivity by forming conductive Ni chains in a polyurethane shape-memory polymer/carbon-black composite. *Applied Physics Letters*, 92, 204101.

Leng JS, Lan X, Liu YJ et al. (2008b). Electrical conductivity of thermoresponsive shape-memory polymer with embedded micron sized Ni powder chains. *Applied Physics Letters*, 92, 014104.

Leng JS, Lv H, Liu Y et al. (2008c). Synergic effect of carbon black and short carbon fibre on shape memory polymer actuation by electricity. *Journal of Applied Physics*, 104, 104917.

Li C, Thostenson ET, and Chou TW (2008). Sensors and actuators based on carbon nanotubes and their composites: a review. *Composites Science and Technology*, 68, 1227–1249.

Li F, Qi L, Yang J et al. (2000). Polyurethane/conducting carbon black composites: structure, electric conductivity, strain recovery behavior, and their relationships. *Journal of Applied Polymer Science*, 75, 68–77.

Liang C, Rogers C, and Malafeew E (1997). Investigation of shape memory polymers and their hybrid composites. *Journal of Intelligent Material Systems and Structures*, 8, 380–386.

Liang JJ, Huang Y, Zhang L et al. (2009). Molecular-level dispersion of graphene into poly(vinyl alcohol) and effective reinforcement of their nanocomposites. *Advanced Functional Materials*, 19, 2297–2302.

Lin Y, Zhou B, Fernando KAS et al. (2003). Polymeric carbon nanocomposites from carbon nanotubes functionalized with matrix polymer. *Macromolecules,* 36, 7199–7204.

Lin Y, Meizani MJ, and Sun YP (2007). Functionalized carbon nanotubes for polymeric nanocomposites. *Journal of Materials Chemistry*, 17, 1143–1148.

Liu Y, Gall K, Dunn M et al. (2004). Thermomechanics of shape memory polymer nanocomposites. *Mechanics of Materials*, 36, 929–940.

Liu C, Qin H, and Mather PT (2007). Review of progress in shape-memory polymers. *Journal of Materials Chemistry*, 17, 1543–1558.

Liu YJ, Lv HB, Lan X et al. (2009). Review of electro-active shape-memory polymer composite. *Composites Science and Technology*, 69, 2064–2068.

Lu H, Liu Y, Gou J et al. (2010). Synergistic effect of carbon nanofibre and carbon nanopaper on shape memory composite. *Applied Physics Letters*, 96, 084102.

Ma PC, Kim JK, and Tang BZ (2007). Effects of silane functionalization on the properties of carbon nanotube/epoxy nanocomposites. *Composites Science and Technology*, 67, 2965–2972.

Marutani E, Yamamoto S, Ninjbadgar T et al. (2004). Surface-initiated atom transfer radical polymerization of methyl methacrylate on magnetite nanoparticles. *Polymer,* 45, 2231–2235.

McClory C, McNally T, Brennan GP et al. (2007). Thermosetting polyurethane multiwalled carbon nanotube composites. *Journal of Applied Polymer Science*, 105, 1003–1011.

Meng QH and Hu JL (2009). A review of shape memory polymer composites and blends. *Composites* A, 40, 1661–1672.

Meng Q, Hu J, and Zhu Y (2007). Shape-memory polyurethane/multiwalled carbon nanotube fibers. *Journal of Applied Polymer Science*, 106, 837–848.

Meng QH, Hu JL, and Mondal S (2008). Thermal sensitive shape recovery and mass transfer properties of polyurethane/modified MWNT composite membranes synthesized via in situ solution pre-polymerization. *Journal of Membrane Science,* 319, 102–110.

Mohr R, Kratz K, Weigel T et al. (2006). Initiation of shape-memory effect by inductive heating of magnetic nanoparticles in thermoplastic polymers. *Proceedings of the National Academy of Sciences of the United States of America*, 103, 3540–3545.

Morgan AB (2006). Flame retarded polymer layered silicate nanocomposites: a review of commercial and open literature systems. *Polymers for Advanced Technologies*, 17, 206–217.

Ni QQ, Zhang CS, Fu YQ et al. (2007). Shape memory effect and mechanical properties of carbon nanotube/shape memory polymer nanocomposites. *Composite Structures*, 81, 176–184.

Ohki T, Ni QQ, Ohsako N et al. (2004). Mechanical and shape memory behavior of composites with shape memory polymer. *Composites A*, 35, 1065–1073.

Paik IH, Goo NS, Jung YC et al. (2006). Development and application of conducting shape memory polyurethane actuators. *Smart Materials and Structures*, 15, 1476–1482.

Pan GH, Huang WM, Ng ZC et al. (2008). The glass transition temperature of polyurethane shape memory polymer reinforced with treated/non-treated attapulgite (palygorskite) clay in dry and wet conditions. *Smart Materials and Structures*, 17, 045007.

Park JS, Chung YC, Lee SD et al. (2008). Shape memory effects of polyurethane block copolymers cross-linked by Celite. *Fibers and Polymers*, 9, 661–666.

Patole AS, Patole SP, Kang H et al. (2010). A facile approach to the fabrication of graphene/polystyrene nanocomposite by in situ microemulsion polymerization. *Journal of Colloid and Interface Science*, 350, 530–537.

Pavlidou S and Papaspyrides CD (2008). A review on polymer-layered silicate nanocomposites. *Progress in Polymer Science*, 33, 1119–1198.

Petrovic ZS, Javni I, Waddon A et al. (2000). Structure and properties of polyurethane–silica nanocomposites. *Journal of Applied Polymer Science*, 76, 133–151.

Pitsillides CM, Joe EK, Wei XB et al. (2003). Selective cell targeting with light-absorbing microparticles and nanoparticles. *Biophysical Journal*, 84, 4023–4032.

Qian D, Dickey EC, Andrews R et al. (2000). Load transfer and deformation mechanisms in carbon nanotube-polystyrene composites. *Applied Physics Letters*, 76, 2868–2870.

Rafiee MA, Rafiee J, Wang Z et al. (2009). Enhanced mechanical properties of nanocomposites at low graphene content. *ACS Nano*, 3, 3884–3890.

Rafiee MA, Rafiee J, Srivastava I et al. (2010a). Fracture and fatigue in graphene nanocomposites. *Small*, 6, 179–183.

Rafiee MA, Rafiee J, Yu ZZ et al. (2010b). Buckling resistant graphene nanocomposites. *Applied Physics Letters*, 95, 223103.

Ramanathan T, Abdala AA, Stankovich S et al. (2008). Functionalized graphene sheets for polymer nanocomposites. *Nature Nanotechnology*, 3, 327–331.

Ratna D and Karger-Kocsis J (2008). Recent advances in shape memory polymers and composites: a review. *Journal of Materials Science*, 43, 254–269.

Razzaq MY and Frormann L (2007). Thermomechanical studies of aluminum nitride filled shape memory polymer composites. *Polymer Composites*, 28, 287–293.

Ren L and Chow GM (2003). Synthesis of NIR-sensitive Au-Au$_2$S nanocolloids for drug delivery. *Materials Science and Engineering C*, 23, 113–116.

Rozenberg BA and Tenne R (2008). Polymer-assisted fabrication of nanoparticles and nanocomposites. *Progress in Polymer Science*, 33, 40–112.

Rusa M, Whitesell JK, and Fox MA (2004). Controlled fabrication of gold/polymer nanocomposites with a highly structured poly(N-acylethylenimine) shell. *Macromolecules*, 8, 2766–2774.

Sahoo NG, Jung YC, and Cho JW (2005). Conducting shape memory polyurethane–polypyrrole composites for an electroactive actuator. *Macromolecular Materials and Engineering*, 290, 1049–1055.

Sahoo NG, Jung YC, Yoo HJ et al. (2006). Effect of functionalized carbon nanotubes on molecular interaction and properties of polyurethane composites. *Macromolecular Chemistry and Physics*, 207, 1773–1780.

Sahoo NG, Jung YC, and Cho JW (2007a). Electroactive shape memory effect of polyurethane composites filled with carbon nanotubes and conducting polymer. *Materials and Manufacturing Processes*, 22, 419–423.

Sahoo NG, Jung YC, Yoo HJ et al. (2007b). Influence of carbon nanotubes and polypyrrole on the thermal, mechanical and electroactive shape-memory properties of polyurethane nanocomposites. *Composites Science and Technology*, 67, 1920–1929.

Sahoo NG, Rana S, Cho JW et al. (2010). Polymer nanocomposites based on functionalized carbon nanotubes. *Progress in Polymer Science*, 35, 837–867.

Sanchez C, Soler-Illia G, Ribot F et al. (2001). Designed hybrid organic–inorganic nanocomposites from functional nanobuilding blocks. *Chemistry of Materials*, 13, 3061–3083.

Sanchez C, Lebeau B, Chaput F et al. (2003). Optical properties of functional hybrid organic–inorganic nanocomposites. *Advanced Materials*, 15, 1969–1994.

Sandler J, Shaffer MSP, Prasse T et al. (1999). Development of a dispersion process for carbon nanotubes in an epoxy matrix and the resulting electrical properties. *Polymer,* 40, 5967–5971.

Sandler JKW, Kirk JE, Kinloch IA et al. (2003). Ultra-low electrical percolation threshold in carbon-nanotube–epoxy composites. *Polymer,* 44, 5893–5899.

Schmidt AM (2006). Electromagnetic activation of shape memory polymer networks containing magnetic nanoparticles. *Macromolecular Rapid Communications,* 27, 1168–1172.

Schmidt G and Malwitz MM (2003). Properties of polymer–nanoparticle composites. *Current Opinions in Colloid and Interface Science,* 8, 103–108.

Schmidt D, Shah D, and Giannelis EP (2002). New advances in polymer/layered silicate nanocomposites. *Current Opinions in Solid State and Materials Science,* 6, 205–212.

Sen R, Zhao B, Perea D et al. (2004). Preparation of single walled carbon nanotube reinforced polystyrene and polyurethane nanofibers and membranes by electrospinning. *Nano Letters,* 4, 459–464.

Shaffer MSP and Windle AH (1999). Fabrication and characterization of carbon nanotube–poly(vinyl alcohol) composites. *Advanced Materials,* 11, 937–941.

Song M and Xia HS (2006). Preparation and characterisation of polyurethane grafted single-walled carbon nanotubes and derived polyurethane nanocomposites. *Journal of Materials Chemistry,* 16, 1843–1851.

Song YS and Youn JR (2005). Influence of dispersion states of carbon nanotubes on physical properties of epoxy nanocomposites. *Carbon,* 43, 1378–1385.

Spange S, Graser A, Muller H et al. (2001). Synthesis of inorganic–organic host–guest hybrid materials by cationic vinyl polymerization within Y zeolites and MCM-41. *Chemistry of Materials,* 10, 3698–3708.

Spinks GM, Mottaghitalab V, Bahrami-Saniani M et al. (2006). Carbon-nanotube-reinforced polyaniline fibers for high-strength artificial muscles. *Advanced Materials,* 18, 637.

Spitalsky Z, Tasis D, Papagelis K et al. (2010). Carbon nanotube-polymer composites: chemistry, processing, mechanical and electrical properties. *Progress in Polymer Science,* 35, 357–401.

Stankovich S, Dikin DA, Dommett D et al. (2006). Graphene-based composite materials. *Nature,* 442, 282.

Tamburri E, Orlanducci S, Terranova ML et al. (2005). Modulation of electrical properties in single-walled carbon nanotube/conducting polymer composites. *Carbon,* 43, 1213–1221.

Terrones M, Botello-Mendez AR, Campos-Delgado J et al. (2010). Graphene and graphite nanoribbons: morphology, properties, synthesis, defects and applications. *Nano Today,* 5, 351–372.

Thostenson ET, Ren Z, and Chou TW (2001). Advances in the science and technology of carbon nanotubes and their composites: a review. *Composites Science and Technology,* 61, 1899–1912.

Tohji K, Takahashi H, Shinoda Y et al. (1997). Purification procedure for single-walled nanotubes. *Journal of Physical Chemistry B,* 101, 1974–1978.

Tonge SR and Tighe BJ (2001). Responsive hydrophobically associating polymers: a review of structure and properties. *Advanced Drug Delivery Reviews,* 53, 109–122.

Utracki LA (2008). Polymeric nanocomposites: compounding and performance. *Journal of Nanoscience and Nanotechnology,* 8, 1582–1596.

Velasco-Santos C, Martinez-Hernandez AL, Fisher FT et al. (2003). Improvement of thermal and mechanical properties of carbon nanotube composites through chemical functionalization. *Chemistry of Materials,* 15, 4470–4475.

Wang Z and Pinnavaia TJ (1998). Nanolayer reinforcement of elastomeric polyurethane. *Chemistry of Materials,* 10, 3769.

Wang TL and Tseng CG (2007). Polymeric carbon nanocomposites from multiwalled carbon nanotubes functionalized with segmented polyurethane. *Journal of Applied Polymer Science*, 105, 1642–1650.

Watts PCP and Hsu WK (2003). Behaviours of embedded carbon nanotubes during film cracking. *Nanotechnology*, 5, L7–L10.

Weigel T, Mohr R, and Lendlein A (2009). Investigation of parameters to achieve temperatures required to initiate the shape-memory effect of magnetic nanocomposites by inductive heating. *Smart Materials and Structures*, 18, 025011.

Wen J and Wilkes GL (1996). Organic/inorganic hybrid network materials by the sol–gel approach. *Chemistry of Materials*, 8, 1667–1681.

Werne TV and Patten TE (2001). Atom transfer radical polymerization from nanoparticles: a tool for the preparation of well defined hybrid nanostructures. *Journal of the American Chemical Society*, 31, 7497–7505.

Xia H and Song M (2005). Preparation and characterization of polyurethane-carbon nanotube composites. *Soft Matter*, 1, 386–394.

Xia H and Song M (2006), Preparation and characterisation of polyurethane grafted single-walled carbon nanotubes and derived polyurethane nanocomposites. *Journal of Materials Chemistry*, 16, 1843–1851.

Xia H, Song M, Jin J et al. (2006). Poly(propylene glycol)-grafted multi-walled carbon nanotube polyurethane. *Macromolecular Chemistry and Physics*, 27, 1945–1952.

Xiao P, Chen J, and Xu XF (2010). Progress in studies on carbon and silicon carbide nanocomposite materials. *Journal of Nanomaterials*, 2010, 1–4.

Xie XL, Mai YW, and Zhou XP (2005). Dispersion and alignment of carbon nanotubes in polymer matrix: a review. *Materials Science and Engineering R Reports*, 49, 89–112.

Xu RJ, Manias E, Snyder AJ et al. (2003). Low permeability biomedical polyurethane nanocomposites. *Journal of Biomedical Materials Research A*, 64A, 114–119.

Xu W, Raychowdhury S, Jiang DD et al. (2008). Dramatic improvements in toughness in poly(lactide-co-glycolide) nanocomposites. *Small*, 4, 662–669.

Xu B, Huang WM, Pei YT et al. (2009). Mechanical properties of attapulgite clay reinforced polyurethane shape-memory nanocomposites. *European Polymer Journal*, 45, 1904–1911.

Xu B, Fu YQ, Ahmad M et al. (2010a). Thermomechanical properties of polystyrene-based shape memory nanocomposites. *Journal of Materials Chemistry*, 20, 3442–3448.

Xu B, Fu YQ, Huang WM et al. (2010b). Thermal-mechanical properties of polyurethane–clay shape memory polymer nanocomposites. *Polymers*, 2, 31–39.

Xu B, Zhang L, Tao SW et al. (2011). Dielectric property of thermo electroactive shape memory polymer nanocomposites. Submitted.

Yakacki CM, Satarkar NS, Gall K et al. (2009). Shape-memory polymer networks with Fe_3O_4 nanoparticles for remote activation. *Journal of Applied Polymer Science*, 112, 3166–3176.

Yamamoto K, Akita S, and Nakayama Y (1998). Orientation and purification of carbon nanotubes using ac electrophoresis. *Journal of Physics D*, 31, L34–L36.

Yang B, Huang WM, Li C et al. (2005a). Effects of moisture on the glass transition temperature of polyurethane shape memory polymer filled with nano-carbon powder. *European Polymer Journal*, 41, 1123–1128.

Yang B, Huang WM, Li C et al. (2005b). Qualitative separation of the effects of carbon nanopowder and moisture on the glass transition temperature of polyurethane shape memory polymer. *Scripta Materialia*, 53, 105–107.

Yang B, Huang WM, Li C et al. (2006). Effects of moisture on the thermomechanical properties of a polyurethane shape memory polymer. *Polymer*, 47, 1348–1356.

Yao KJ, Song M, Hourston DJ et al. (2002). Polymer/layered clay nanocomposites 2: polyurethane nanocomposites. *Polymer*, 43, 1017–1020.

Yoo HJ, Jung YC, Sahoo NG et al. (2006). Polyurethane-carbon nanotube nanocomposites prepared by *in situ* polymerization with electroactive shape memory. *Journal of Macromolecular Science B*, 45, 441–451.

Yu Y and Ikeda T (2005). Photodeformable polymers: a new kind of promising smart material for micro- and nano-applications. *Macromolecular Chemistry and Physics*, 206, 1705–1708.

Zhang WD, Shen L, Phang IY et al. (2004). Carbon nanotubes reinforced nylon-6 composite prepared by simple melt-compounding. *Macromolecules*, 37, 256–259.

Zhang CS, Ni QQ, Fu SY et al. (2007). Electromagnetic interference shielding effect of nanocomposites with carbon nanotube and shape memory polymer. *Composites Science and Technology*, 67, 2973–2980.

Zhang HB, Zheng WG, Yan Q et al. (2010). Electrically conductive polyethylene terephthalate–graphene nanocomposites prepared by melt compounding. *Polymer*, 51, 1191–1196.

Zhou QY, Wang SX, Fan XW et al. (2002). Living anionic surface-initiated polymerization (LASIP) of a polymer on silica nanoparticles. *Langmuir*, 8, 3324–3331.

Zhu J, Morgan AB, Lamelas FJ et al. (2001). Fire properties of polystyrene clay nanocomposites. *Chemistry of Materials*, 13, 3774–3780.

Zou H, Wu SS, and Shen J (2008). Polymer/silica nanocomposites: preparation, characterization, properties, and applications. *Chemical Reviews*, 108, 3893–3957.

8 Porous Polyurethane Shape Memory Polymers

8.1 INTRODUCTION

Porous polymers are very important in many fields, for example, in tissue engineering they can be applied as scaffolds for cellular attachments and tissue development (Thompson et al. 2000). Traditionally, many approaches can develop pores in a polymer. Typical methods include solvent casting, particulate leaching, phase separation, emulsion freezing, carbon dioxide expansion, and combinations of these methods. However, the common agents used to develop the porous or foaming structures in polyurethanes are organic solvents that may remain in the pores. The residues of these agents may be harmful to cells and tissues (Mooney et al. 1996). Therefore, these foams may not be appropriate for biomedical applications.

As discussed in Chapters 3 and 5, moisture exerts a great influence on polyurethane shape memory polymers (SMPs). Their glass transition temperatures can be reduced dramatically after immersion in water (Yang et al. 2004). This phenomenon allows the development of SMP devices that can be actuated by water (moisture-responsive) instead of by heat (thermo-responsive). Furthermore, recovery of a piece of SMP follows a pre-determined sequence; the material is programmable (Huang et al. 2005). Water absorbed in polyurethane SMPs may cause bubbles during hot molding or upon further heating to higher temperatures. The bubbles are not necessarily problems. In fact, they may be used as a simple approach for producing SMP foams. From an application view, we must ensure the sizes of the bubbles and/or their porosities under control or eliminate the bubbles if they are unwanted.

In the production of open cellular polyurethane SMP foams, a concept known as cold hibernated elastic memory (CHEM) was proposed for space-bound structural applications (Sokolowski et al. 1999; Figure 8.1). As shown in Figure 8.2, up to 400% inelastic strain may be recovered upon heating. These foams are intended to deploy structures in a simple manner at extremely low cost and require far less stowage space in comparison with other deployment mechanisms (Sokolowski and Tan 2007).

We may extend the CHEM concept into a variety of new applications, for example, a foldable vehicle that has two shapes. The vehicle is shaped for packing to reduce the occupied space, then assumes full size during operation, just like a *Transformers* toy. Figure 8.3 reveals the shape recovery sequence of a compacted toy car inside a hot chamber. After cooling back to room temperature (below the T_g of the SMP), the foam becomes stiff and strong enough to carry a load. Passing an electrical current

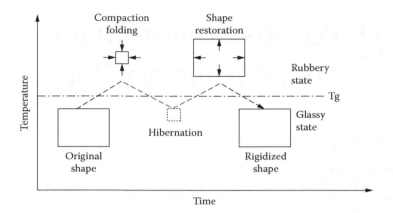

FIGURE 8.1 CHEM processing cycle. (Reprinted from Tey SJ, Huang WM, and Sokolowski WM. *Smart Materials and Structures*, 10, 321–325, 2001. With permission.)

through the embedded high-resistance wires inside the SMP foam is an alternative to heating it (Watt et al. 2001).

Significant shape recovery also allows quick molding into a particular shape (Huang et al. 2006). Thus, an SMP foam easily takes on the shape, for example, of the ear canal of a patient as an aid to shaping a hearing aid. Another potential application of SMP foam is tagging. Tags made of SMP foam, if strong enough, should be better than those made of SMP solids, because SMP foam can be easily compressed by 90% and fits compactly into objects of various dimensions. (Huang et al. 2006)

The polyurethane SMP foam from SMP Technologies, Japan, is biocompatible as reported by Sokolowski et al. (2007). According to Nardo et al. (2009), plasma sterilization is applicable to this polyurethane SMP foam. Based on the CHEM concept, such materials have been used for aneurysm treatment (Ortega et al. 2007, Nardo et al. 2009), endovascular interventions (Metcalfe et al. 2003), and other minimally

FIGURE 8.2 Left: 83% compressed polyurethane SMP foam. Right: foam after shape recovery by heating. (Reprinted from Huang WM. In *Shape-Memory Polymers and Multifunctional Composites*, Taylor & Francis, Boca Raton, 2010, pp. 333–363. With permission.)

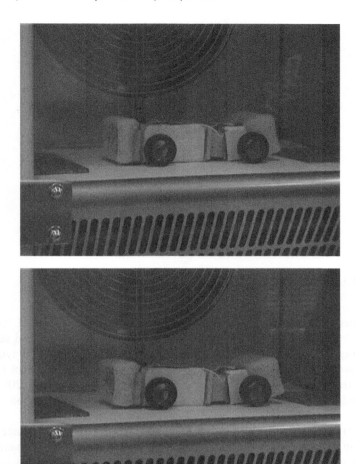

FIGURE 8.3 Shape recovery of compacted toy car inside hot chamber.

invasive surgical procedures. Laser heating SMP foam to treat stroke patients was also proposed (Maitland et al. 2007, Small et al. 2009). Open-cell polyurethane-based SMP foams have been produced with conventional Poisson's and auxetic (negative Poisson's) ratios as reported by Bianchi et al. (2010).

Organoclay fillers have been added to porous SMP foams to improve their mechanical properties (Simkevitz and Naguib 2010). Inductive heating has been proposed to remotely activate nano magnetite-reinforced SMP foams (Buckley et al. 2006, Vialle et al. 2009).

Recently, thermomechanical studies have been conducted on various types of SMP foams (Chung and Park 2010, Domeier et al. 2010, Di Prima et al. 2007, 2010a, 2010b). However, the pioneer work on systematic characterization of polyurethane

FIGURE 8.3   (Continued)

SMP foams under various conditions should be attributed to Prof. H. Tobushi's group (Tobushi et al. 2000, 2001b, 2004). Watt et al. (2001) reported results of their testing the work capability and maximum actuation forces of polyurethane CHEM foams. Models of strain recovery of SMP foam under different temperature and stress conditions were also proposed. Di Prima et al. (2010a, 2010c) used macro- and meso-scale constitutive models to simulate the deformation of epoxy SMP foam.

However, most investigations and simulations target instant performance without considering storage factors. In real engineering applications, CHEM foam will be packed and later heated to recover its shape. To be of real commercial interest, a foam must be subjected to thorough investigation of the effects of long-term storage below its glass transition temperature T_g (cold hibernation).

This chapter first presents a systematic investigation of the formation and control of the bubbles generated by hot molding or heat treatment of polyurethane SMPs. We then study the thermomechanical behavior of SMP foam, including the influences of long-term storage. Unless otherwise stated, the SMPs discussed hereinafter were obtained from SMP Technologies, Japan.

8.2   WATER AS FOAMING AGENT FOR POROUS SMPs

Chemical, water, or gas blowing agents (Lee et al. 2007, Maitland et al. 2007, Domeier et al. 2010, Simkevitz and Naguib 2010) and microballoons (Li and John 2008) are conventional steps used to produce polymer foams. It is well known that moisture may cause trapping of air bubbles inside polymers during production. Traditionally, this was considered a problem, but this phenomenon now serves as a novel non-toxic technique for mass production of porous SMPs with adjustable pore sizes and porosities.

TABLE 8.1
Processing Configuration of Haake Extruder

Channel	1	2	3	4	5	RPM
Temperature (°C)	165	175	185	195	205	20

8.2.1 MATERIALS AND SAMPLE PREPARATION

Ether-based polyurethane SMP (MM5530, T_g about 55°C) in pellet form was used in this study. For good quality control, the SMP pellets were pre-dried in a vacuum oven at 80°C for 12 hours. Two different processes were used to prepare two different types of materials.

Pure SMP — Two trays were placed inside a controllable humidifier at a high relative humidity (RH) of about 80% at 21 ± 1°C. Each tray contained 250 g of SMP pellets. The weight of one tray was tracked to determine the exact amount of moisture absorption and the SMP in the other tray was used in the real fabrication.

With sodium chloride — As in Haugen et al. (2004), sodium chloride powder (98.5% purity, ~50 µm diameter) was used. The powder was pre-dried in a vacuum oven at 80°C for 12 hours for moisture removal. After that, SMP pellets were mixed with the powder at different weight fractions and then extruded into wires of 5 mm diameter by a Haake Rheocord 90 (with five heating channels). Table 8.1 shows setup details of the extruder for the mixing process. Subsequently, the SMP wires filled with sodium chloride were cut into ~3 mm long pellets using Pelletizer Postex. The pellets were returned to the humidifier following the same procedure used to prepare pure SMPs for moisture absorption. SMP sheets ~1.0 mm thick were fabricated from different pellets by hot pressure molding at 200°C for 15 minutes using Teflon molds.

8.2.2 RESULTS AND DISCUSSION

As an example, the increment in weight (percent) of a sample without sodium chloride against storage time is shown in Figure 8.4, revealing an approximate linear relationship after storage for 3 hours or more (Table 8.2). After 140 hours in a humidifier, the weight increased by 2.5%.

The morphologies of SMP sheets were obtained using a digital camera (CV-M50IR, JAI Corporation). Image processing software (Aphelion) was used for threshold segmentation and subsequent analysis. Figure 8.5 presents typical images of SMP sheets without sodium chloride. Generally speaking, bubbles at the millimeter scale did not develop evenly but their distribution is reasonably uniform. Figure 8.6 shows two typical images of SMP sheets containing sodium chloride. No remarkable differences in porosity are revealed between samples filled with 10 and 20% (weight fraction) sodium chloride. Furthermore, the bubbles are ~0.3 to 0.4 mm in diameter—much smaller than those without sodium chloride, but they are more evenly developed. The porosities and mean diameters of bubbles in the samples without sodium chloride were plotted against storage time as shown in Figures 8.7 and 8.8, respectively.

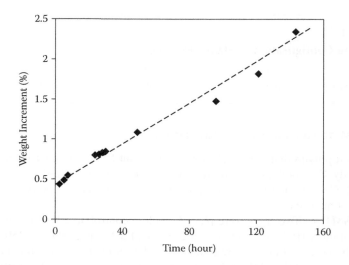

FIGURE 8.4   Relationship of weight increment versus storage time in sample without sodium chloride. (Reprinted from Huang WM, Yang B, Wooi LH et al. In *Materials Science Research Horizons*, Nova Science Publishers, New York, 2007, pp. 235–250. With permission.)

Figure 8.7 reveals the relationship between porosity and storage time and indicates that the influence of moisture absorption on porosity is not significant. Figure 8.8 reveals that bubble sizes decrease slightly with the increase in storage time, indicating that the influence of moisture fraction on bubble size was small. These results demonstrate the simplicity but not-so-precise nature of this technique in terms of porosity and pore size. However, excellent shape recovery was noted in all samples. We should bear these factors in mind in real applications.

8.3   FORMATION OF BUBBLES BY HEAT TREATMENT

According to Huang et al. (2005), moisture absorbed in this polyurethane SMP is in the form of free water and bound water. The ratios of the free, bound, and total absorbed water were calculated as functions of immersion time in room-temperature water (refer to Chapter 3 for details). As reported, bound water significantly reduced the T_g in an almost linear manner, while the effect of free water was negligible.

TABLE 8.2
Effects of Storage on Porosity and Pore Diameter

Storage Time (Hours)	Porosity (%)	Pore Diameter (mm)
3	15	2.00
30	21	2.54
49	23	2.87

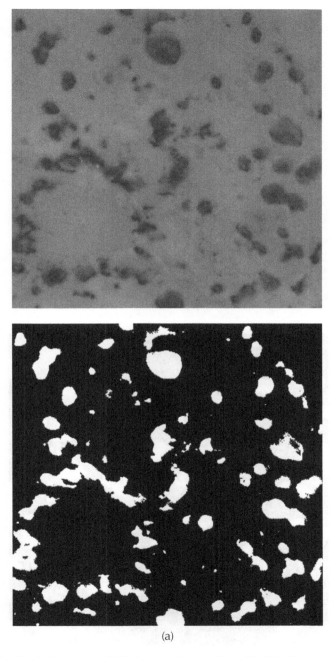

(a)

FIGURE 8.5 Typical images of SMP sheets without sodium chloride. Top: images of digital camera (gray); bottom: after processing (monochrome). (a) 3 hours: porosity 15%; pore diameter 2.0 mm. (Reprinted from Huang WM, Yang B, Wooi LH et al. In *Materials Science Research Horizons*, Nova Science Publishers, New York, 2007, pp. 235–250. With permission.)

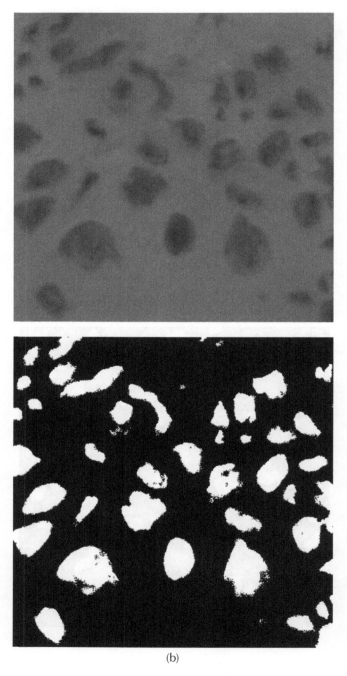

(b)

FIGURE 8.5 (b) 30 hours: porosity 21%; pore diameter 2.54 mm.

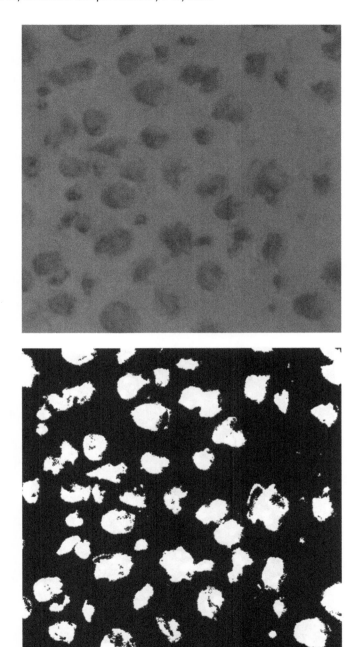

(c)

FIGURE 8.5   (c) 49 hours: porosity 23%; pore diameter 2.87 mm.

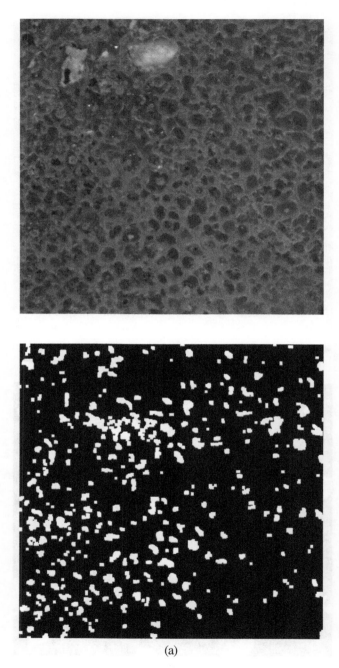

(a)

FIGURE 8.6 Typical images of SMP sheets with (a) 10% sodium chloride. Top: images of digital camera (gray); bottom: after processing (monochrome). (Reprinted from Huang WM, Yang B, Wooi LH et al. In *Materials Science Research Horizons*, Nova Science Publishers, New York, 2007, pp. 235–250. With permission.)

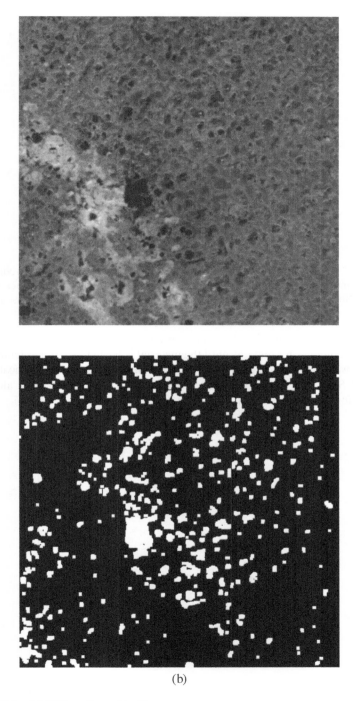

(b)

FIGURE 8.6   (b) 20% sodium chloride.

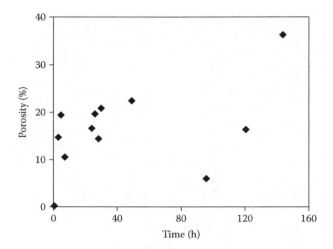

FIGURE 8.7 Porosity versus storage time relationship of SMP sheets without sodium chloride. (Reprinted from Huang WM, Yang B, Wooi LH et al. In *Materials Science Research Horizons*, Nova Science Publishers, New York, 2007, pp. 235–250. With permission.)

Further investigation revealed that all free water may be eliminated upon heating to 120°C; the bound water remains until the material is heated to a higher temperature. The high water absorption ability of this SMP indicates that one can heat a piece of soaked SMP over 100°C to produce bubbles—heat treatment as an alternative approach for bubble generation.

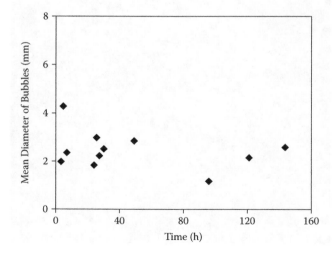

FIGURE 8.8 Mean diameter of bubbles versus storage time relationship of SMP sheets without sodium chloride. (Reprinted from Huang WM, Yang B, Wooi LH et al. In *Materials Science Research Horizons*, Nova Science Publishers, New York, 2007, pp. 235–250. With permission.)

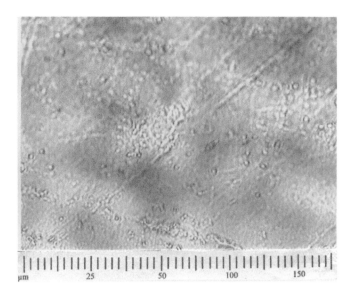

FIGURE 8.9 Optical microscope image of an as-fabricated sample. Scale bar unit = μm. (Reprinted from Huang WM, Yang B, Wooi LH et al. In *Materials Science Research Horizons*, Nova Science Publishers, New York, 2007, pp. 235–250. With permission.)

8.3.1 SAMPLE PREPARATION AND BUBBLE FORMATION

The same SMP pellets described in Section 8.2.1 were used. We followed the procedure suggested by SMP Technologies to prepare SMP sheets about 1 mm thick by hot pressing at 200°C using Teflon molds. Figure 8.9 is an optical microscope image of the morphology of a typical as-fabricated sample. At this scale, no bubbles can be observed. Subsequently, the SMP sheets were immersed into room-temperature water (about 22°C) for different durations. After that, two types of tests were carried out to study the influence of heating temperature and immersion times.

In the first type of test, the immersion time in water was fixed at 2 hours. After soaking, SMP samples were heated to five temperatures (110, 120, 130, 140, and 150°C) for 3 minutes, followed by quenching in room-temperature water. The quenching is intended to prevent significant shape recovery and maintain bubble size. Figure 8.10 reveals many closed-cell bubbles in the samples. It is clear that with increases in heating temperature, the size of bubbles increase remarkably, particularly above 120°C, resulting from the evaporation of bound water above 120°C and further softening of the SMP above 120°C.

In the second type of test, the SMP samples were immersed into room-temperature water for 1, 2, 6, 12, 24, and 48 hours, then heated to 120°C for 3 minutes, followed by quenching in room-temperature water. Because most bound water remains at 120°C, free water and its evaporation are the only players behind these bubbles. The morphologies of the tested samples presented in Figure 8.11 reveal that at a lower

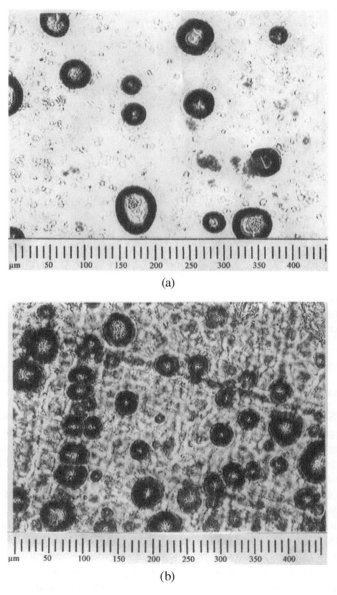

(a)

(b)

FIGURE 8.10 Morphologies of samples soaked in water for 2 hours, then heated to (a) 110°C, (b) 120°C, (c) 130°C, (d) 140°C, (e) 150°C, followed by room-temperature water quenching. (Reprinted from Huang WM, Yang B, Wooi LH et al. In *Materials Science Research Horizons*, Nova Science Publishers, New York, 2007, pp. 235–250. With permission.)

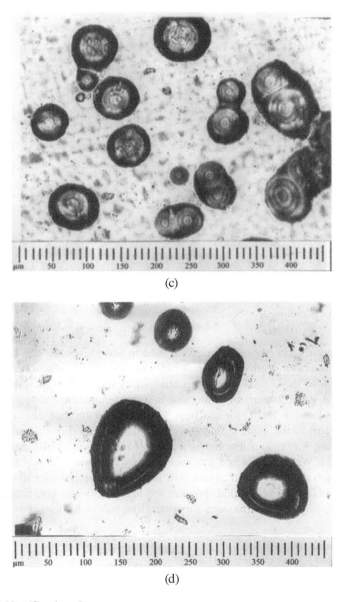

(c)

(d)

FIGURE 8.10 (Continued)

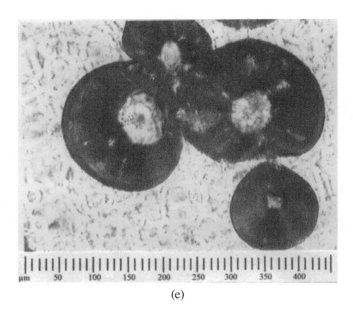

(e)

FIGURE 8.10 (Continued)

water absorption ratio (shorter immersion time in water), bubble size is smaller. A prolonged immersion results in bigger bubbles and eventually semi-open bubbles.

In addition to these tests, we conducted two more tests to study the effects of heating speed. After 2 hours of immersion in room-temperature water, one piece of sample was heated to 120°C for less than 3 minutes and another was heated gradually (for 1 hour) to the same temperature. Bubbles appeared in the quickly heated sample (Figure 8.12a) and few bubbles appeared in the gradually heated sample (Figure 8.12b). When heated at a low speed, water evaporates slowly and manages to pass through the nano-sized channels or gaps inside the polymer to escape. As a result, no bubbles can be formed. We can see that both water content and heating procedure are important parameters for controlling bubble formation and size.

8.3.2 TUNING BUBBLE SIZES

A series of tests were performed to investigate possible approaches to alter the size of bubbles by means of heat treatment after formation.

8.3.2.1 Bubble Size Adjustment

A piece of SMP sample was soaked in room-temperature water for 2 hours, then heated to 105°C for 3 minutes and quenched in room-temperature water. As expected, many bubbles formed, as shown in Figure 8.13a. We focus on the three bubbles circled in Figure 8.13. In the next step, the sample was instantly heated to 110, 120, and 150°C in a step-by-step manner. Figure 8.13b shows no apparent change in these bubbles after heating to 110°C. However, they were significantly bigger after heating

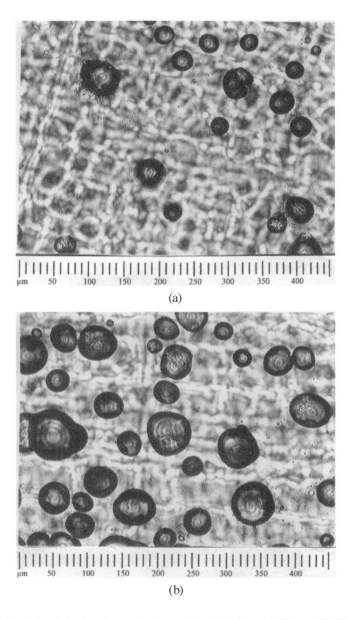

(a)

(b)

FIGURE 8.11 Morphologies of samples after soaking for 1 hour (a), 2 hours (b), 6 hours (c), 12 hours (d), 24 hours (e), and 48 hours (f) followed by heating to 120°C and quenching in room-temperature water. (Reprinted from Huang WM, Yang B, Wooi LH et al. In *Materials Science Research Horizons*, Nova Science Publishers, New York, 2007, pp. 235–250. With permission.)

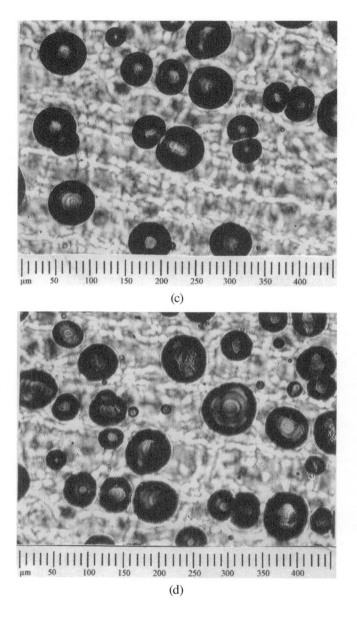

(c)

(d)

FIGURE 8.11   (Continued)

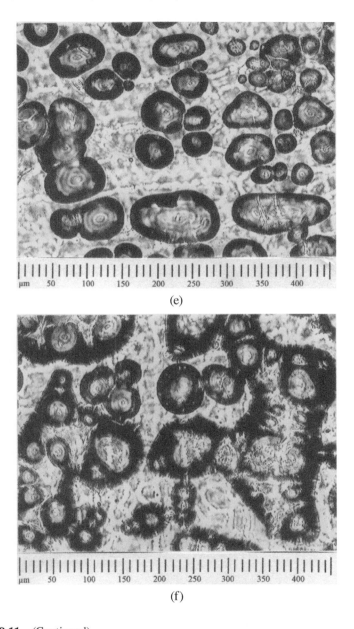

FIGURE 8.11 (Continued)

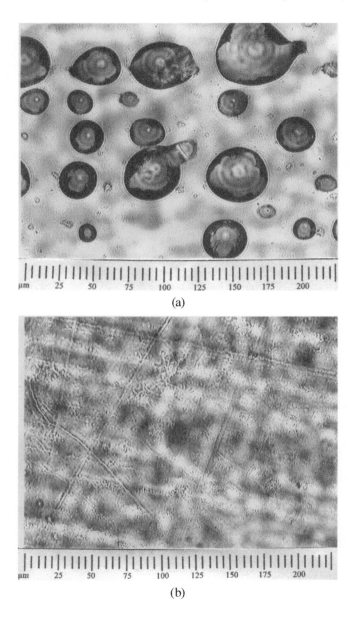

(a)

(b)

FIGURE 8.12 Heating to 120°C (a) instantly (<3 minutes); (b) gradually (1 hour). (Reprinted from Huang WM, Yang B, Wooi LH et al. In *Materials Science Research Horizons*, Nova Science Publishers, New York, 2007, pp. 235–250. With permission.)

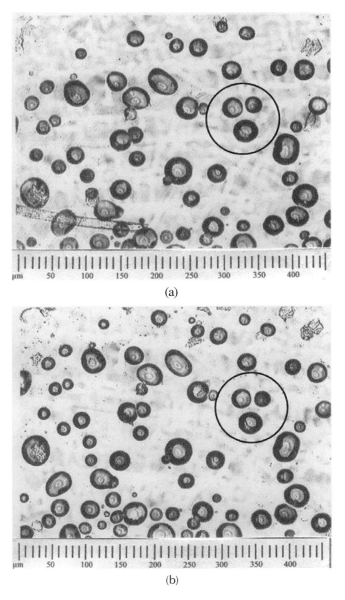

(a)

(b)

FIGURE 8.13 Evolution of bubbles from soaking in water for 2 hours and heating to 105°C (a), 110°C (b), 120°C (c), and 150°C (d). (Reprinted from Huang WM, Yang B, Wooi LH et al. In *Materials Science Research Horizons*, Nova Science Publishers, New York, 2007, pp. 235–250. With permission.)

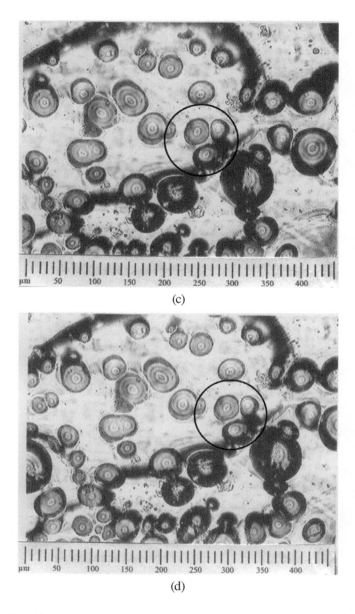

FIGURE 8.13   (Continued)

to 120°C. Upon further heating to 150°C, bubble size did not change significantly. Thus, 120°C appears to be a critical temperature that, based on a previous study (Chapter 3), distinguishes the evaporation of free and bound water. Clearly, the increase in bubble size upon heating over 120°C is largely due to bound water.

We also studied the effects of water absorption–heating cycles. A piece of sample was soaked in room-temperature water for 1 hour and then heated to 120°C for initial bubble generation. After re-soaking in room-temperature water for 1 hour and re-heating to 120°C, no apparent change in bubble size can be observed. In another test, the sample was re-soaked in water for 48 hours and then heated to 120°C. Again, no dramatic size or shape variations of bubbles were noted. In the last test, the heating temperature was increased to 150°C in the second round of the absorption–heating cycle (re-soaking time = 1 hour). Figure 8.14 reveals a significant increase in bubble size.

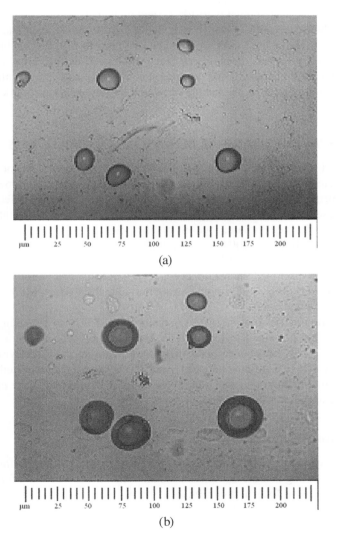

FIGURE 8.14 (a) Initial bubbles formed at 120°C after soaking 1 hour. (b) Same sample re-soaked 1 hour before heating to 150°C. (Reprinted from Huang WM, Yang B, Wooi LH et al. In *Materials Science Research Horizons*, Nova Science Publishers, New York, 2007, pp. 235–250. With permission.)

8.3.2.2 Reversible Bubbles

SMPs have a characteristic ability to recover large and seemingly plastic deformations. This ability may be utilized to produce bubbles that reduce their sizes. If proven feasible, this reversible bubble phenomenon may be used for drug release as reported by Gall et al. (2004), but via a much simpler and more cost-effective fabrication process.

As in earlier studies, a piece of SMP sample was soaked in room-temperature water for 2 hours, heated quickly to 105°C, and then quenched in room-temperature water for bubble generation (Figure 8.15a). Subsequently, the same sample was heated to 80°C—well above its T_g. As we can see in Figure 8.15b, the bubble becomes smaller. A laser beam has the ability to heat a small, well-defined area, and we used one to heat a piece of soaked SMP to generate a single bubble at a specific location. As revealed in Figure 8.16, the bubble generated by laser heating shrank upon heating to 80°C.

8.4   THERMOMECHANICAL BEHAVIORS OF SMP FOAMS

A number of SMP foams are commercially available at present. The most successfully marketed forms may be the polyurethane SMP foam series developed by Dr. S. Hayashi (formerly with Mitsubishi Heavy Industries, now with SMP Technologies, Japan). These foams have a T_g range from –70°C to 70°C. Experimental investigation, although limited to date, focused on characterizing their thermomechanical behaviors. Tobushi et al. (2004) investigated the influence of shape-holding conditions on the shape recovery of MF5520.

This section discussed the results of four types of tests: compression, free recovery constrained cooling, and gripping, using the same SMP foam (MF5520) Tobushi et al. (2004) studied.

Thermogravimetric analysis (TGA) results for MF5520 at a heating speed of 10°C/minute are presented in Figure 8.17. The decomposition temperature of this SMP foam is 200°C and above. Cyclic differential scanning calorimetryr (DSC) was also conducted at 10°C/minute between 0 and 120°C. Results are reported in Section 8.6.1. Note that unless otherwise stated, the strain and stress cited are engineering strain and engineering stress, respectively.

8.4.1   Sample Preparation and Experimental Setup

Samples were cut to 30 × 30 × 45 mm or 30 × 30 × 15 mm size for testing. Uniaxial compression testing is more conventional than uniaxial tensile testing for characterizing foam because foam is more likely to carry compressive loading rather than tensile loading. For shape memory foam, two scenarios are of practical interest. One is recovery without constraint i.e., the free recovery test in which the shape recovery ratio is the key concern. The other scenario is recovery with fixed sample height. This is the constrained recovery test for analyzing recovery stress.

An Instron 5569 instrument with a temperature control chamber was used for the compression, constrained cooling, and gripping tests. An aluminum rod was

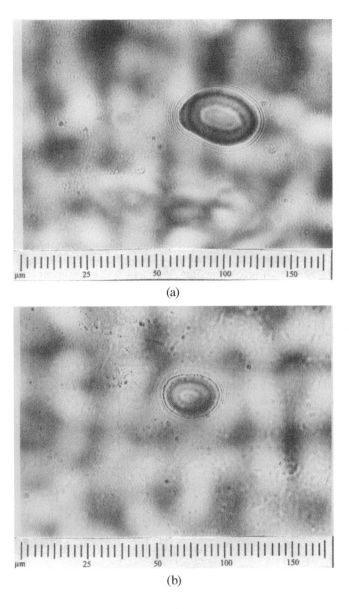

(a)

(b)

FIGURE 8.15 (a) Sample soaked 2 hours initially, heated to 102°C for bubble generation. (b) Subsequently heated to 80°C. (Reprinted from Huang WM, Yang B, Wooi LH et al. In *Materials Science Research Horizons*, Nova Science Publishers, New York, 2007, pp. 235–250. With permission.)

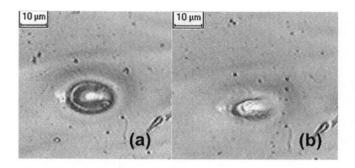

FIGURE 8.16   Single bubble generated by laser (a) and its shrinkage after heating (b).

extended into the temperature-controlled chamber to compress the SMP foam as shown in Figure 8.18. An ESPEC SH-240 temperature and humidity chamber was used for the free recovery test. The deformation of the SMP foam was measured by a Keyence VH-6300 high-resolution digital microscope. For convenience, the details of the gripping test will be discussed later.

8.4.2   EXPERIMENTS AND RESULTS

8.4.2.1   Compression Test

The SMP samples (45 mm in height) were heated to 83°C, about 20°C above their nominal T_g, and held for 5 minutes to achieve uniform temperature distribution. The samples were then compressed to 50, 75, 80, and 93.4% strain at a strain rate of 3.7 × 10^{-3}/s. Figure 8.19 shows compression to 50% and 93.4% and cooling back to room temperature (about 20°C).

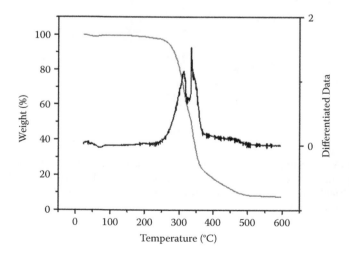

FIGURE 8.17   TGA result for MF5520.

(a)

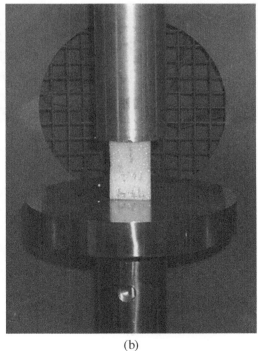

(b)

FIGURE 8.18 Experimental setup. (a) Overall view. (b) Zoom-in view. (Reprinted from Huang WM, Lee CW, and Teo HP. *Journal of Intelligent Material Systems and Structures*, 17, 753–760, 2006. With permission.)

(a)

(b)

(c)

FIGURE 8.19 Compression of SMP foam at 83°C and after cooling back to room temperature. (a) At 50% compression. (b) At 93.4% compression. (c) Compressed sample. (Reprinted from Huang WM, Lee CW, and Teo HP, *Journal of Intelligent Material Systems and Structures*, 17, 753–760, 2006. With permission.)

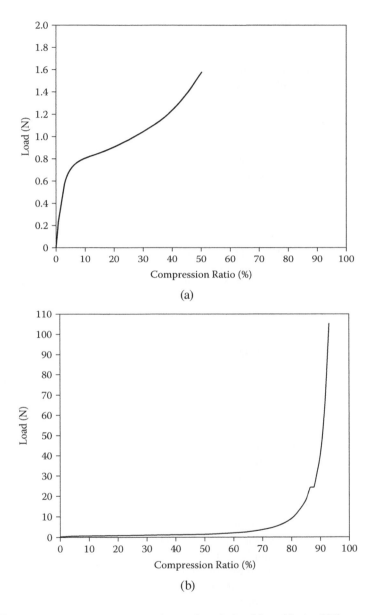

FIGURE 8.20 Force-versus-compression ratio relationships. (a) At 50% compression. (b) At 93.4% compression. (Reprinted from Huang WM, Lee CW, and Teo HP. *Journal of Intelligent Material Systems and Structures*, 17, 753–760, 2006. With permission.)

Figure 8.20 shows the load-versus-compression ratio relationships at maximum compressions of 50 and 93.4%, respectively. Except for an apparent kink shown in Figure 8.20b around 25 N, the curves are reasonably smooth. However, a closer look at Figure 8.19 reveals that even at 50% compression the sample is already in an arrowhead (<) shape due to buckling—an issue that will be discussed in Section 8.5.

8.4.2.2 Free Recovery Test

Each sample (45 mm in height) was heated to 83°C and held for 10 minutes. The samples were then subjected to a constant load (dead weight) of 20 N, corresponding to ~83% compression (Figure 8.20b). Subsequently, the samples were air cooled for a day under a dead weight.

In the free recovery test, the dead weight was removed and the samples were heated to eight different temperatures (40, 50, 53, 55, 58, 60, 63, and 70°C). The exact temperature (measured by a thermocouple) and height (in percent of original sample height) against time are plotted in Figure 8.21. In Figure 8.22, the final height against setup temperature is summarized for comparison. Below ~53°C, no visible shape recovery occurs. Shape recovery appears between 53 and 62°C. Also, around 57°C, the shape recovery exceeded 95% and a full recovery was seen at 65°C and above.

8.4.2.3 Constrained Cooling Test

This test determines the evolution of the recovery force and stress in pre-compressed SMP samples resulting from constrained cooling. The testing procedure (Huang et al. 2006) is

1. Heat sample to 83°C.
2. Compress sample to 50, 75, 80, 90, or 93.4%.
3. Hold until recovery force is stabilized.
4. Re-set chamber temperature to 27°C and record recovery stress and temperature during cooling.

Figure 8.23 shows the evolution of the recovery stress upon cooling at different compressive pre-strains. As shown, the recovery force decreases upon cooling in a monotonic manner in all samples. It is gradual above 55°C, sharp from 55 to ~ 35°C, and then stable below ~35°C. Based on Figures 8.22 and 8.23, we can conclude that, similar to shape memory alloys (SMAs), this SMP exhibits a hysteresis of ~15°C but we cannot clearly see any hysteresis in the DSC result (see Section 8.6.1).

Figure 8.24 shows the maximum (at 80°C) and minimum (at 27°C) recovery stresses against the compression ratios. Figure 8.25 shows the percent decrease of recovery stress plotted against the compression ratio relationship, in which the maximum stress (at 80°C) of each sample is taken as the reference for calculation. According to Figure 8.24, the recovery stress does not increase proportionally to the compression ratio. The maximum recovery stress starts to increase dramatically

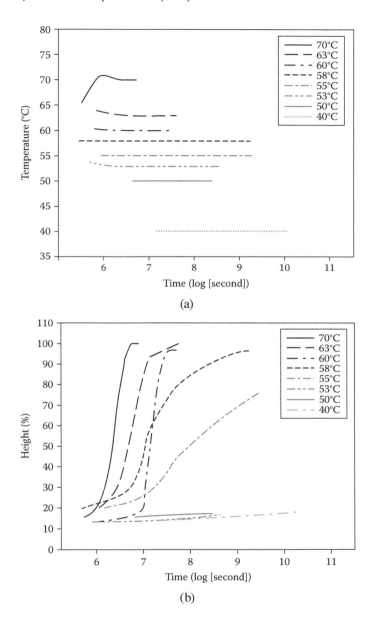

FIGURE 8.21 Results of free recovery test. (a) Time versus temperature. (b) Time versus height. (Reprinted from Huang WM, Lee CW, and Teo HP. *Journal of Intelligent Material Systems and Structures*, 17, 753–760, 2006. With permission.)

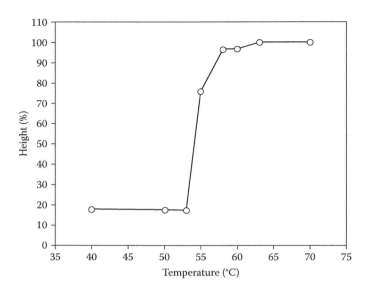

FIGURE 8.22 Temperature-versus-height relationship in free recovery test. (Reprinted from Huang WM, Lee CW, and Teo HP. *Journal of Intelligent Material Systems and Structures*, 17, 753–760, 2006. With permission.)

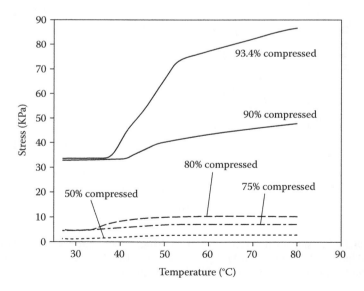

FIGURE 8.23 Evolution of compressive stress upon cooling.

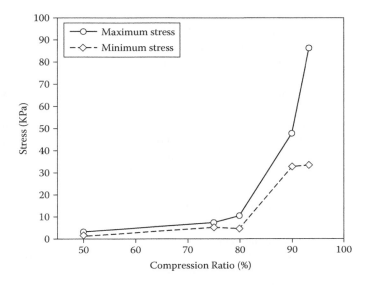

FIGURE 8.24 Maximum and minimum compressive stresses against compression ratios.

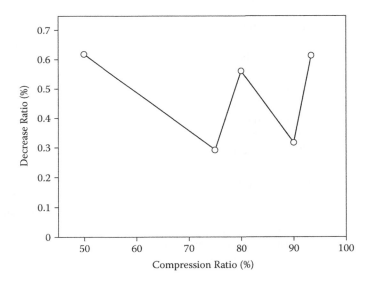

FIGURE 8.25 Decrease of recovery stress (%) upon cooling. Maximum stress serves as reference.

at a pre-compression over 80%, similar to the trend revealed by the stress-against-compression ratio curve in Figure 8.20b. Furthermore, the minimum recovery stress becomes stable when the compression ratio is over 90%.

As for the decrease in percentage in the recovery stress upon cooling, we can see that the decrease is between 30 and 60%. The remaining compressive stress is not insignificant, although the final temperature is very close to room temperature. Tobushi et al. (2001a, 2001b) reported that in an 80% compressed sample (comparable to our pre-compression ratio), the compressive stress eventually diminished after cooling back to room temperature. The possible reasons are (1) the samples we tested were bent into arrowhead (<) shapes during compression; and (2) the actual T_g values of our samples were lower due to moisture effects.

8.4.2.4 Gripping and Shape Recovery Test

Cracks may appear after severe impact loading and cause a significant decrease in strength. Embedded shape memory alloy wires can be utilized for self-healing of materials mainly by means of crack closure (Files and Olson 1997, Kirkby et al. 2008). Recently, SMPs, particularly foams, have also been proposed for self-healing and also for shape recovery (Li and John 2008).

We conducted gripping and shape recovery tests for two reasons. One was to find the gripping forces of SMP foams and rods of three different diameters (6, 8, and 10 mm). Gripping force is a measure of wear during cyclic penetration of a rod. The second reason was to determine the shape recovery ability of the SMP foam after quasi-static penetration loading. The real-life application is achieving self-healing or closure of holes produced by the penetration of a sharp object. Figure 8.26 presents the details of the testing rigs.

For the gripping tests, a rod was pre-penetrated half-way through the SMP foam at 90°C (Figure 8.27a). After that, a pushing force was applied to the flat end of the rod at a speed of 10 mm/minute to a maximum displacement of 15 mm (Figure 8.27b shows the experimental setup). The displacement and pushing force were recorded for further analysis. The same test was repeated 12 times on the same sample through the same hole.

Figure 8.28 presents the gripping force versus displacement (up to 1 mm) curves of three rods in the first, second, third, and twelfth tests. In all tests, the initial gripping force increased almost linearly with displacement before a peak was reached. After that, the gripping force dropped and gradually became stable. The peak indicates the transition point of friction from static to dynamic. For a better view, Figure 8.29 shows the relationships between maximum gripping force and the number of tests:

- A larger rod is always associated with a higher gripping force.
- The maximum gripping force drops dramatically in the second test, then gradually becomes stable upon further testing due to wear caused by the contact between foam surface and rod.
- The gripping force in the 10 mm rod case decreases continuously even in the twelfth test. In contrast, the gripping force in the other two cases became stable after only a few tests.

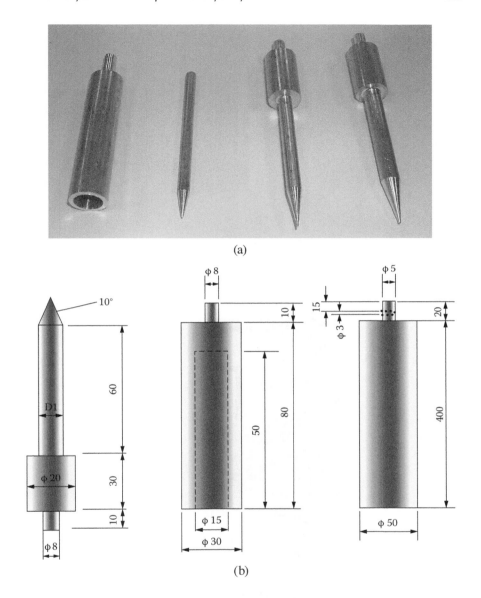

FIGURE 8.26 Rigs for gripping test. (a) Testing rigs. (b) Details of dimensions (mm). (Reprinted from Huang WM, Lee CW, and Teo HP. *Journal of Intelligent Material Systems and Structures*, 17, 753–760, 2006. With permission.)

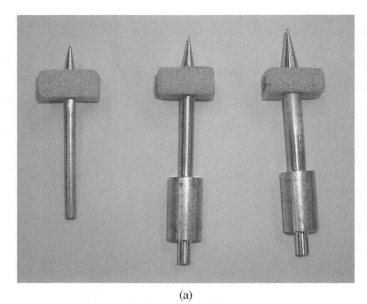

(a)

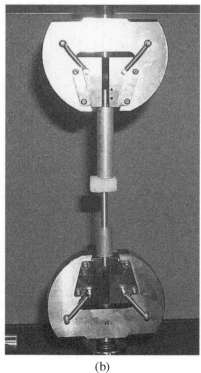

(b)

FIGURE 8.27 Samples with rods before gripping test (a) and setup of gripping test (b). (Reprinted from Huang WM, Lee CW, and Teo HP. *Journal of Intelligent Material Systems and Structures*, 17, 753–760, 2006. With permission.)

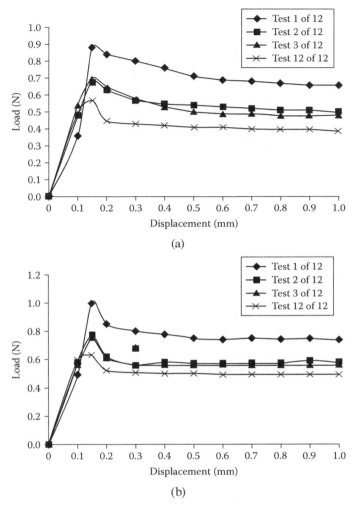

FIGURE 8.28 Gripping force versus displacement curves for (a) 6 mm rod, (b) 8 mm rod, and (c) 10 mm rod. (Reprinted from Huang WM, Lee CW, and Teo HP. *Journal of Intelligent Material Systems and Structures*, 17, 753–760, 2006. With permission.)

The friction causes foam to gradually flake off as evidenced by the permanent deformation shown in Figure 8.30. Tearing is more significant when a larger-sized rod is used. Figure 8.31 shows displacement corresponding to the maximum gripping force in each test against the testing number. Despite fluctuations, the general trend is that the displacement at the maximum gripping force depends only on the diameter of the rod. A larger diameter is accompanied by a larger displacement at the point where maximum gripping force occurs. While the displacements in the 6 and 8 mm rods are very close (between 0.1 and 0.2 mm), the displacement in the 10 mm rod case is higher at 0.2 to 0.3 mm.

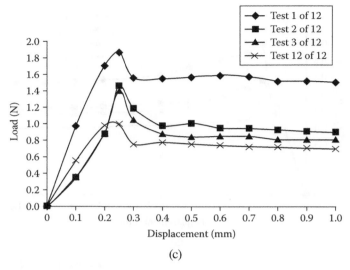

(c)

FIGURE 8.28 (Continued)

8.5 YIELD SURFACE OF FOAM

Although foam is mainly intended for compression loading due to its great contraction ability, it is often deformed under multiaxial loading, for instance, due to boundary constraints in panels made of foam. Thus, understanding the yield start point (yield surface) of foam in a multiaxial stress state is important in engineering applications (Evans et al. 1998, Gibson 2000). Well-documented experiments in the

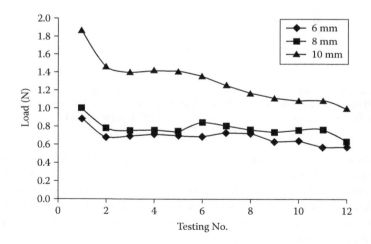

FIGURE 8.29 Maximum gripping force versus testing number relationship. (Reprinted from Huang WM, Lee CW, and Teo HP. *Journal of Intelligent Material Systems and Structures*, 17, 753–760, 2006. With permission.)

(a)

(b)

FIGURE 8.30 Recovered shape (a) after 1 test and (b) after 12 tests. (Reprinted from Huang WM, Lee CW, and Teo HP. *Journal of Intelligent Material Systems and Structures*, 17, 753–760, 2006. With permission.)

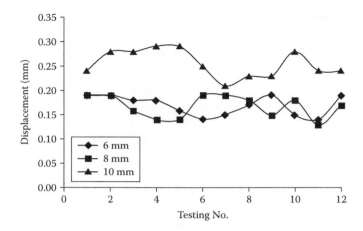

FIGURE 8.31 Displacement at maximum gripping force versus testing number relationship. (Reprinted from Huang WM, Lee CW, and Teo HP. *Journal of Intelligent Material Systems and Structures*, 17, 753–760, 2006. With permission.)

literature reveal that, based on content, production method, and relative density, a foam may yield or fail via three mechanisms: elastic buckling, plastic yielding, and brittle crushing or fracture (Triantafillou et al. 1989). These mechanisms apply to SMP foams at high or low temperatures.

The three major foam types are honeycomb, open cell, and closed cell. We focus on the open- and closed-cell types because current SMP foams are of these two types. For simplicity, we assume that the original foam is uniform and isotropic, i.e., non-textured. Both theoretical and phenomenological approaches have been used to determine the yield surface of foam. The theoretical approach is extremely tedious. The phenomenological approach requires significant experimental effort before data fitting. As an example, the model proposed by Gibson et al. (1989) requires many parameters that are difficult to obtain from non-honeycomb foam. A simple, practical, and applicable approach to estimating the failure surface of a non-textured foam is badly needed.

8.5.1 Framework

Huang (1999) proposed a framework to determine the yield surfaces of non-textured polycrystal shape memory alloys in which the yield (phase transformation start) criterion for a single crystal can be expressed as:

$$\operatorname*{Max}_{1 \leq l \leq m}\left(\boldsymbol{\Sigma} : \boldsymbol{\varepsilon}_l^p\right) = K \tag{8.1}$$

where K, the driving force, is a constant. $\boldsymbol{\Sigma}$ and $\boldsymbol{\varepsilon}_l^p$ $(l = 1, 2, \ldots m)$ are the applied stress and strain associated with the phase transformation, respectively; m is the total number of possible transformation (yield) systems. We consider yield of foam due to a particular type of mechanism as transformation as well. For convenience, we normalize the yield stress by the yield stress of uniaxial tension. Given a polycrystal with n grains, the transformation strain of grain k in the global coordinate system is:

$$\mathbf{E}_{kl}^{p} = \mathbf{R}_{k}^{T}\mathbf{\varepsilon}_{l}^{p}\mathbf{R}_{k}; \quad (k = 1, 2, \ldots n) \tag{8.2}$$

$\mathbf{R}_{k}$ is the rotation matrix for the transition from the local coordinate system attached by the austenite lattice structure of grain k to the global coordinate system. Two schemes have been proposed to find the corresponding yield surface (Huang 1999). For a given stress state $\mathbf{\Sigma}^{0}$ in the first scheme, the yield stress $\mathbf{\Sigma}_{\text{Max}}^{p}$ relative to that of uniaxial tension is expressed as:

$$\mathbf{\Sigma}_{\text{Max}}^{p} = \frac{\mathbf{\Sigma}^{0}}{\alpha_{\text{Max}}} \tag{8.3}$$

where

$$\alpha_{\text{max}} = \frac{\underset{1 \leq k \leq n}{\text{Max}}\left(K_{\text{Max}}^{k,0}\right)}{\underset{1 \leq k \leq n}{\text{Max}}\left(K_{\text{Max}}^{k,\text{tension}}\right)} \tag{8.4}$$

and

$$K_{\text{Max}}^{k,0} = \underset{1 \leq l \leq m}{\text{Max}}\left(\mathbf{\Sigma}^{0} : \mathbf{E}_{kl}^{p}\right) \tag{8.5}$$

$$K_{\text{Max}}^{k,\text{tension}} = \underset{1 \leq l \leq m}{\text{Max}}\left(\mathbf{\Sigma}^{\text{tension}} : \mathbf{E}_{kl}^{p}\right) \tag{8.6}$$

$\mathbf{\Sigma}^{\text{tension}}$ stands for the stress state of uniaxial tension. This is a maximum scheme because the start of yield depends only on one grain that produces the maximum K, which is analogous to having all grains in series. Interaction among grains is ignored. We can prove that the yield criterion of the maximum scheme (Huang and Zhu 2002) is equivalent to:

$$\sigma_{1}\varepsilon_{1}^{p} + \sigma_{2}\varepsilon_{2}^{p} + \sigma_{3}\varepsilon_{3}^{p} = K \tag{8.7}$$

where σ_{1}, σ_{2}, and σ_{3} $(\sigma_{1} \geq \sigma_{2} \geq \sigma_{3})$ are the principal stresses and $\varepsilon_{1}^{p}, \varepsilon_{2}^{p}$, and ε_{3}^{p} $(\varepsilon_{1}^{p} \geq \varepsilon_{2}^{p} \geq \varepsilon_{3}^{p})$ are the principal transformation strains. The resulting yield surface is shown in Figure 8.32. As pointed out in Huang and Gao (2004), any strain with the principal strain in a format of α $(1, 0, -1)$, where α is a constant, results in the exact Tresca yield surface.

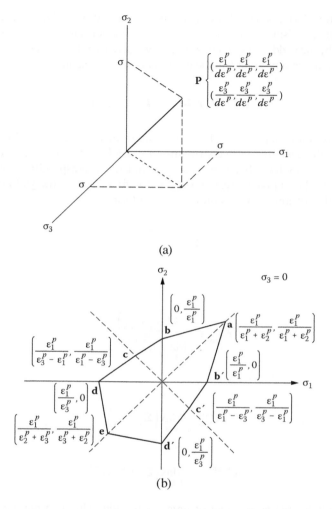

FIGURE 8.32 Apex (a) and yield surface in $(\sigma_1 - \sigma_2, \sigma_3 = 0)$ plane (b) resulting from maximum scheme. (Reprinted from Huang WM and Zhu JJ. *Mechanics of Materials*, 34, 547–561, 2002. With permission.)

In the second scheme, the interaction among grains is counted by means of simple averaging over all grains:

$$\Sigma^P_{Ave} = \frac{\Sigma^0}{\alpha_{Ave}} \tag{8.8}$$

where

$$\alpha_{Ave} = \frac{\sum\limits_{k=1}^{n} K_{Max}^{k,0}}{\sum\limits_{k=1}^{n} K_{Max}^{k,\text{tension}}}$$

(8.9)

This is analogous to putting all grains in parallel. Note that if $\mathbf{R}_k$ is expressed by three Euler angles ($_1$, ϕ, and $_2$), the orientation element dg (Bunge 1982) is given by

$$dg = \frac{1}{8\pi^2}\sin\phi d\,_1 d\phi d\,_2$$

(8.10)

Thus, Equation (8.9) may be rewritten as

$$\alpha_{Ave} = \frac{\int\limits_{0}^{2\pi}\int\limits_{0}^{\pi}\int\limits_{0}^{2\pi} K_{Max}^{k,\ 0}\sin\phi d\,_1 d\phi d\,_2}{\int\limits_{0}^{2\pi}\int\limits_{0}^{\pi}\int\limits_{0}^{2\pi} K_{Max}^{k,\ \text{tension}}\sin\phi d\,_1 d\phi d\,_2}$$

(8.11)

Following the second (average) scheme, Huang and Gao (2004) studied the strains that precisely produce the von Mises surface. Figure 8.33 shows these principal strains in the ($\varepsilon_{xx}, \varepsilon_{yy}, \varepsilon_{zz}$) space. As we can see, not all strains with a principal strain in the format of α (1, 0, −1) result in the exact Tresca yield surface. In the ($\varepsilon_{xx}, \varepsilon_{yy}, \varepsilon_{zz}$)

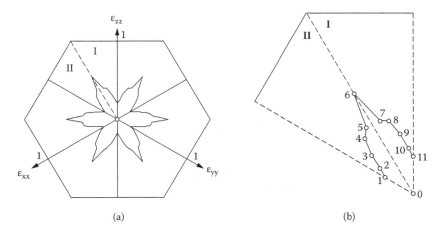

(a)   (b)

FIGURE 8.33   Strain meeting von Mises criterion in ($\varepsilon_{xx}, \varepsilon_{yy}, \varepsilon_{zz}$) space. (a) Overall view. (b) Details in zones I and II. (Reprinted from Huang WM and Gao XY. *Philosophical Magazine Letters*, 84, 625–629, 2004. With permission.

space, these strains form a continuous flower-like curve plus a separate single point at the center. However, they are discontinuous in the $(\varepsilon_{xx}, \varepsilon_{yy}, \varepsilon_{zz}, \varepsilon_{xy}, \varepsilon_{yz}, \varepsilon_{zx})$ space (Huang and Gao 2004).

This framework can be naturally extended to non-crystalline materials, in particular to non-textured ones. For estimation, it may be reasonably accurate to consider "plastic" strain in principal strain format for any non-crystalline material. Assume that we have the following normalized typical principal "plastic" strains:

$$\begin{bmatrix} 1 & & \\ & a & \\ & & -1 \end{bmatrix} ; \begin{bmatrix} 1 & & \\ & b & \\ & & b \end{bmatrix} ; \begin{bmatrix} c & & \\ & c & \\ & & -1 \end{bmatrix} \quad (8.12)$$

where $-1 < a < 1$, $-1 < b < 1$, and $-1 < c < 1$. We may call them shearing, tension, and compression mechanisms, respectively; if $a \approx 0$, the first one is pure shearing; while in the second and third, the dominant strains are 1 and -1, respectively. The real strain differs from the normalized strain by a scale factor H. In real materials, multiple failure mechanisms may co-exist. Hence, the real failure surface of a material may be obtained by superposition. However, the size of each failure surface must be re-scaled accordingly.

One case in which the three mechanisms co-exist is schematically illustrated in Figure 8.34. The real failure surface may be somewhere between the thick solid line (maximum scheme) and the thick dashed line (average scheme), i.e., *a-b-c-d-e-b-a*

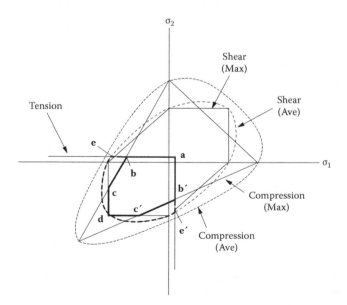

FIGURE 8.34 Combinations of different mechanisms. (Reprinted from Huang WM. *International Journal of Mechanical Sciences*, 45, 1531–1540, 2003. With permission.)

and a-b'-c'-d-e'-b'-a. Ideally, the actual size of each surface can be decided by only one test if the principal plastic strain is known.

8.5.2 APPLICATIONS IN FOAMS

Foam may be treated as a particular type of non-crystalline material. Elastic buckling and brittle crushing may be considered similar to compression mechanisms, plastic yielding as a shearing mechanism, and brittle fracture as a tension mechanism. Many experiments (Gibson et al. 1989, Triantafillou et al. 1989, Gioux et al. 2000) noted that the failure surfaces of many foams can be estimated by straight lines in some specified stress planes. Consequently, as an approximation, it should be reasonable to consider only the maximum scheme for foams in these particular planes.

Figure 8.35 illustrates the possible failure surfaces upon biaxial loading in the $[\sigma_1 - \sigma_2, \ \sigma_3 = 0]$ plane and axisymmetric triaxial loading in the $[S-T]$ or $[\sigma_1 - \sigma_2(=\sigma_3)]$ plane due to shearing, tension, and compression mechanisms. In Figure 8.36, the failure surfaces of two polymer foams (H100 and H200) in the $(S-T)$ plane are estimated against those of the experimental results of Deshpande and Fleck (2000). In Triantafillou et al. (1989) and Gibson et al. (1989), the principal stress criterion was used for polymer foam. This is similar to the case if we take $a = b = c = 0$ in the normalized typical principal "plastic" strains in Equation (8.12). Because the failure of foam most likely initializes from a small local area—from one or a few cells—the situation is the same as the maximum scheme, thus the application of the maximum scheme for foam may be justified.

For verification purposes, we should obtain the failure surface in one plane by data fitting and then use the parameters from data fitting to predict the yield surface in another plane. This is the best way to check whether a proposed approach is reliable and applicable in real engineering practice. Gioux et al. (2000) reported both failure surfaces of an aluminum foam in the $(\sigma_1 - \sigma_2, \ \sigma_3 = 0)$ plane and $(S-T)$ plane. Figure 8.37a presents the fitted failure surface data for the $(S-T)$ plane using the above three failure mechanisms. Subsequently, the identical set of parameters in the normalized typical principal "plastic" strains was used to predict the failure surface in the $(\sigma_1 - \sigma_2, \ \sigma_3 = 0)$ plane. The results are shown in Figure 8.37b along with experimental results for comparison. Good agreement is observed, proving the reliability of this approach and verifying its feasibility in real engineering practice.

Because we have used the simplest plastic strains for all mechanisms, in which $a = b = c = 0$, three specified tests are required to determine the size of each surface, after which the complete failure surface of a foam can be constructed. The recommended testing stress states are $\sigma_1 = \sigma_2 = \sigma_3 > 0$, $\sigma_1 = \sigma_2 = \sigma_3 < 0$, and $\sigma_1 = -\sigma_2 = -\sigma_3$ (or $\sigma_1 = -\sigma_2, \ \sigma_3 = 0$) for tension, compression, and shearing mechanisms, respectively.

The failure surface of foam in the 3-D principal stress space is illustrated in Figure 8.38. Clearly, the 3-D failure surface is a slanted hexagonal prism (Tresca yield surface associated with shearing mechanism) with two triangular pyramids at the ends, representing failure due to the tension and compression mechanisms.

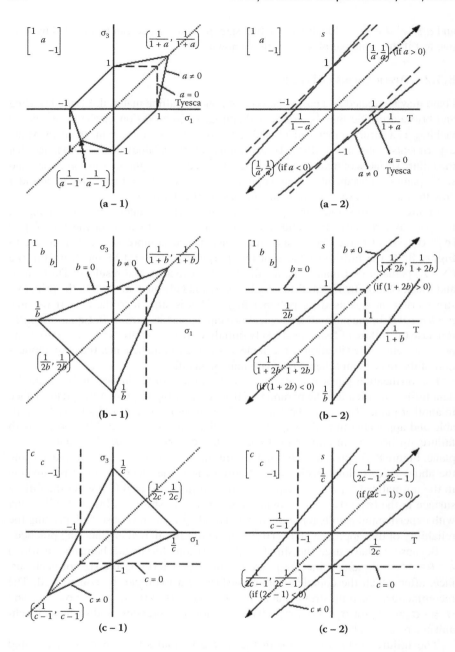

FIGURE 8.35 Failure surfaces due to (a) shearing, (b) tension, and (c) compression mechanisms in $(\sigma_1 - \sigma_2, \sigma_3 = 0)$ plane and $(S - T)$ plane. (Reprinted from Huang WM. *International Journal of Mechanical Sciences*, 45, 1531–1540, 2003. With permission.)

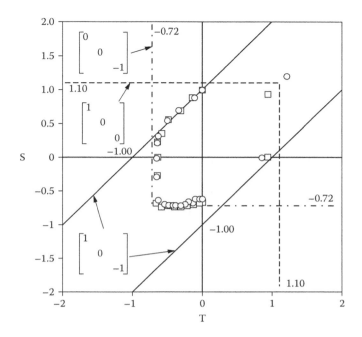

FIGURE 8.36   Failure surfaces of H100 and H200 polymer foams. Experimental data (symbols) from Deshpande and Fleck (2000). (Reprinted from Huang WM. *International Journal of Mechanical Sciences*, 45, 1531–1540, 2003. With permission.)

8.6   INFLUENCE OF STORAGE ON POLYURETHANE SMP FOAMS

Tobushi et al. (2006) investigated the influence of strain-holding conditions on shape recovery and secondary-shape forming in a 0.25 mm thick thin-film polyurethane SMP (MM6520). Tobushi et al. (2004) also studied the influence of shape-holding conditions on the shape recovery of polyurethane SMP foam (MF5520). From an engineering view, it is necessary to characterize the thermomechanical behavior of SMP foams after hibernation to investigate their long-term stability and reliability in real engineering applications, for instance, in space missions (Sokolowski and Tan 2007).

The effects of long-term room-temperature storage on the maximum recovery force in constrained recovery and strain recovery against a constant load were investigated on an open-cell polyurethane SMP foam (MF5520). The same foam was tested by Watt AM, Pellegrino S, and Sokolowski WM [(2001). Thermomechanical properties of a shape memory polymer foam (unpublished data)] and Tobushi et al. (2004). An Instron 5500R with a temperature-controlled chamber was used for characterization. All specimens were prepared in the same size (30 × 30 × 50 mm). The *in situ* temperature of the foam was monitored by a built-in probe. As described in Section 8.4.2, two aluminum rods were extended into the hot chamber to apply pressure along the longitudinal direction of the foam. Each sample was tested following compression, hibernation, and recovery.

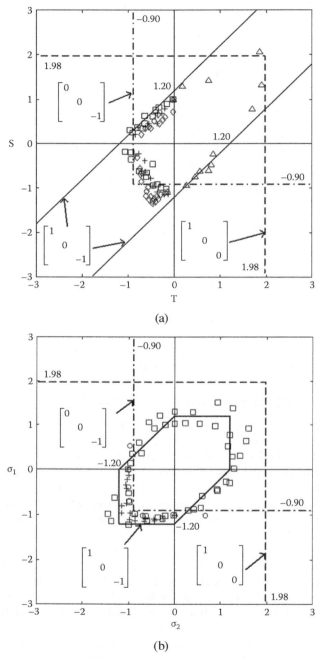

FIGURE 8.37 Failure surfaces upon axisymmetric triaxial loading (a) and biaxial loading (b). Experimental data (symbols) from Gioux et al. (2000). (Reprinted from Huang WM. *International Journal of Mechanical Sciences*, 45, 1531–1540, 2003. With permission.)

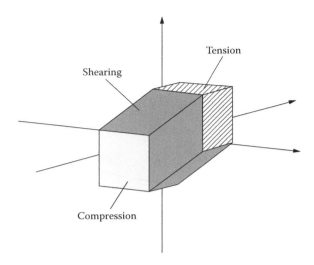

FIGURE 8.38 Yield surfaces of foams (3-D view) in principal stress space. (Reprinted from Huang WM. *International Journal of Mechanical Sciences*, 45, 1531–1540, 2003. With permission.)

8.6.1 PRE-COMPRESSION AND HIBERNATION

First, the foam was heated to 83°C (~30°C above its T_g) in a hot chamber and held for 10 minutes at 83°C for temperature stabilization. After that, the foam with original length of 50 mm was compressed along the longitudinal direction at a strain rate of 3.3×10^{-3}/s to final lengths of 10 mm and 3.3 mm, corresponding to compressive strains of 80 and 93.4%, respectively. A typical strain-versus-stress relationship (to 80% strain) is shown in Figure 8.39.

If pre-compressed foam was stored at about 30°C without constraints, it would return to its original length in ~4 days as opposed to the time reported by Hayashi et al. (1995). However, Watt AM, Pellegrino S, and Sokolowski WM [(2001). Thermomechanical properties of a shape memory polymer foam (unpublished data)] noted that the T_g of the MF5520 they used was only 40 to 50°C. Cyclic DSC results for our MF5520 foam shown in Figure 8.40 reveal that the actual T_g of the foam in heating shifts toward higher temperatures during thermal cycling. This is apparently the result of moisture as discussed in Chapter 3. For convenience, mechanical constraints to prevent shape recovery during hibernation for periods up to 2 months were applied.

8.6.2 RECOVERY TESTS

After different periods of hibernation, SMP foam was placed between two aluminum rods for two types of recovery tests.

8.6.2.1 Constrained Recovery

In constrained recovery testing, the length of pre-compressed foam was fixed during gradual heating to 83°C in a hot chamber. The exact temperatures and recovery

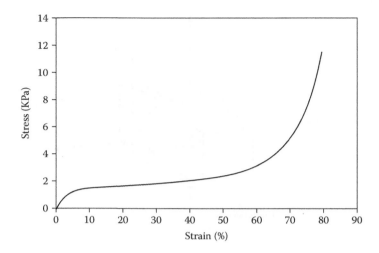

FIGURE 8.39 Typical strain-versus-stress curve of SMP foam (compressed at 83°C). (Reprinted from Tey SJ, Huang WM, and Sokolowski WM. *Smart Materials and Structures*, 10, 321–325, 2001. With permission.)

forces were recorded against time. Due to the highly non-uniform nature of the foam in heating, it is more reasonable to present the experimental results in terms of testing time against recovery stress for comparison. Figure 8.41 shows the relationships of recovery stress and testing time for different hibernation periods after 80 and 93.4% pre-strains.

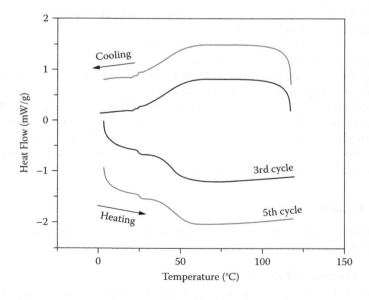

FIGURE 8.40 Cyclic DSC results for polyurethane SMP foam (MF5520).

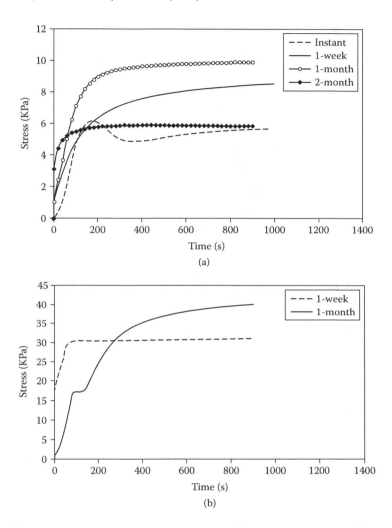

FIGURE 8.41 Comparison of testing time versus recovery stress relationships of SMP foam after different hibernation periods with (a) 80% and (b) 93.4% pre-compression strains. (Reprinted from Tey SJ, Huang WM, and Sokolowski WM. *Smart Materials and Structures*, 10, 321–325, 2001. With permission.)

8.6.2.2 Recovery against Constant Load

Pre-compressed foam was placed atop the lower aluminum rod. The position of the upper rod was adjusted to apply a fixed initial (constant) load. After that, the specimen was heated to 83°C while the applied load was automatically adjusted to hold the constant load. The temperature and expansion during heating were recorded against testing time. For the reason stated above, we plotted recovery stress against testing time instead of temperature.

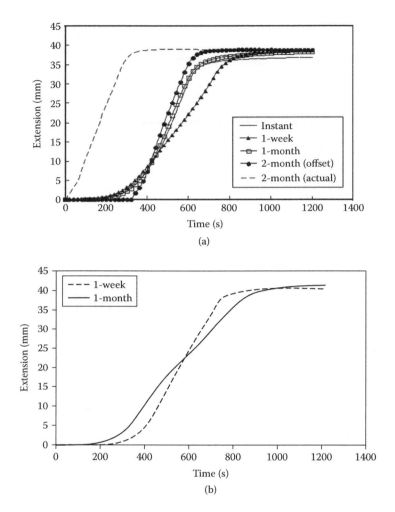

FIGURE 8.42 Comparison of extension-versus-testing time relationships of SMP foam after different hibernation periods with (a) 80% and (b) 93.4% pre-compression strains. (Reprinted from Tey SJ, Huang WM, and Sokolowski WM. *Smart Materials and Structures*, 10, 321–325, 2001. With permission.)

Figure 8.42 shows the extension-versus-testing time relationships for different hibernation periods after 80 and 93.4% pre-strains. The applied constant load was 1 N. Note that the maximum recovery stress in 80% pre-strained foam varied from 6 to 10 kPa after different hibernation periods. Based on these pre-strain and storage conditions, the SMP foam can expand in length from 10 mm to ~38 mm upon heating. This is equivalent to a 380% expansion against a constant load of 1 N. For the foam with 93.4% pre-strain and up to 1 month of hibernation, the maximum recovery

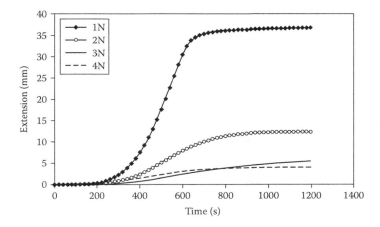

FIGURE 8.43 Extension-versus-testing time curves of SMP foams against various constant loads (without hibernation). (Reprinted from Tey SJ, Huang WM, and Sokolowski WM. *Smart Materials and Structures*, 10, 321–325, 2001. With permission.)

stress ranged from 32.5 to 40 kPa—a few times higher than that achieved in the 80% pre-strained foam. These samples can expand from 3.3 to ~42 mm—equivalent to a 1273% expansion against a constant force of 1 N. Figure 8.43 shows the extension-versus-testing time curves of 80% pre-compressed foams upon heating without hibernation against different constant loads. After another round of heating without any constraint (free recovery test), complete shape recovery was observed in all samples.

We can conclude that the MF5520 polyurethane SMP foam can maintain its shape memory effect even after storage in a compacted state up to 2 months. Complete strain recovery is achievable after hibernation up to 2 months.

8.7 SUMMARY

In the first part of this chapter, two approaches are presented to produce porous SMPs using water as a non-harmful agent. We demonstrated the possibility of controlling sizes and numbers of pores by varying moisture ratios and heating procedures. It is possible to further modify (increase or decrease) the size of pores. Shrinkable pores that utilize the shape memory features of SMPs are more attractive because they can be used for resizable micro bubbles and even channels. Lasers are useful for achieving a single pore with controlled geometric dimensions and properties.

In the second part of this chapter, we studied the thermomechanical behaviors of polyurethane SMP foam (MF5520) under various conditions relevant to engineering applications, including the influence of long-term storage, which is an important concern. The moisture effect in this foam was also investigated. We also discussed a possible approach to obtain the yield surface of a foam under a complex stress state that serves as a useful tool for mechanical modeling of SMP foam.

ACKNOWLEDGMENTS

We sincerely appreciate the help and contributions of M.X. Lee, K.C. Chan, and Drs. N.T. Nguyen and C.C. Wang in preparing this chapter. C. Tang helped compile the material.

REFERENCES

Bianchi M, Scarpa F, and Smith CW (2010). Shape memory behaviour in auxetic foams: mechanical properties. *Acta Materialia*, 58, 858–865.

Buckley PR, McKinley GH, Wilson TS et al. (2006). Inductively heated shape memory polymer for the magnetic actuation of medical devices. *IEEE Transactions on Biomedical Engineering*, 53, 2075–2083.

Bunge HJ (1982). *Texture Analysis in Materials Science: Mathematical Methods*, Butterworths, London.

Chung SE and Park CH (2010). The thermoresponsive shape memory characteristics of polyurethane foam. *Journal of Applied Polymer Science*, 117, 2265–2271.

Deshpande VS and Fleck NA (2000). Isotropic constitutive models for metallic foams. *Journal of the Mechanics and Physics of Solids*, 48, 1253–1283.

Di Prima MA, Lesniewski M, Gall K et al. (2007). Thermomechanical behavior of epoxy shape memory polymer foams. *Smart Materials and Structures*, 16, 2330–2340.

Di Prima M, Gall K, McDowell DL et al. (2010a). Deformation of epoxy shape memory polymer foam I: experiments and macroscale constitutive modeling. *Mechanics of Materials*, 42, 304–314.

Di Prima MA, Gall K, McDowell DL et al. (2010b). Cyclic compression behavior of epoxy shape memory polymer foam. *Mechanics of Materials*, 42, 405–416.

Di Prima MA, Gall K, McDowell DL et al. (2010c). Deformation of epoxy shape memory polymer foam II: mesoscale modeling and simulation. *Mechanics of Materials*, 42, 315–325.

Domeier L, Nissen A, Goods S et al. (2010). Thermomechanical characterization of thermoset urethane shape-memory polymer foams. *Journal of Applied Polymer Science*, 115, 3217–3229.

Evans AG, Hutchinson JW, and Ashby MF (1998). Multifunctionality of cellular metal systems. *Progress in Materials Science*, 43, 171–221.

Files B and Olson GB (1997). Terminator 3: biomimetic self-healing alloy composite. In *Proceedings of Second International Conference on Shape Memory and Superelastic Technologies,* California.

Gall K, Kreiner P, Turner D et al. (2004). Shape-memory polymers for microelectromechanical systems. *Journal of Microelectromechanical Systems*, 13, 472–483.

Gibson LJ (2000). Mechanical behavior of metallic foams. *Annual Review of Materials Science,* 30, 191–227.

Gibson LJ, Ashby MF, Zhang J et al. (1989). Failure surfaces for cellular materials under multiaxial loads I: modelling. *International Journal of Mechanical Sciences*, 31, 635–663.

Gioux G, McCormack TM, and Gibson LJ (2000). Failure of aluminum foams under multiaxial loads. *International Journal of Mechanical Sciences*, 42, 1097–1117.

Haugen H, Ried V, Brunner M et al. (2004). Water as foaming agent for open cell polyurethane structures. *Journal of Materials Science: Materials in Medicine*, 15, 343–346.

Hayashi S, Kondo S, Kapadia P et al. (1995). Room-temperature-functional shape-memory polymers. *Plastics Engineering*, 51, 29–31.

Huang W (1999). "Yield" surfaces of shape memory alloys and their applications. *Acta Materialia*, 47, 2769–2776.

Huang WM (2003). A simple approach to estimate failure surface of polymer and aluminum foams under multiaxial loads. *International Journal of Mechanical Sciences*, 45, 1531–1540.

Huang WM (2010). Novel applications and future of shape memory polymers. In *Shape-Memory Polymers and Multifunctional Composites*, Leng J and Du S, Eds. Taylor & Francis, Boca Raton, pp. 333–363.

Huang WM and Gao XY (2004). Tresca and von Mises yield criteria: a view from strain space. *Philosophical Magazine Letters*, 84, 625–629.

Huang WM and Zhu JJ (2002). To predict the behavior of shape memory alloys under proportional load. *Mechanics of Materials*, 34, 547–561.

Huang WM, Lee CW, and Teo HP (2006). Thermomechanical behavior of a polyurethane shape memory polymer foam. *Journal of Intelligent Material Systems and Structures*, 17, 753–760.

Huang WM, Yang B, An L et al. (2005). Water-driven programmable polyurethane shape memory polymer: demonstration and mechanism. *Applied Physics Letters,* 86, 114105.

Huang WM, Yang B, Wooi LH et al. (2007). Formation and adjustment of bubbles in a polyurethane shape memory polymer. In *Materials Science Research Horizons*, Glick HP, Ed. Nova Science Publishers, New York, pp. 235–250.

Kirkby EL, Rule JD, Michaud VL et al. (2008). Embedded shape-memory alloy wires for improved performance of self-healing polymers. *Advanced Functional Materials*, 18, 2253–2260.

Lee SH, Jang MK, Kim SH et al. (2007). Shape memory effects of molded flexible polyurethane foam. *Smart Materials and Structures*, 16, 2486–2491.

Li GQ and John M (2008). A self-healing smart syntactic foam under multiple impacts. *Composites Science and Technology*, 68, 3337–3343.

Maitland DJ, Small W, Ortega JM et al. (2007). Prototype laser-activated shape memory polymer foam device for embolic treatment of aneurysms. *Journal of Biomedical Optics*, 12, 030504.

Metcalfe A, Desfaits AC, Salazkin I et al. (2003). Cold hibernated elastic memory foams for endovascular interventions. *Biomaterials*, 24, 491–497.

Mooney DJ, Baldwin DF, Suh NP et al. (1996). Novel approach to fabricate porous sponges of poly(d,l-lactic-co-glycolic acid) without the use of organic solvents. *Biomaterials*, 17, 1417–1422.

Nardo LD, Alberti R, Cigada A et al. (2009), Shape memory polymer foams for cerebral aneurysm reparation: effects of plasma sterilization on physical properties and cytocompatibility. *Acta Biomaterialia*, 5, 1508–1518.

Ortega J, Maitland D, Wilson T et al. (2007). Vascular dynamics of a shape memory polymer foam aneurysm treatment technique. *Annals of Biomedical Engineering*, 35, 1870–1884.

Simkevitz SL and Naguib HE (2010). Fabrication and analysis of porous shape memory polymer and nanocomposites. *High Performance Polymers*, 22, 159–183.

Small W, Gjersing E, Herberg JL et al. (2009). Magnetic resonance flow velocity and temperature mapping of a shape memory polymer foam device. *Biomedical Engineering Online*, 8, 42.

Sokolowski WM and Tan SC (2007). Advanced self-deployable structures for space applications. *Journal of Spacecraft and Rockets*, 44, 750–754.

Sokolowski WM, Chmielewski AB, Hayashi S et al. (1999). Cold hibernated elastic memory (CHEM) self-deployable structures: smart structures and materials. *Electroactive Polymer Actuators and Devices*, 3669, 179–185.

Sokolowski W, Metcalfe A, Hayashi S et al. (2007). Medical applications of shape memory polymers. *Biomedical Materials*, 2, S23–S27.

Tey SJ, Huang WM, and Sokolowski WM (2001). Influence of long-term storage in cold hibernation on strain recovery and recovery stress of polyurethane shape memory polymer foam. *Smart Materials and Structures*, 10, 321–325.

Thompson RC, Shung AK, Yaszemski MJ et al. (2000). *The Principles of Tissue Engineering*, Academic Press, San Diego.

Tobushi H, Okumura K, Endo M et al. (2000). Thermomechanical properties of polyurethane-shape memory polymer foam. *Journal of Advanced Science*, 12, 281–286.

Tobushi H, Okumura K, Endo M et al. (2001a). Thermomechanical properties of polyurethane-shape memory polymer foam. *Journal of Intelligent Material Systems and Structures*, 12, 283–287.

Tobushi H, Okumura K, Endo M et al. (2001b). Strain fixity and recovery of polyurethane shape memory polymer foam. *Transactions of the Materials Research Society of Japan*, 26, 351–354.

Tobushi H, Matsui R, Hayashi S et al. (2004). The influence of shape-holding conditions on shape recovery of polyurethane-shape memory polymer foams. *Smart Materials and Structures*, 13, 881–887.

Tobushi H, Hayashi S, Hoshio K et al. (2006). Influence of strain-holding conditions on shape recovery and secondary-shape forming in polyurethane-shape memory polymer. *Smart Materials and Structures*, 15, 1033–1038.

Triantafillou TC, Zhang J, Shercliff TL et al. (1989). Failure surfaces for cellular materials under multiaxial loads II: comparison of models with experiment. *International Journal of Mechanical Sciences*, 31, 665–678.

Vialle G, Di Prima M, Hocking E et al. (2009). Remote activation of nanomagnetite reinforced shape memory polymer foam. *Smart Materials and Structures*, 18, 115014.

Yang B, Huang WM, Li C et al. (2004). On the effects of moisture in a polyurethane shape memory polymer. *Smart Materials and Structures*, 13, 191–195.

9 Shape Memory Effects at Micro and Nano Scales

9.1 INTRODUCTION

Shape memory polymers (SMPs) have the ability to recover from a significant pre-deformation (on the order of 100% strain) upon exposure to the right stimulus (Mather et al. 2009, Huang et al. 2010a). A variety of applications of SMPs have been explored, from morphing wings in aerospace engineering to micro devices for minimally invasive surgery in the biomedical field (Leng and Du 2010). Figure 9.1 shows the formation of micro-sized dimples or protrusions arrayed atop the wing of a toy airplane made of SMP. Such surface-patterned wings can reduce drag, in particular for unmanned aircraft (Mahmood et al. 2001). In comparison with their metallic shape memory alloy (SMA) counterparts (Miyazaki et al. 2009, Fu et al. 2004, Huang 2002), SMPs exhibit higher levels of recoverable strain and are much easier to produce into different sizes and shapes by traditional polymer processing techniques at a much lower cost (Gunes and Jana 2008).

Thermo-responsive SMPs, particularly polyurethane-based materials, have been investigated extensively for many years (Mather et al. 2009, Leng and Du 2010, Gunes and Jana 2008). The thermoplastic polyurethane SMP originally invented by Dr. S. Hayashi (1990) is both thermo-responsive as originally designed and also moisture-responsive (Yang et al. 2004, Huang et al. 2010b). The moisture-responsive feature is induced by the moisture absorbed by the polymer that causes the glass transition temperature T_g (also the shape recovery temperature) to drop significantly (Yang et al. 2006).

The shape memory effect (SME) at nano scale in thermo-responsive bulk SMPs has been demonstrated (Nelson et al. 2005, Wornyo et al. 2007) and proposed for data storage (Altebaeumer et al. 2008), tunable nanowrinkles (Fu et al. 2009), and other uses. SMP nano fibers, nano thin films, and nano protrusions have been produced by electrospinning, water float casting (Sun and Huang 2010), and indentation–polishing–heating (IPH; Liu et al. 2007a). Yang et al. (2005) revealed that within a certain range of particle sizes, the shape of a nanoparticle becomes naturally ellipsoidal and assumes a significant aspect ratio. These studies shed light on the possible uses of SMPs to make micro devices and micro machines. Such tiny machines could be designed for *in situ* testing and characterization of micro and nano structures such as studying the molecular forces, mechano-transitions, and other fundamental biological processes of DNA in living cells (Liedl et al. 2010).

Tiny machines made of SMP have additional advantages. They can be packed into compact size and then delivered into microbes and even cells via surgery, after which they recover their original shape. The procedure is very similar to the

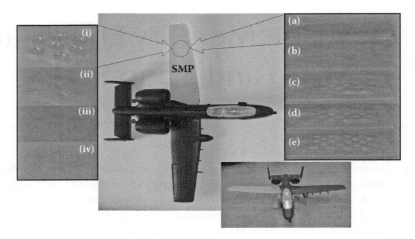

FIGURE 9.1 Formation of arrays of dimples (i to iv) and protrusions (a to e) atop wing made of SMP.

minimally invasive surgery in common use now, but at a far smaller level. We may eventually realize the idea that "the material is the machine" (Bhattacharya and James 2005).

To realize such targets, we must know the sizes and constraints of SMPs that will allow them to maintain their SMEs. This chapter presents a systematic study of SME dynamics in micro- and nano-scaled polyurethane SMPs and micro and nano patterning atop SMPs.

9.2 SMP THIN WIRES

The SMPs used in this study were MM5520 pellets from SMP Technologies, Japan. The fabrication procedure (Figure 9.2) is as follows: (1) melt about 50 pieces of MM5520 pellets; (2) gently draw a fine thread from a melting SMP using a pair of forceps until the thread breaks; (3) note the small spring found at the end of the broken thread. Depending on the pulling speed and temperature, differently sized and right- and left-hand springs have been fabricated (Huang et al. 2010c). Figure 9.3 presents a typical spring with a coil diameter ~100 μm and a wire diameter ~25 μm. Half of the spring is right-handed and the other half is left-handed.

The spiral-shaped thread (spring) may result from buckling at the moment the thread breaks. As we know, during pulling a temperature gradient is present across the cross-sectional area of the thread. For a thinner thread, the gradient is larger, in particular in polymers that are poor thermal conductors. The surface of the thread is in direct contact with the air and at a relatively lower temperature. It thus becomes cold and stiff (forming a kind of elastic layer), while the core of the thread remains hot and soft (as a substrate). While the core can easily deform in a quasi-plastic manner, the surface layer is more elastically stretched. Immediately after the thread breaks, the core material contracts less than the surface and the surface (elastic layer) buckles when a critical condition is reached. Such buckling may appear as wrinkles

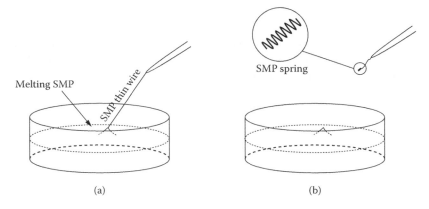

(a)   (b)

FIGURE 9.2   Procedure for producing micro-sized polyurethane SMP spring.

only on the surface area (Chapter 10 covers details about this topic). Under significant contraction force, the thread twists and a spiral spring forms. SMP springs are elastic at room temperature. According to Wahl (1963), when stretched within the linear elastic range, the extension of a linear elastic spring δ can be estimated from

$$\delta = \frac{4R^3 n}{a^4 G} P \qquad (9.1)$$

where n is the number of coils, a is the radius of the spring wire, and R is the mean coil radius.

For a more precise analysis, we must consider the pitch angle γ, the curvature of the spring, and tension–torsion coupling. In practice, no spiral spring can twist

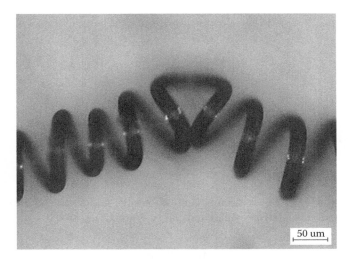

FIGURE 9.3   Typical micro-sized SMP spring. Length of scale bar is 50 μm.

freely upon uniaxial stretching due to the constraint from clamps and/or holding points.

As a classic solution by Ancker and Goodier (1958), the total wrap-up angle Φ and the total extension δ under a torsional moment about the spring axis M and a tension–compression load along the spring axis P can be expressed as

$$\Phi = a_{11}M + a_{12}P \tag{9.2}$$

$$\delta = a_{21}M + a_{22}P \tag{9.3}$$

where

$$a_{11} = \frac{8Rn}{a^4E}\left[1 + \frac{(3 - 7\upsilon - 20\upsilon^2 - 8\upsilon^3)}{48(1+\upsilon)}\left(\frac{a}{R}\right)^2 + (1+\upsilon)p^2\right] \tag{9.4}$$

$$a_{22} = \frac{4R^3n}{a^4G}\left[1 - \frac{3}{16}\left(\frac{a}{R}\right)^2 + \frac{3+\upsilon}{2(1+\upsilon)}p^2\right] \tag{9.5}$$

Because $E = 2(1+\upsilon)G$, where E is the Young's modulus, G is the shear modulus, and n is Possion's ratio, the other two coefficients are

$$a_{21} = a_{12} = \frac{8R^2n}{a^4E}p\nu \tag{9.6}$$

In Equations (9.4) through (9.6), p is the spring pitch [$= \tan(\gamma)$], a can be considered a constant, and p and R can be calculated based on the assumption that the wire length is always constant:

$$R^2(1 + p^2) = C \tag{9.7}$$

and

$$(2\pi R)^2 + (\bar{h} + \delta/n)^2 = 4\pi^2C \tag{9.8}$$

where C is a constant, and

$$\bar{h} = 2\pi\bar{R}\tan(\bar{\gamma}) \tag{9.9}$$

where a macron sign ($^-$) over a component denotes the initial value of a freestanding spring. If rotation is relatively small and can be ignored, i.e., $\Phi = 0$:

$$M = -\frac{a_{12}}{a_{11}}p \tag{9.10}$$

The total elongation can be expressed:

$$\delta = \left(-\frac{a_{12}a_{21}}{a_{11}} + a_{22} \right) P \tag{9.11}$$

Substituting a_{11}, a_{12}, a_{21}, and a_{22} into Equation (9.11) results in

$$\delta = (1 + \xi_1 + \xi_2) \frac{4R^3 n}{a^4 G} P \tag{9.12}$$

where

$$\xi_1 = -\frac{3}{16} \left(\frac{a}{R} \right)^2 + \frac{3 + \upsilon}{2(1 + \upsilon)} p^2 \tag{9.13}$$

and

$$\xi_2 = -\frac{p^2 \upsilon^2}{1 + \upsilon + (3 - 7\upsilon - 20\upsilon^2 - 8\upsilon^3)\left(\dfrac{a}{R} \right)^2 + (1 + \upsilon^2)p^2} \tag{9.14}$$

Note that ξ_1 is the contribution of the spring or coil geometry and ξ_2 is caused by M, the torsional moment due to constraints at the ends. Although the above equations are classical, they are very precise even for nano-sized springs such as the spiral-shaped multi-wall carbon nanotubes studied by Huang (2005) as long as the material remains within its linear elastic range.

The SME of a micro spring is demonstrated in Figure 9.4. The spring was pre-stretched at 90°C, well above its T_g glass transition temperature of ~55°C. Figure 9.4a shows a freestanding spring after cooling back to room temperature (~22°C) with the stretched length held during cooling. Subsequently, the pre-stretched freestanding spring was placed atop a hotplate for thermally induced free recovery. As shown in Figure 9.4 (b through h), upon heating, the spring gradually recovers its original spiral shape.

9.3 SMP MICRO BEADS

As-received MM5520 pellets (SMP Technologies, Japan) were dissolved in dimethylformamide (DMF) to achieve an SMP concentration of 20 wt.% in a Petri dish at room temperature (T_o, ~23°C). A single string of carbon fiber ~7 μm in diameter (Figure 9.5, left inset) was extracted from a carbon fiber bundle and placed into the solution with both ends attached to copper connectors. The length of fiber between the connectors was fixed at 40 mm. Subsequently, a constant DC voltage of 23 V and current of 0.0027 A were applied on the carbon fiber through the

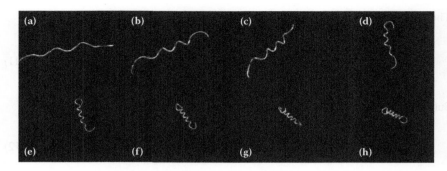

FIGURE 9.4   Shape recovery of micro SMP spring upon heating.

connectors for different time periods. Finally, the carbon fiber was removed and dried in air at room temperature.

Figure 9.5 presents one typical experimental result: many micro beads are formed on the string of carbon fiber. A zoom-in view reveals that a bead has a maximum wall thickness of ~2.7 µm. Figure 9.6a shows that upon Joule heating the bead grows bigger and longer and eventually becomes a wire after a prolonged heating time. For a better view, we plot the geometrical dimensions of beads against heating time in Figure 9.6b. Standard deviations are also indicated. Figure 9.6c defines the geometrical dimensions (length and diameter). Clearly, the sizes and shapes of the beads can be determined by controlling the Joule heating time. The variations in geometrical dimensions at the same heating times were small.

It is difficult to measure the exact temperature of a string of carbon fiber during Joule heating because the string is very thin. However, it is not difficult to estimate

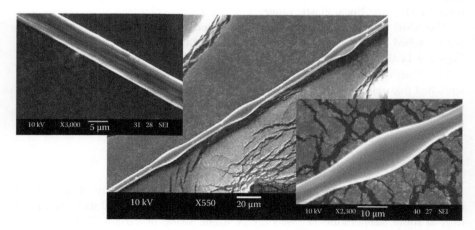

FIGURE 9.5   SEM images of single string of carbon fiber (left inset), SMP micro beads on carbon fiber string (middle), and zoom-in view of micro bead (right inset). The wall thickness of the bead in the right inset is ~2.7 µm. (Reprinted from Sun L and Huang WM. *Materials and Design*, 31, 2684–2689, 2010. With permission.)

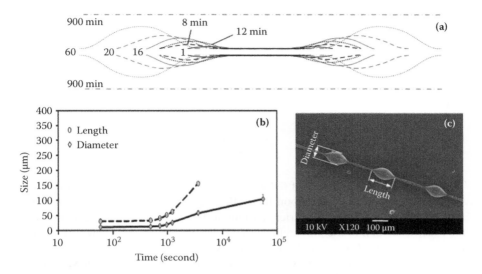

FIGURE 9.6 Evolution of geometrical dimensions of micro beads at different Joule heating times. (a) Superimposition of typical beads (and eventually wire) produced at different heating times for easy comparison. (b) Geometrical dimensions against heating time relationships. (c) Geometrical dimensions.

temperature by applying a classic solution. Consider a piece of conductive wire (length L, radius r, thermal conductivity k) that is Joule heated by passing a constant electrical current of I. The temperature distribution within the length direction of thin wire at a steady state can be obtained by solving the following equation (Carslaw and Jaeger 1959):

$$\frac{\partial^2 T}{\partial x^2} - \frac{hl}{Ak}(T - T_o) = -\frac{RI^2}{kV} \tag{9.15}$$

with boundary conditions:

$$T(0) = T(L) = T_o \tag{9.16}$$

A is the cross-sectional area of the wire, l is the perimeter of the cross-section, V is its volume, R is its resistance, and h is the conventional heat-transfer coefficient. If a wire is very long ($L > 1000\ r$), the temperature within the wire (except the parts near the ends) is almost uniform, and can be calculated (An et al. 2008) as

$$T = \frac{2B\left(e^{\frac{mL}{2}} - e^{\frac{-mL}{2}}\right)}{m^2(e^{mL} - e^{-mL})} + T_o - \frac{B}{m^2} \tag{9.17}$$

where

$$m = 2\sqrt{\frac{h}{kd}} \tag{9.18}$$

and

$$B = -\frac{4RI^2}{\pi^2 kd^2} \tag{9.19}$$

The d represents the diameter of the wire. Figure 9.7 shows the T versus h relationship of the current case. A carbon fiber (40 mm long and 7 μm in diameter) was Joule heated by passing an electrical current of 0.0027 A. Although we do not have the exact value of h of the polyurethane SMP–DMF solution, we know that the h of water ranges from 500 to 1000 W/m°C (Carslaw and Jaeger 1959). We can expect the exact value of the solution to be close to the lower limit of water, if not lower. Hence, given that T_o is ~23°C, the carbon wire should be over 80°C upon Joule heating. As reported (Leng et al. 2008a, 2008b), a standard procedure for curing this polyurethane SMP from its DMF solution is to heat the solution at 80°C for 24 hours. Thus, in our Joule-heating case, only the part of the solution very near the carbon fiber was heated for solidification, while the rest was still in liquid form.

Because the temperature of the middle part of the carbon fiber is more or less uniform, beads form first instead of a uniform wire (tube) because of the surface

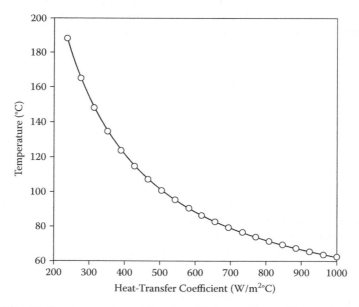

FIGURE 9.7 Heating temperature as function of heat transfer coefficient of solution. (k of carbon fiber is 125 W/m°C, resistivity of carbon fiber is 14.16×10^{-5} Ω m [measured]).

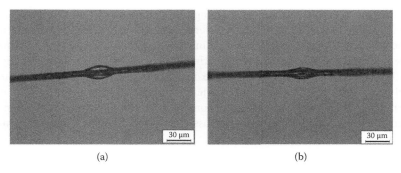

(a)   (b)

FIGURE 9.8 SME of micro bead (optical microscope images). (a) After compression (flattened). (b) After heating for shape recovery. Length of scale bar is 30 μm.

tension among semi-solid SMP and the solid carbon fibers in the liquid SMP–DMF solution. The observed SMP beads were similar to the droplets used in the microbond pull-out technique to measure the fiber–resin interfacial shear strength (Miller et al. 1991). Droplet-shaped resin on fiber is also the result of surface tension among the semi-solid resin, solid fiber, and the air, without Joule heating, so that the exact shapes and sizes of droplets are not easily controllable unless the materials used in the experiment are pre-determined.

The SMEs of micro beads and wires were characterized. Figure 9.8a shows a micro bead after pre-compression (flattening). As we can see, the flattened bead is 41 μm long and has a diameter (maximum width) of 19 μm. After heating (Figure 9.8b), the bead length is 40 μm and 15.5 μm is the maximum width. The change in width is significant (over 20% reduction). Because the string of carbon fiber is 7 μm in diameter, the wall thickness of the original bead is only 4.25 μm in this particular bead. After excluding the width of carbon fiber, the reduction of SMP wall thickness is over 40%.

Figure 9.9 demonstrates the shape recovery of a 100 μm diameter SMP wire. The original wire was slightly curved. At 70°C, the wire was straightened and cooled for 5 minutes in a refrigerator. After removal from the refrigerator, it was left in the air

(b) At 70°C straightened and cooled
in fridge for 5 minutes

(a) Original shape   (c) Left in air   (d) After heating (70°C)

FIGURE 9.9 Demonstration of shape memory phenomenon of 100 μm diameter SMP wire.

for 45 minutes and no visible shape change was observed. Upon heating to 70°C, the wire instantly curved. It was also observed that the shape recovery temperature can be reduced after immersing the wire in room-temperature water for 45 minutes. Shape recovery could be achieved by heating to 37°C—18°C lower than the original T_g of this SMP (Sun and Huang 2010). The decrease in shape recovery temperature is largely due to water absorbed upon immersion of the wire into water (Yang et al. 2006). This confirms the moisture-responsive (in addition to thermo-responsive) character of the material and also the possibility of a gradient T_g for programmed shape recovery (Huang et al. 2005a).

9.4 SMP THIN AND ULTRATHIN FILMS

In this section, we present two techniques, namely water float casting and spin coating, to fabricate thin and ultrathin polyurethane SMP films. The SMEs of the films resulting from heating are also demonstrated.

9.4.1 WATER FLOAT CASTING

The spread phenomenon and subsequent formation of thin films on water surfaces are combined as a standard technique known as the water float casting to fabricate thin and ultrathin films (Adamson and Gast 1997). Two grams of MM5520 pellets were added to 50 mL of DMF (corresponding to a 4.0 wt.% concentration of SMP) and continuously stirred until the pellets were completely dissolved. After a drop of SMP solution (Figure 9.10a) was deposited on the water surface, it spread quickly and spontaneously. After the DMF solvent completely evaporated, a thin film was left on the water surface (Figure 9.10b).

The spreading of the SMP solution on the water surface can be explained. The surface tension of DMF is 37.10 mN/m (at 22°C), significantly lower than that of water (72.80 mN/m at 22°C). The difference in surface tension is large enough to overcome the interfacial tension between the two liquids, giving rise to spontaneous spreading of the SMP. In addition, DMF is a water-miscible solvent. During spreading, the two liquid phases are in contact to form an interfacial layer, hence mutual diffusion between water and the solvent in the polymer solution occurs. As the polymer solution spreads on the surface of the water, a quick solvent–non-solvent exchange occurs between DMF and water, and the polymer chains aggregate and then precipitate during the phase-inversion process, resulting in the formation of a thin film. The main factors affecting the thickness and uniformity of the resultant polymer films include

- Viscosity of polymer solution
- Solubility of polymer in solvent
- Surface thermodynamic properties (e.g., surface tension) of solvent
- Miscibility of solvent and non-solvent (e.g., water)
- Densities of spreading solution and water

Polyurethane SMP thin films up to 300 nm thick have been fabricated using this method. Figure 9.11 reveals the shape recovery process of a piece of 300 nm thick

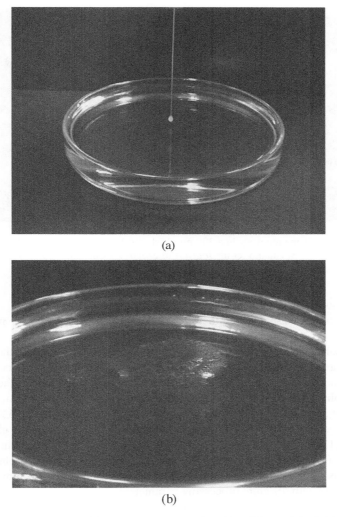

(a)

(b)

FIGURE 9.10 Deposition of droplet atop water surface (a) and formation of thin film (b).

SMP film upon heating and straightening from its curved shape. Full shape recovery was observed.

9.4.2 SPIN COATING

Spin coating is a procedure to fabricate uniform polymer thin films atop a flat substrate. This traditional technique has been used for decades. Standard spin coating steps are

- A certain amount of solution (polymer dissolved in solvent) is deposited atop the center of a flat substrate.

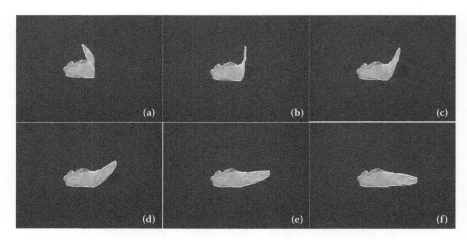

FIGURE 9.11 Shape recovery in a 300 nm thick SMP film upon heating. (Reprinted from Huang WM, Yang B, Zhao Y et al. *Journal of Materials Chemistry*, 20, 3367–3381, 2010. With permission.)

- The substrate is spun at a certain speed for a specific period.
- The solution is spread over the substrate by the centrifugal force of spinning.

When the solvent is completely evaporated, a polymer thin film is left atop the substrate. Many factors affect the thickness and quality of the resultant thin films including the nature of the polymer (viscosity, evaporation speed, concentration, surface tension, etc.) and spin-coating parameters (spin speed, acceleration, fume exhaust, spin time, deceleration, and amount of polymer deposited). We dissolved MM5520 in DMF with a concentration of 1 wt.%. After adding MM5520 pellets into a glass bottle containing DMF solvent, the bottle was put into an orbital shaking bath at a shaking speed of 150 rpm and 60°C for 24 hours to achieve a homogeneous solution. Spin coating was done atop a piece of flat silicon wafer (N doped, 100 mm diameter, 450 μm thick) with different processing parameters.

Figure 9.12 plots the average film thickness against the average surface roughness of the obtained SMP films. Despite an average thickness close to 50 nm, the surface roughness of these films was still unsatisfactory. More efforts are required to achieve a smooth ultrathin film.

A closer look reveals a number of protrusions atop the films (Figure 9.13), which cause the film surface to appear uneven; high surface roughness was observed.

We carried out an instrumented indentation (micro hardness) test with a flat conical diamond tip (200 μm in diameter) to flatten a protrusion at room temperature (Figure 9.14a and b). The applied loading and unloading rate was 1600 mN/minute, approach speed was 50%/minute, pause time was 60.0 seconds, and maximum force was 800 mN. After heating to 100°C for 5 minutes, the protrusion recovered its original shape (Figure 9.14c). A careful comparison of the cross-sections at different stages reveals that a cone-shaped protrusion (about 210 nm high and 5 μm in diameter at the base) had an excellent SME.

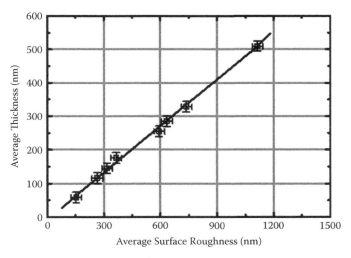

FIGURE 9.12 Average surface roughness versus average thickness of ultrathin SMP films.

To verify the SME in ultrathin films, a pyramid-shaped Berkovich indenter was used to test instrumented indentation at a loading and unloading rate of 1000 mN/min, approach speed of 50%/minute, pause time of 60.0 seconds, and maximum force of 500 mN. After indentation, the thin film was placed into a hot chamber at 100°C for 5 minutes for thermally induced shape recovery. Figure 9.15 presents the experimental results atop a 170 nm thick SMP ultrathin film. After indentation, an indent ~70 nm deep was obtained. Bear in mind that the thickness of the film is only 170 nm. After heating, we can barely see a hint

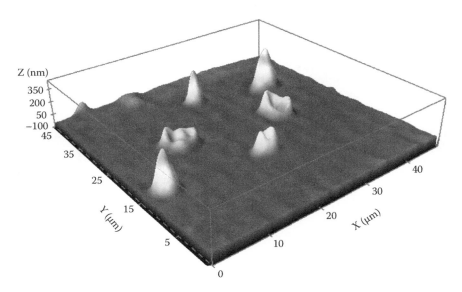

FIGURE 9.13 Typical protrusions atop a piece of spin-coated SMP thin film.

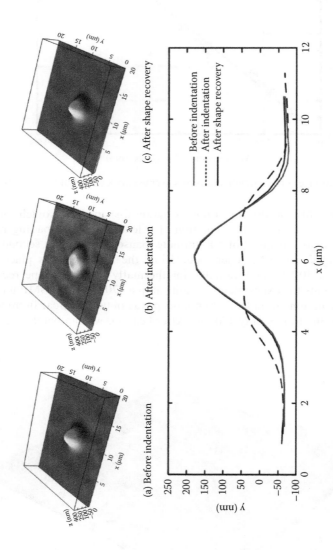

FIGURE 9.14 SME demonstrated by indentation test (using flat-end indenter). Top: three-dimensional surface scanning images. Bottom: cross-sectional view.

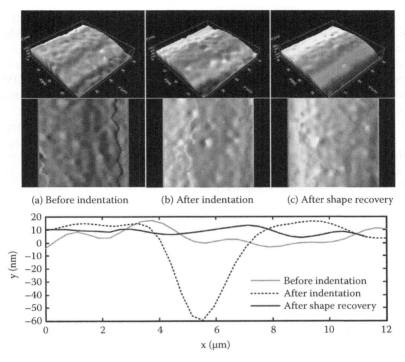

(a) Before indentation　　(b) After indentation　　(c) After shape recovery

FIGURE 9.15 SME demonstrated by indentation test (using Berkovich diamond indenter) in 170 nm thick ultrathin film. Top: three- and two-dimensional surface scanning images. Bottom: cross-sectional view. (Reprinted from Huang WM, Ding Z, Wang CC et al. *Materials Today*, 13, 54–61, 2010. With permission.)

of the indent atop the ultrathin film. The excellent SME of this SMP ultrathin film was confirmed.

9.5 SURFACE PATTERNING ATOP SHAPE MEMORY POLYMERS

In nature, many types of surface features at micro and nano scales significantly enhance surface-related properties of biological organisms. Inspired by these properties, artificial micro- and nano-scale surface features have been developed and applied in engineering applications for self-cleaning, reduced friction and drag, and other functions. This section demonstrates a few techniques for micro- and nano-sized patterning atop SMPs. Instead of using polyurethane SMPs, we used a polystyrene polymer from Cornerstone Research Group (CRG) for this study.

The as-received SMP sheet was about 4 mm thick. Long rectangular or square samples were cut from the sheet for various tests. The samples were carefully polished to an average surface roughness of about 26 nm (Ra). Figure 9.16 presents the results of differential scanning calorimetry (DSC), thermogravimetry analysis (TGA), and dynamic mechanical analysis (DMA). DSC with a Perkin-Elmer sub-ambient instrument was conducted between 20 and 200°C at a heating and cooling rate of 10°C/minute. As shown in Figure 9.16a, the T_g of this polymer is at about 64°C, very close to the value stated by CRG (62°C). TGA testing (TGA 2950, TA Instruments) was conducted from room temperature (~20°C) to 600°C at a heating rate of 10°C/minute. Figure 9.16b reveals that the decomposition temperature of this polymer is around 300°C. The DMA result (Figure 9.16c, Perkin-Elmer DMA-7) shows that upon heating, its storage modulus decreased sharply until the temperature exceeded 100°C.

Figure 9.17 presents a typical stress-versus-strain curve of a long rectangular-shaped sample in uniaxial (in-plane) stretching at 80°C, followed by unloading after cooling back to room temperature (23°C). Figure 9.18 shows the typical stress-ver-sus-strain curves of square-shaped samples in compression (out of plane) at 80 and 23°C, respectively. Similarly, the tests were carried out at 80°C and the samples

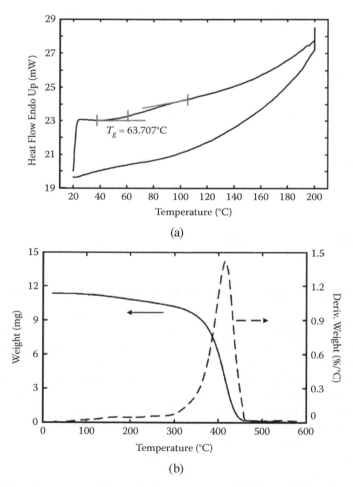

(a)

(b)

FIGURE 9.16 Thermal characterization of polystyrene SMP. (a) DSC results. (b) TGA results. (c) DMA results.

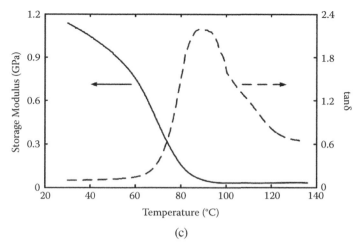

(c)

FIGURE 9.16 (Continued)

were cooled down to room temperature and unloaded. In all experiments, the applied strain rate was 10^{-3}/second.

9.5.1 BUTTERFLY-LIKE FEATURE

Incidentally, surface scanning using a WYKO interferometer on a 50% stretched (at 80°C) sample revealed that there were many butterfly-like features on the pre-polished surface (Liu et al. 2007b). The body of the butterfly was a protrusion (peak)

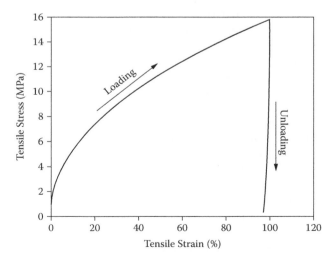

FIGURE 9.17 Stress-versus-strain relationship in uniaxial stretching (in-plane) at 80°C and unloading after cooling back to room temperature (23°C). (Reprinted from Liu N, Huang WM, Phee SJ et al. *Smart Materials and Structures,* 17, 057001, 2008. With permission.)

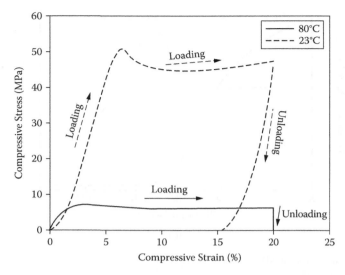

FIGURE 9.18 Stress-versus-strain relationships in compression (out-of-plane) at two different temperatures and subsequently unloading at room temperature (23°C). (Reprinted from Liu N, Huang WM, Phee SJ et al. *Smart Materials and Structures,* 17, 057001, 2008. With permission.)

and the two wings were troughs. Interestingly, the bodies of all butterflies were more or less parallel to the direction of pre-stretching. Further observation reveals that all bodies were apparently split into two parts, suggesting that something underneath was broken so that a sharp valley resulted.

Subsequently, the sample was slightly polished and then heated to 100°C for 1 minute in a laboratory oven (Grieve). Butterfly-like features still can be observed everywhere, as shown in Figure 9.19. Body protrusions and wing troughs are shown. A closer look reveals that the butterfly bodies are more or less perpendicular to the direction of pre-stretching, a 90-degree switch from the previous position.

The sizes of butterflies (in-plane) ranged from about 6 to 90 mm. Figure 9.20 reveals the approximate relationship between butterfly size and total height (height of peak plus depth of trough). As butterfly size increases, the total height also increases in a somewhat linear fashion.

An underlying mechanism modified from the one proposed in Liu et al. (2007b) is discussed below. Refer to Figure 1.4 in Chapter 1. SMPs are made of elastic and transition segments. While the elastic segment is always hard within the temperature range of our interest, the transition segment becomes soft in the presence of the right stimulus (heat in the case of thermo-responsive SMPs). Based on the size ranges of these micro butterflies, the activity should be at micro scale, well above the size of a single molecular chain of an elastic segment.

According to both DSC and DMA results (Figure 9.16a and c), at 80°C, this SMP is not yet fully in its rubber-like state so that some of the transition segment (along with the elastic segment it contains) is still hard, as illustrated in Figure 9.21a. Upon stretching, the softened transition segment in the polymer is flexible and

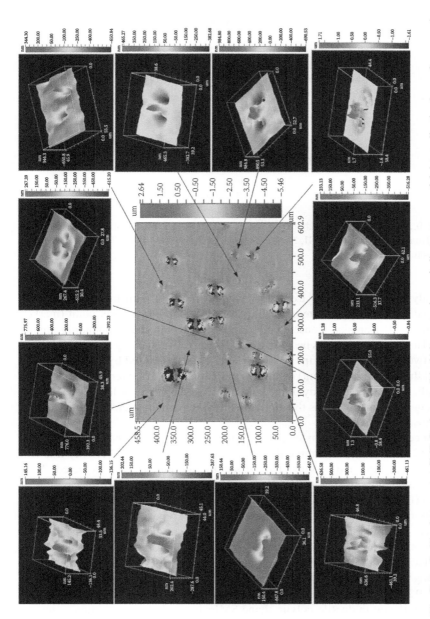

FIGURE 9.19 Butterfly-like feature atop SMP. Dark spots (black) in center image indicate unsuccessful data collection by surface scanning machine at this scan scale. Zoom-in views provide details of each butterfly-like feature. (Reprinted from Huang WM, Su JF, Hong MH et al. *Scripta Materialia*, 53, 1055–1057, 2005. With permission.)

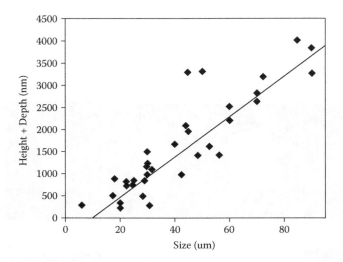

FIGURE 9.20 In-plane size of butterfly versus vertical height relationship (Reprinted from Liu N, Huang WM, and Phee SJ. *Surface Review and Letters*, 14, 1187–1190, 2007. With permission.)

can easily accommodate the extension, while a piece of hard transition segment is gradually straightened. After stretching to a certain point, a weak point (marked in Figure 9.21b) breaks. Consequently, the stress within the surrounding area is released and recovery occurs because the temperature is relatively high (80°C), producing two peaks (due to shape recovery) and a trough (caused by the broken hard piece of the transition segment) as shown in Figure 9.21c. A butterfly-like feature is the result. Obviously, the body of the butterfly should be more or less parallel to the stretching direction, as this is the direction of the maximum stretching and thus the point most likely to break.

After cooling back to room temperature, these features should be largely retained. Figure 9.21d shows top and three-dimensional views of a typical butterfly-like feature after stretching and cooling. They reveal the morphologies as expected. After subsequent slight polishing, the peaks are removed and troughs become shallow or even flat (Figure 9.21e).

Upon re-heating above T_g, the sample largely recovers its original shape. However, the material at the previous peak positions has been removed and a two-bottomed trough is formed. At the previous trough position, peak(s) can be found due to little or no polishing. A new butterfly with its body parallel to the stretching direction (90-degree angle switch from the previous direction) results (Figure 9.21f), exactly as we observed (Figure 9.21g). The butterfly phenomenon is less significant in ductile polyurethane SMPs because their transition segments are still able to deform remarkably well below the T_g (see Figure 1.16a in Chapter 1). The transition segment of this polystyrene SMP is brittle when it is in the low-temperature hard state.

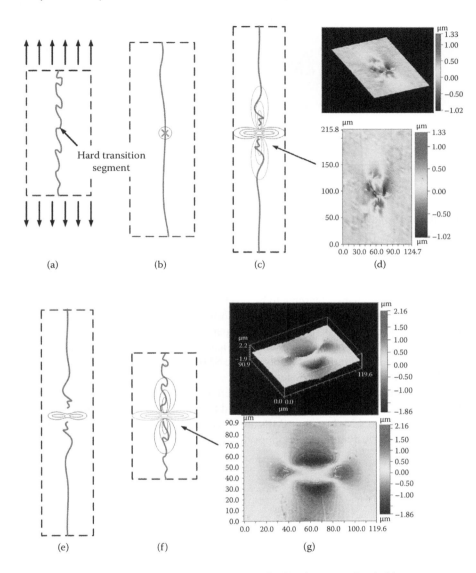

FIGURE 9.21   Mechanism of formation of butterfly-like feature and switching.

9.5.2   PATTERNING BY INDENTATION, POLISHING, AND HEATING (IPH)

The butterfly-like feature is a kind of natural product—difficult to control in terms of exact size, shape, and distribution. To achieve precise patterning, we sought a technique enabling us to control the location, shape, and size of features effectively. We previously demonstrated a technique for micro and nano patterning atop SMPs in three steps, namely indentation, polishing, and heating (IPH; Liu et al. 2007a). The technique was first proposed to produce protrusive features atop shape memory

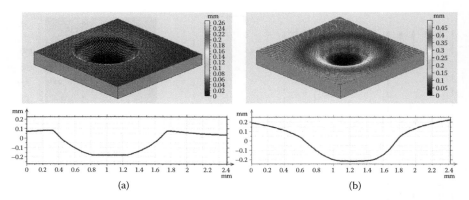

FIGURE 9.22 Indents generated at low (a) and high (b) temperatures (using spherical indenter).

alloys (SMAs; Zhang et al. 2006), but it is more applicable for patterning atop SMPs because they exhibit much higher shape recovery strain than SMAs (Huang 2002), allowing variously shaped features to be produced.

Unlike SMAs (Su et al. 2007), SMPs normally do not have two-way SMEs, and the resulting pattern is more stable and permanent. Figure 9.22 reveals typical indents produced by a spherical indenting device at high (rubber state) and low (glass state) temperatures atop the same polystyrene SMP. The cross-sectional views show that the indents produced at low temperatures are "pile-up," and those produced at high temperatures are "sink-in." This is different from SMAs in which the indents produced at low temperatures are sink-in, and those produced at high temperatures are pile-up (Figure 9.23) because SMPs are soft at high temperatures and hard at low temperatures; SMA dynamics are opposite.

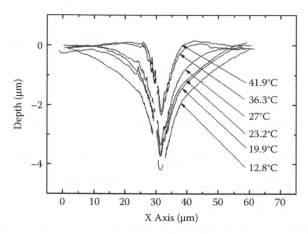

FIGURE 9.23 Typical indents produced at different temperatures atop NiTi SMA. (Reprinted from Huang WM, Su JF, Hong MH et al. *Scripta Materialia*, 53, 1055–1057, 2005. With permission.)

It should be pointed out that the full shape recovery of SMAs is almost impossible to achieve because the strain involved during indentation is normally beyond their transition strain range, producing remarkable permanent deformation. Indents atop SMAs change their shape upon heating (Huang et al. 2005b, Su et al. 2007), i.e., all indents become pile-ups. After that, the indents are shallow at high temperatures but deep at low temperatures, showing the two-way SME. Upon heating of SMPs, the indents produced at low or high temperatures almost totally disappear because SMPs have much higher recoverable strain. The IPH technique is detailed in Figure 9.24.

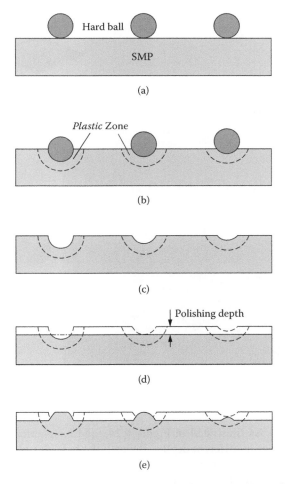

FIGURE 9.24 Formation of protrusive bumps atop SMP by IPH. (a) Original SMP. (b) Indentation to different depths using hard spherical balls. (c) After indentation. (d) After polishing. (e) Different shaped bumps formed after heating (left: flat top; right: round top). (f) Finite element simulation and typical three-dimensional experimental results for comparison. ((a) to (e) Reprinted from Liu N, Xie Q, Huang WM et al. *Journal of Micromechanics and Microengineering*, 18, 027001, 2008. With permission. (f) Reprinted from Huang WM, Zhao Y, Sun L et al. *Proceedings of SPIE*, 7493, 749337, 2009. With permission.)

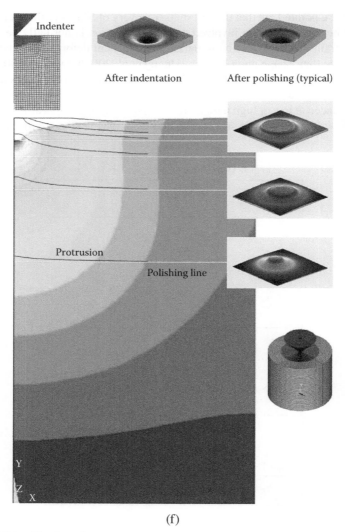

(f)

FIGURE 9.24 (Continued)

Indentation — As illustrated in Figure 9.24a and b, spherical hard balls were used for indentations at different depths. Beneath the balls are *plastic zones* within which significant deformation occurs. After removal of the balls indents with different depths were obtained as shown in Figure 9.24c.

Polishing — The SMP was polished slightly so that two indents disappeared (one over-polished, the other polished only to remove the indent). The third is less polished so that we still can see a shallow indent but the depth of the indent is shallower (Figure 9.24d).

Heating — After polishing, the SMP was heated for shape recovery, resulting in differently sized and shaped bumps (Figure 9.24e). The one with the over-polished

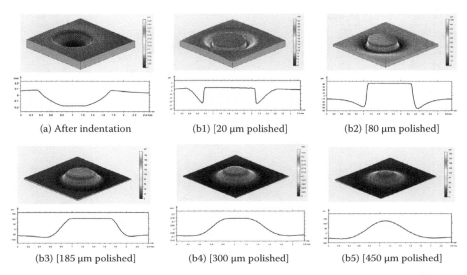

(a) After indentation   (b1) [20 μm polished]   (b2) [80 μm polished]

(b3) [185 μm polished]   (b4) [300 μm polished]   (b5) [450 μm polished]

FIGURE 9.25 Various protrusive shapes realized using a macrospherical indenter. (a) After indentation at room temperature. (b) After polishing to different depths and heating for shape recovery (polishing depths indicated individually). (Reprinted from Liu N, Huang WM, Phee SJ et al. *Smart Materials and Structures*, 16, N47–N50, 2007. With permission.)

indent is a crown-shaped low protrusion. The polished indent is a crown-shaped high protrusion, and the less-polished indent resulted in a flat-topped protrusion.

The mechanism for different sized and shaped protrusions is further illustrated in Figure 9.24f. A finite element method (ANSYS) was applied to simulate the deformation beneath the indents and the results were then compared with typical experimental results. Figure 9.25 reveals the details of experiments in which a macrospherical indenter was used for indentation at room temperature (resulting indents were pile-up) and the polishing depths varied. The results ranged from circular trench (only the pile-up part was polished away) to flat tops and crown tops. All these features are isotropic (in-plane).

When the SMP was pre-stretched in one (in-plane) direction, after IPH, the elliptically shaped features obtained as shown in Figure 9.26 were highly anisotropic (in-plane). Figure 9.27 shows the comparison of indents and protrusions (after polishing and heating) in an experiment with a piece of 50% pre-stretched (at 80°C) SMP. Such highly anisotropic (elliptical) features are difficult to achieve by conventional techniques and impossible in SMAs even after IPH because their recoverable strain is far more limited (below 10%).

Using different shaped and sized indenters, nano-sized and different shapes were achieved (Liu et al. 2007a). Figure 9.28a presents a protrusion 339 nm high. Figure 9.28b shows a 3.4 μm high pyramid-shaped feature produced via a Berkovich nanoindenter.

As demonstrated, the IPH approach can produce protrusions of various sizes and shapes. However, for practical engineering applications, we must extend the concept to the production of arrays of features over large areas. Instead of using one indenter

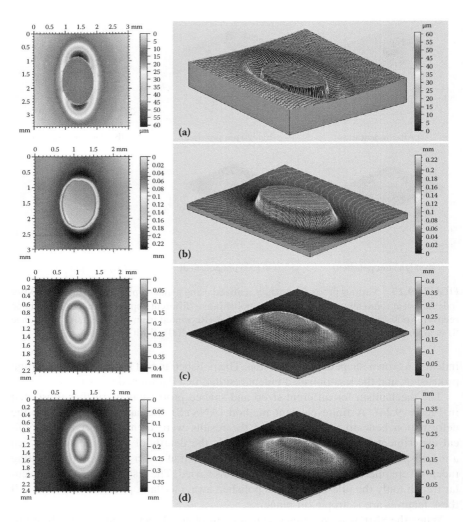

FIGURE 9.26 Elliptically shaped protrusions in 50% pre-stretched (at 80°C) samples. Left column: top view. Right column: three-dimensional view.

or a single hard ball to produce a single indent—a very time-consuming process—we packed an array of hard balls atop a piece of SMP and compressed the balls simultaneously. Figure 9.29 shows a typical depth-versus-force curve from compressing an array of 1 mm diameter well-packed steel balls atop a piece of SMP. A typical array of indents produced is shown in Figure 9.30a. After polishing and heating, an array of protrusions appeared. As seen in Figure 9.30b, the protrusions are flat topped because the SMP was less polished. The protrusions in Figure 9.30c are crown shaped because the sample was polished until the indents were not observable.

Micro- and nano-sized protrusion arrays were developed with smaller-sized balls. The result shown in Figure 9.31 was produced by 0.1 mm diameter soda lime glass balls, while the Figure 9.32 result was from 30.1 μm diameter soda lime glass

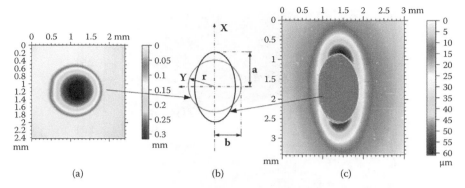

(a) (b) (c)

FIGURE 9.27 (a) Indent after polishing. (b) Comparison of indent and protrusion. (c) Protrusion after heating.

balls. Both sizes of balls were from Duke Scientific Corporation, USA. In addition, an elliptically shaped protrusion array (with a height of only a few micrometers) was demonstrated by Huang et al. (2009b). Note that tremendous time and effort are required to pack very small balls atop SMPs. In addition, the distribution pattern of the resulting array of features is limited to hexagons.

9.5.3 Laser-Assisted Patterning

To overcome the limitations and problems of the IPH method, we proposed an alternative approach for patterning atop a piece of SMP using a laser (Liu et al. 2008b). The procedure is illustrated in Figure 9.33. In the first step, a piece of SMP is polished to obtain a very smooth surface, compressed at a high temperature, and then

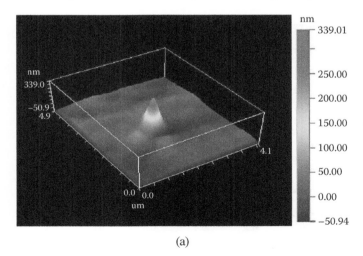

(a)

FIGURE 9.28 Nano sized (a) and pyramid shaped (b) protrusive features. (Reprinted from Liu N, Huang WM, Phee SJ et al. *Smart Materials and Structures*, 16, N47–N50, 2007 and Liu N, Huang WM, Phee SJ et al. *Smart Materials and Structures*, 17, 057001, 2008. With permission.)

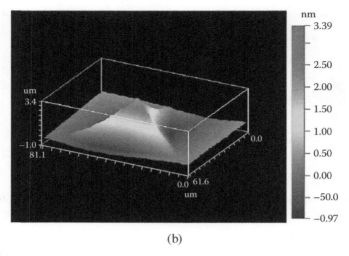

(b)

FIGURE 9.28   (Continued)

cooled back to room temperature. A laser is used to heat a small local area of the pre-compressed SMP. The temperature within the small heated area increases and a bump is generated.

A laser can heat a very small area at a very high speed (He et al. 2004). Arrays of dots and even straight lines and curves can be produced by programming the motion and controlling the activation time of the laser. One can also place a micro-lens array as a mask right atop the pre-compressed SMP, so that a single laser shot will produce an array of dots instantly, as shown in Figure 9.34. The dots

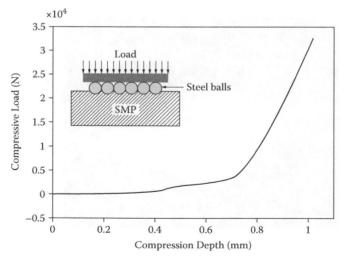

FIGURE 9.29 Typical indentation depth-versus-compressive load relationship. Inset: experimental setup. (Reprinted from Huang WM, Zhao Y, Sun L et al. *Proceedings of SPIE*, 7493, 749337, 2009. With permission.)

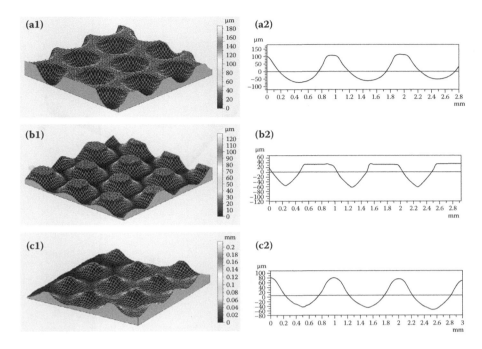

FIGURE 9.30 Typical results. (a) After indentation. (b) After slight polishing and heating. (c) After deep polishing and heating. Left column: three-dimensional view. Right column: cross-sectional view.

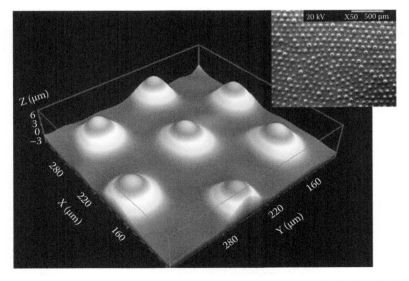

FIGURE 9.31 Micro-sized protrusion array. Inset: SEM image (top view). (Reprinted from Liu N, Xie Q, Huang WM et al. *Journal of Micromechanics and Microengineering*, 18, 027001, 2008. With permission.)

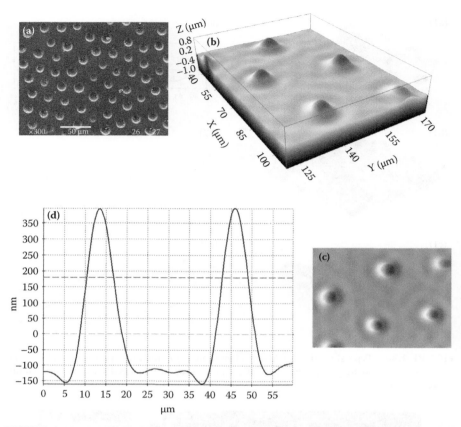

FIGURE 9.32 Nano-sized protrusion array. (a) SEM image of large area. (b) Three-dimensional surface scanning image (zoom-in on protrusions). (c) Top view of (b). (d) Cross-sectional view. (Reprinted from Huang WM, Liu N, Lan X et al. *Materials Science Forum*, 614, 243–248, 2009. With permission.)

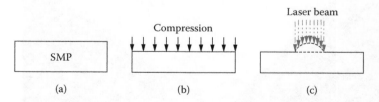

FIGURE 9.33 Formation of protrusive bump atop SMP using laser. (a) Original sample. (b) Pre-compressed. (c) Bump formed after laser heating. (Reprinted from Liu N, Xie Q, Huang WM et al. *Journal of Micromechanics and Microengineering*, 18, 027001, 2008. With permission.)

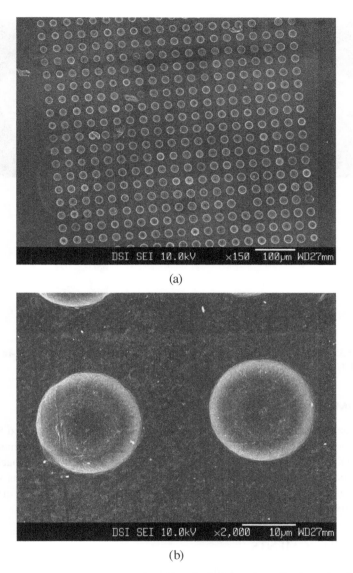

(a)

(b)

FIGURE 9.34 (a) Array of dots produced by single laser shot. (b) Zoom-in view of two dots. (Reprinted from Zhao Y, Cai M, Huang WM et al. In *Proceedings of EPD Congress: Characterization of Minerals, Metals and Materials*, Hagni A. et al., Eds., pp. 167–174, 2010. With permission.)

are ~15 μm in diameter and the gaps between dots are ~10 μm. Based on the conditions (SMP pre-compression amount, laser type and power, dimensions of microlens array, etc.), the size and distribution of dots can be adjusted easily. Figure 9.35 presents an array of protrusions that are so densely packed that there is no flat surface among the dots.

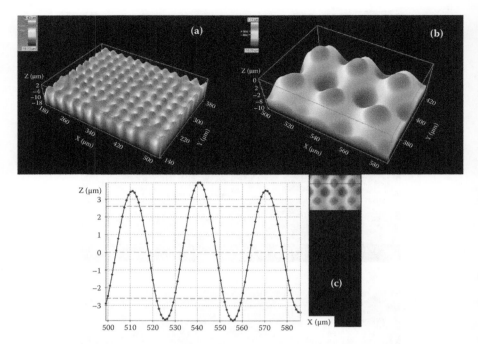

FIGURE 9.35   Densely packed array of protrusions. (a) Overall view. (b) Zoom-in view over 6 (3 × 2) dots. (c) Cross-sectional view.

9.6   SUMMARY

We proposed several novel approaches to fabricate micro- and nano-sized SMP features and confirmed that SMEs persist even in submicro-sized polyurethane SMPs. In addition, surface patterns atop SMP may be achieved by two approaches: IPH and laser heating. Although the IPH can produce protrusions of various sizes and shapes, the laser method is more convenient for producing protrusion arrays in a cost-effective manner due to its high speed.

ACKNOWLEDGMENTS

Some of the experiments were conducted by Y. Zhao, J.Q. Lin, Z.G. Chan, J.F. Royce, K.C. Cen, J. Khusbir, L. Chia, K. Daniel, L. Chew, N. Liu, and T.T.T. Vu. We also thank Dr. Joan Q. Xie of DSI for granting us permission to use laser equipment for our experiments, Prof. L. Li for valuable discussions, and Dr. C.C. Wang for his help in compiling and editing this chapter.

REFERENCES

Adamson AW and Gast AP (1997). *Physical Chemistry of Surfaces*. 6th Ed. Wiley, New York.
Altebaeumer T, Gotsmann B, Pozidis H et al. (2008). Nanoscale shape-memory function in highly cross-linked polymers. *Nano Letters*, 8, 4398–4403.

An L, Huang WM, Fu YQ et al. (2008). A note on size effect in actuating NiTi shape memory alloys by electrical current. *Materials and Design*, 29, 1432–1437.

Ancker CJ and Goodier JN (1958). Pitch and curvature corrections for helical springs. *Journal of Applied Mechanics*, 25, 466–470.

Bhattacharya K and James RD (2005). The material is the machine. *Science,* 307, 53–54.

Carslaw HS and Jaeger JC (1959). *Conduction of Heat in Solids.* Oxford University Press, Oxford.

Fu Y, Du H, Huang WM et al. (2004). TiNi-based thin films in MEMS applications: a review. *Sensors and Actuators A*, 112, 395–408.

Fu CC, Grimes A, Long M et al. (2009). Tunable nanowrinkles on shape memory polymer sheets. *Advanced Materials*, 21, 1–5.

Gunes IS and Jana SC (2008). Shape memory polymers and their nanocomposites: a review of science and technology of new multifunctional materials. *Journal of Nanoscience and Nanotechnology*, 8, 1616–1637.

Hayashi S (1990). Technical report on preliminary investigation of shape memory polymers. Research and Development Center, Mitsubishi Heavy Industries, Nagoya, Japan.

He Q, Hong MH, Huang WM et al. (2004). CO_2 laser annealing of sputtering deposited NiTi shape memory thin films. *Journal of Micromechanics and Microengineering*, 14, 950–956.

Huang WM (2002). On the selection of shape memory alloys for actuators. *Materials and Design*, 23, 11–19.

Huang WM (2005). Mechanics of coiled nanotubes in uniaxial tension. *Materials Science and Engineering A*, 408, 136–140.

Huang WM, Yang B, An L et al. (2005a). Water-driven programmable polyurethane shape memory polymer: demonstration and mechanism. *Applied Physics Letters*, 86, 114105.

Huang WM, Su JF, Hong MH et al. (2005b). Pile-up and sink-in in micro-indentation of a NiTi shape-memory alloy. *Scripta Materialia*, 53, 1055–1057.

Huang WM, Zhao Y, Sun L et al. (2009a). Wrinkling atop shape memory polymer with patterned surface. *Proceedings of SPIE*, 7493, 749337.

Huang WM, Liu N, Lan X et al. (2009b). Formation of protrusive micro/nano patterns atop shape memory polymers. *Materials Science Forum*, 614, 243–248.

Huang WM, Ding Z, Wang CC et al. (2010a). Shape memory materials. *Materials Today*, 13, 54–61.

Huang WM, Yang B, Zhao Y et al. (2010b). Thermo-moisture responsive polyurethane shape-memory polymer and composites: a review. *Journal of Materials Chemistry*, 20, 3367–3381.

Huang WM, Fu YQ, and Zhao Y (2010c). High performance polyurethane shape-memory polymer and composites. *PU Magazine*, 7, 155–160.

Leng J, Du S, Eds. (2010). *Shape-Memory Polymers and Multifunctional Composites.* Taylor & Francis, Boca Raton, FL.

Leng JS, Huang WM, Lan X et al. (2008a). Significantly reducing electrical resistivity by forming conductive Ni chains in a polyurethane shape-memory polymer/carbon-black composite. *Applied Physics Letters*, 92, 204101.

Leng JS, Lan X, Liu YJ et al. (2008b). Electrical conductivity of thermo-responsive shape-memory polymer with embedded micron sized Ni powder chains. *Applied Physics Letters*, 92, 014104.

Liedl T, Hogberg B, Tytell J et al. (2010). Self-assembly of three-dimensional prestressed tensegrity structures from DNA. *Nature Nanotechnology*, 5, 520–524.

Liu N, Huang WM, Phee SJ et al. (2007a). A generic approach for producing various protrusive shapes on different size scales using shape-memory polymer. *Smart Materials and Structures*, 16, N47–N50.

Liu N, Huang WM, and Phee SJ (2007b). A secret garden of micro butterflies: phenomenon and mechanism. *Surface Review and Letters*, 14, 1187–1190.

Liu N, Huang WM, Phee SJ et al. (2008a). Formation of micro protrusions atop thermo-responsive shape memory polymer. *Smart Materials and Structures*, 17, 057001.

Liu N, Xie Q, Huang WM et al. (2008b). Formation of micro protrusion array atop shape memory polymer. *Journal of Micromechanics and Microengineering*, 18, 027001.

Mahmood GI, Hill ML, Nelson DL et al. (2001). Local heat transfer and flow structure on and above a dimpled surface in a channel. *Journal of Turbomachinery*, 123, 115–123.

Mather PT, Luo X, and Rousseau IA (2009). Shape memory polymer research. *Annual Review of Materials Research*, 39, 445–471.

Miller B, Gaur U, and Hirt DE (1991). Measurement and mechanical aspects of the micro-bond pull-out technique for obtaining fiber/resin interfacial shear strength. *Composites Science and Technology*, 42, 207–219.

Miyazaki S, Fu YQ, and Huang WM, Eds. (2009). *Thin Film Shape Memory Alloys: Fundamentals and Device Applications*. Cambridge University Press, Cambridge.

Nelson BA, King WP, and Gall K (2005). Shape recovery of nanoscale imprints in a thermoset "shape memory" polymer. *Applied Physics Letters*, 86, 103108.

Su JF, Huang WM, and Hong MH (2007). Indentation and two-way shape-memory in a NiTi polycrystalline shape-memory alloy. *Smart Materials and Structures*, 16, S137–144.

Sun L and Huang WM (2010). Thermo/moisture responsive shape-memory polymer for possible surgery/operation inside living cells in future. *Materials and Design*, 31, 2684–2689.

Wahl AM (1963). *Mechanical Springs*. McGraw-Hill, New York.

Wornyo E, Gall K, Yang FZ et al. (2007). Nanoindentation of shape memory polymer networks. *Polymer*, 48, 3213–3225.

Yang B, Huang WM, Li C et al. (2004). On the effects of moisture in a polyurethane shape memory polymer. *Smart Materials and Structures*, 13, 191–195.

Yang Z, Huck WTS, Clarke SM et al. (2005). Shape-memory nanoparticles from inherently non-spherical polymer colloids. *Nature Materials*, 4, 486–490.

Yang B, Huang WM, Li C et al. (2006). Effect of moisture on the thermomechanical properties of a polyurethane shape memory polymer. *Polymers*, 47, 1348–1356.

Zhang YJ, Cheng YT, and Grummon DS (2006). Shape memory surfaces. *Applied Physics Letters*, 89, 041912.

Zhao Y, Cai M, Huang WM et al. (2010). Patterning atop shape memory polymers and their characterization. In *Proceedings of EPD Congress: Characterization of Minerals, Metals and Materials*, Hagni A. et al., Eds., pp. 167–174.

10 Wrinkling atop Shape Memory Polymers

10.1 INTRODUCTION

The first systematic investigation of the wrinkling phenomenon in a thin elastic film deposited atop a polymer was probably reported by Bowden et al. (1998); polydimethylsiloxane (PDMS) was used as the substrate. This wrinkling phenomenon, resulting from the buckling of the film due to a strain mismatch between a polymer substrate and elastic thin film atop it, attracts great interest from many researchers. The system may be the most convenient and easiest approach to achieving surface features with different patterns for many engineering applications. Various types of polymers and techniques have utilized the wrinkling phenomenon since then. For example, Chung et al. (2009) deposited polystyrene (PS) dissolved in toluene atop a silicon wafer by spin casting. After UV oxidation of the surface layer, the toluene vaporized and wrinkles were produced atop the PS film.

Chung et al. (2007) also investigated the wetting behavior atop patterned surfaces with anisotropic wrinkles. The conclusion was that droplets exhibit obvious anisotropic wetting behavior governed by the roughness aspect ratio. This work demonstrates the possible uses of directional and spatial variations of surface properties for controlled wetting, adhesion, and friction applications (Chung et al. 2007).

Koo et al. (2010) used patterned surfaces to enhance the light extraction efficiency of organic light-emitting diodes (OLEDs). The patterned surface was fabricated by spontaneous wrinkling. The authors reported that the wrinkled surface can at least double the outcoupling of waveguide modes and can be applied to white OLEDs with broad spectra on flexible substrates. A tunable phase grating was created with controllable surface patterns generated from the mechanical instability of a glassy polymer film atop a pre-stressed silicone sheet. The obtained grating was controlled by the compressive strain and stress in the system and the film thickness. This tunable phase grating can be used to tune the intensity of a coherent beam (Harrison et al. 2004).

As reported in Chan and Crosby (2006), after selective oxidation of the surface layer, buckling instantly occurred and thus microlenses were produced. This method is simpler and more economic than traditional ones such as lithography and surface-tension-driven techniques.

Well-documented experimental results reveal that the wavelengths of wrinkles can be altered by controlling the mechanical properties and thickness of the system. If the stress and strain for wrinkling can be adjusted, different wrinkle patterns can be achieved. To describe the wrinkling behavior, several theoretical frameworks

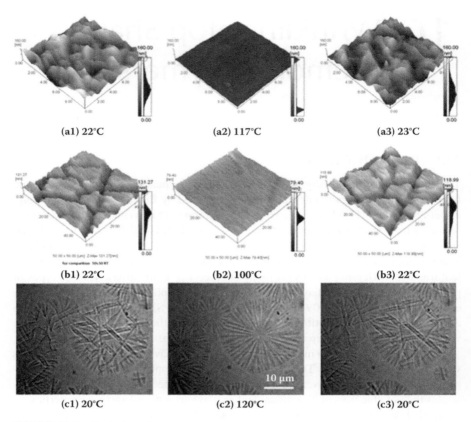

(a1) 22°C   (a2) 117°C   (a3) 23°C

(b1) 22°C   (b2) 100°C   (b3) 22°C

(c1) 20°C   (c2) 120°C   (c3) 20°C

FIGURE 10.1 Evolution of surface morphologies atop NiTi SMA thin films during thermal cycling. (a) Surface relief. (b) Overall wrinkling in fully crystallized thin film. (c) Partial wrinkling in partially crystallized thin film. (Reprinted from Wu MJ, Huang WM, Fu YQ et al. *Journal of Applied Physics*, 105 (3): 033517. 2009. With permission.)

were developed. Huang and Suo (2002) developed a theoretical model based on the differential governing equations for buckling of sheets; Cerda and Mahadevan (2003) developed a theory based on the strain energy method; Li et al. (2010) conducted a general theoretical analysis for wrinkling atop a planar substrate that resulted in two kinds of wrinkling modes: shallow substrate and deep substrate.

As for wrinkling atop shape memory materials (SMMs), Wu et al. (2009) reported the evolution of three types of surface morphologies atop sputter-deposited NiTi-based shape memory thin films upon thermal cycling. As presented in Figure 10.1, in addition to surface relief due to the martensitic (reversible) transformation, the wrinkles in fully and partially crystallized thin films are also reversible, but exhibit different patterns.

Unlike the wrinkles atop shape memory alloys (SMAs) that may be reversible, wrinkles atop shape memory polymers (SMPs) are more permanent and can be easily manipulated into various patterns due to their significant recoverable strain (see Figure 1.3 in

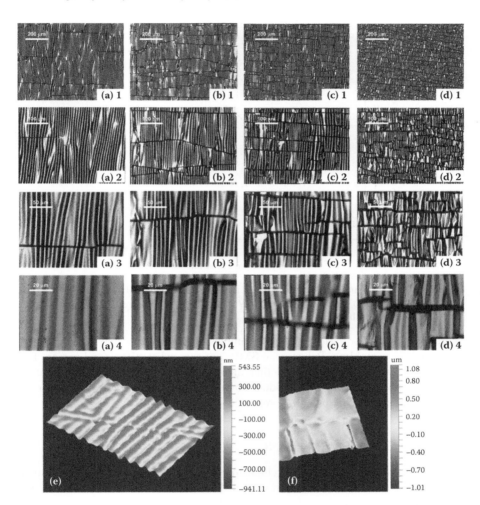

FIGURE 10.2   Wrinkles and cracks atop pre-stretched SMP (coated with 50 nm thick nickel): (a) 5% pre-stretched; (b) 10% pre-stretched; (c) 20% pre-stretched; (d) 40% pre-stretched; (e) typical crack line (in three dimensions); and (f) zoom-in view. ((a)–(d) reprinted from Huang WM, Liu N, Lan X et al. *Materials Science Forum*, 614, 243–248, 2009. With permission.)

Chapter 1). However, it is important to avoid cracks resulting from excessive expansion due to the effect of Poisson's ratio during shape recovery, particularly when a brittle metallic thin film is used. In fact, according to Figure 10.2, the widths of all cracks with different pre-strains are about the same. Figure 10.3 reveals that the number of crack lines is about linearly proportional to the maximum pre-strain.

This chapter starts with a summary of the basic theory behind wrinkling atop polymers and approaches for investigating this phenomenon. Next is a systematic study of various kinds of wrinkles atop SMPs. Polyurethane SMPs (from SMP Technologies, Japan) and polystyrene SMPs (from Cornerstone Research Group, USA) were used for this study. Polyurethane SMP is a thermo-plastic (allowing us

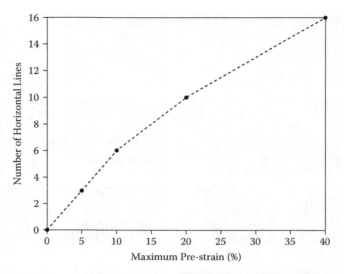

FIGURE 10.3 Maximum pre-strain versus average number of horizontal lines (cracks). (Reprinted from Zhao Y, Cai M, Huang WM et al. EPD Congress, Washington, DC, pp. 167–174, 2010. With permission.)

to prepare tiny and/or variously shaped samples) that is both thermo- and moisture-responsive for wrinkling. Polystyrene SMPs are somewhat stiffer (and present less difficulty in polishing). We found that thermally induced and moisture induced wrinkles in polyurethane SMPs produce about the same configurations. Hence, only thermally induced wrinkles will be reported in this chapter.

10.2 THEORY OF WRINKLING

Figure 10.4 illustrates the basic mechanism of wrinkling of a stiff thin layer atop a soft substrate that exhibits highly nonlinear behavior and can be described by the Föppl-von Kármán equations (Landau and Lifshitz 1986) as

$$\frac{Eh^3}{12(1-v^2)} \nabla^2 w - h \frac{\partial}{\partial x_\beta}\left(\sigma_{\alpha\beta}\frac{\partial w}{\partial x_\alpha}\right) = P$$

$$\frac{\partial \sigma_{\alpha\beta}}{\partial x_\beta} = 0$$

(10.1)

where E is Young's modulus, $\sigma_{\alpha\beta}$ is the stress tensor, h is the thickness of the thin layer, w is out-of-plane deflection, v is Poisson's ratio, P is the external normal force per unit area of the plate, and Δ is a two-dimensional Laplacian. However, these nonlinear partial differential equations are too difficult to be solved analytically. Only some one-dimensional cases can be solved and presented in an analytical form (Volynskii et al. 2000). Semi-analytical methods and numerical simulations are normally applied to quantify the wrinkling behavior of a thin layer under an external force (Allen 1969, Bowden et al. 1998, Cerda and Mahadevan 2003).

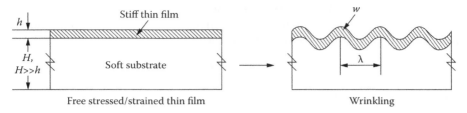

FIGURE 10.4 Wrinkling phenomenon.

10.2.1 SEMI-ANALYTICAL METHOD

Theoretical studies of wrinkling in a sandwich structure can be traced back to the 1940s (Gough et al. 1940, Wan 1947, Goodier and Neou 1951). A summary of the historical development can be found in Allen (1969). Later, Cerda and Mahadevan (2003) derived a general theory for wrinkling studies using a semi-analytical method.

The semi-analytical method focuses on the energies due to in-plane strain (stretching), out-of-plane strain (bending), and the work of external in-plane stresses. Upon wrinkling, the work done by in-plane stress is transformed into the wrinkling energy of the thin film and the elastic energy of the soft substrate. This method can solve the critical conditions (critical stress σ_c and critical wavelength λ_c) for wrinkling.

Strain energy of wrinkled thin film — The strain energy in a wrinkled thin film U_f is composed of stretching energy from the in-plane strain and bending energy from the out-of-plane deformation (Allen 1969, Cerda and Mahadevan 2003) and can be presented as

$$U_f = \int_A \left(\frac{h}{2} L_{ijkl} \dot{E}_{kl} \dot{E}_{ij} + \frac{h^3}{24} L_{ijkl} \dot{K}_{kl} \dot{K}_{ij} \right) dA \qquad (10.2)$$

where h is the thickness of the thin film, L_{ijkl} is the fourth-order modulus tensor, $\dot{E}_{ij}$ is the increment in stretch strain, $\dot{K}_{ij}$ is the increment in bending strain, and A is the area where wrinkling occurs.

Wrinkling energy of substrate — According to Winkler's hypothesis (Allen 1969), the influence of the soft substrate is equivalent to a group of springs with a stiffness of k (further discussed later in this section). Therefore, part of the energy is stored in the substrate during wrinkling. Assume that the thin film (which may look like a shell if the surface of substrate is not smooth) is always well bonded to the substrate, and the displacement of the thin film or shell also represents the displacement of the top surface of the substrate. Hence, the wrinkling energy of the substrate, U_s, can be expressed as

$$U_s = \frac{k}{2} \int_A w^2 dA \qquad (10.3)$$

where w is the out-of-plane deflection.

Work done by in-plane stress — This work, W, can be presented in terms of membrane stress σ_{ij} in the thin film and variation in its length Δl due to the out-of-plane deflection. Δl (Allen 1969) can be presented as

$$l = \frac{1}{2} w_{,i}^2 dx_i \tag{10.4}$$

Subsequently, the work done by membrane force can be worked out as

$$W = \frac{h}{2} \int_A \sigma_{ij} w_{,i} w_{,j}\, dA \tag{10.5}$$

where σ_{ij} is the in-plane stress tensor,

$$w_{,i} = \frac{\partial w}{\partial x_i} \quad \text{and} \quad w_{,j} = \frac{\partial w}{\partial x_j}.$$

Winkler's hypothesis — This is a classical method for studying the effects of a soft substrate on the mechanical deformation of a bilayer or sandwich structure. A parameter k is used to represent the effective stiffness of the substrate that depends on the mechanical properties of the material and characteristic depth of the deformation. It is expressed (Allen 1969, Li et al. 2010) as

$$k = \frac{f(\lambda/l_p)E_s}{l_p(1 - v_s^2)} \tag{10.6}$$

where E_s and v_s are the Young's modulus and Poisson's ratio of the substrate, respectively,

$$\bar{E}_s = \frac{E_s}{1 - v_s^2}.$$

$f(\lambda/l_p) \sim \lambda^2/l_p^2$ is a dimensionless function that depends on the geometry of the system, and l_p is the characteristic depth of the deformed substrate. Figure 10.5 shows two kinds of substrate models in which the effective stiffness is different. For the shallow substrate model ($h \ll H$), l_p is close to the characteristic thickness of the substrate H, so that the stiffness of the substrate is given as

$$k \sim \frac{\bar{E}_s \lambda^2}{H^3}$$

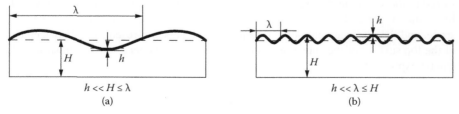

FIGURE 10.5   (a) Shallow substrate model. (b) Deep substrate model.

(Bowden et al. 1998, Cerda and Mahadevan 2003). For the deep substrate model $(h << \lambda \leq H)$, l_p is close to the characteristic wavelength of the wrinkles on the substrate surface, and the stiffness of the substrate is

$$k \sim \frac{\bar{E}_s}{\lambda}$$

(Cerda and Mahadevan 2003).

Wrinkling criterion — Wrinkles will appear when the work done by the thin film stress is larger than the sum of the wrinkling energy of the thin film and the elastic energy of the soft substrate. Like the result of Hutchinson and Neale (1985) for buckling of thin sheets or shells, the critical condition for wrinkling of thin films atop a soft substrate may be expressed as

$$W = U_f + U_s \tag{10.7}$$

Substituting Equations (10.1) through (10.6) into Equation (10.7), the wrinkling criterion can be derived as

$$U_f + U_s - W = C \begin{pmatrix} w_0 \\ u_0 \\ v_0 \end{pmatrix} [M](w_0 \quad u_0 \quad v_0) = 0 \tag{10.8}$$

C is a constant (non-zero); w_0, u_0, and v_0 are amplitudes of the out-of-plane deflection w and the in-plane deformation (u and v), respectively; and

$$[M] = \begin{pmatrix} M_{11} & M_{12} & M_{13} \\ M_{21} & M_{22} & M_{23} \\ M_{31} & M_{32} & M_{33} \end{pmatrix}$$

For an elastic thin film atop a flat substrate:

$$M_{11} = \frac{2\pi^4 h^3}{3}\left(\frac{L_{1111}}{\lambda_1^4} + \frac{L_{2222}}{\lambda_2^4} + \frac{2L_{1122}}{\lambda_1^2\lambda_2^2} + \frac{4L_{1212}}{\lambda_1^2\lambda_2^2} \right) - 2\pi^2 h\left(\frac{\sigma_1}{\lambda_1^2} + \frac{\sigma_2}{\lambda_2^2} \right) + \frac{k}{2}$$

$$M_{22} = 2\pi^2 h\left(\frac{L_{1111}}{\lambda_1^2} + \frac{L_{1212}}{\lambda_2^2} \right) + \frac{k}{2}$$

$$M_{33} = 2\pi^2 h\left(\frac{L_{1212}}{\lambda_1^2} + \frac{L_{2222}}{\lambda_2^2} \right) + \frac{k}{2} \tag{10.9}$$

$$M_{12} = M_{21} = 0$$

$$M_{13} = M_{31} = 0$$

$$M_{23} = M_{32} = 2\pi^2 h\left(\frac{L_{1122}}{\lambda_1\lambda_2} + \frac{L_{1212}}{\lambda_1\lambda_2} \right)$$

where λ_1 and λ_2 are the wrinkle wavelengths along two principal axes, respectively, and σ_1 and σ_2 are the corresponding in-plane stresses along the principal axes, respectively. The fourth-order tensor (L_{ijkl}) denotes the instantaneous modulus L_{ij}. For an isotropic elastic thin film, the elastic moduli (non-zero components) are

$$L_{1111} = L_{11} = \frac{E_f}{1 - v_f^2}$$

$$L_{2222} = L_{22} = \frac{E_f}{1 - v_f^2}$$

$$L_{1122} = L_{12} = \frac{v_f E_f}{1 - v_f^2}$$

$$L_{1212} = L_{44} = \frac{E_f}{2(1 + v_f)}$$

where E_f and v_f are the Young's modulus and Poisson's ratio of the thin film, respectively, and

$$\bar{E}_f = \frac{E_f}{1 - v_f^2}.$$

Because w_0, u_0, and v_0 are not all zero, the wrinkling criterion can be obtained from

$$\det[M] = \begin{vmatrix} M_{11} & M_{12} & M_{13} \\ M_{21} & M_{22} & M_{23} \\ M_{31} & M_{32} & M_{33} \end{vmatrix} = 0 \tag{10.10}$$

Case study: wrinkling of elastic thin film atop flat substrate — Assume that a system is under uniform biaxial stress condition. For a flat substrate, h is the thickness of the thin film, H is the thickness of the substrate, $\lambda_1 = \lambda_2 = \lambda$, $\sigma_1 = \sigma_2 = \sigma_c$, $w \neq 0$, and $u = v = 0$. M is obtained as

$$M_{11} = \frac{8\pi^4 h^3 \bar{E}_f}{3\lambda^4} + \frac{k}{2} - \frac{4\pi^2 h \sigma_c}{\lambda^2}$$

$$M_{22} = M_{33} = \frac{\pi^2 h (3 - v_f) \bar{E}_f}{\lambda^2}$$

$$M_{12} = M_{21} = M_{13} = M_{31} = 0 \tag{10.11}$$

$$M_{23} = M_{32} = \frac{\pi^2 h (1 + v_f) \bar{E}_f}{\lambda^2}$$

Hence, the wrinkling criterion, det[M] = 0, can be worked out as

$$\det[M] = \begin{vmatrix} M_{11} & 0 & 0 \\ 0 & M_{22} & M_{23} \\ 0 & M_{32} & M_{33} \end{vmatrix} = M_{11}(M_{22}M_{33} - M_{23}M_{32}) = 0 \quad (10.12)$$

Because $(M_{22}M_{33} - M_{23}M_{32}) \neq 0$, we can see that $M_{11} = 0$ according to Equation (10.11). The criterion for wrinkling of a thin film atop a flat substrate is thus obtained as

$$\sigma_c = \frac{2\pi^2 h^2 \bar{E}_f}{3\lambda^2} + \frac{\lambda^2 k}{8\pi^2 h} \quad (10.13)$$

where the effective stiffness of the substrate can be determined according to the substrate mode (shallow mode or deep mode). The critical wavelength λ_c can be derived from Equation (10.13) when $d\sigma_c/d\lambda = 0$. For the shallow mode ($h \ll H \leq \lambda$), substituting $k = \pi \bar{E}_s \lambda^2/H^3$ into Equation (10.13), the critical wavelength and critical stress are obtained:

$$\lambda_c = (2\pi h H)^{1/2}\left(\frac{\bar{E}_f}{3\bar{E}_s}\right)^{1/6} ; \quad \sigma_c = \frac{\pi h}{2H}(3\bar{E}_s)^{1/3}\bar{E}_f^{2/3} \quad (10.14)$$

Similarly, for the deep mode ($h \ll \lambda \leq H$), substituting $k = \pi \bar{E}_s/\lambda$ into Equation (10.13), the critical wavelength and critical stress are obtained:

$$\lambda_c = 2\pi h\left(\frac{4\bar{E}_f}{3\bar{E}_s}\right)^{1/3} ; \quad \sigma_c = \frac{1}{4}(3\bar{E}_s)^{2/3}(4\bar{E}_f)^{1/3} \quad (10.15)$$

Equations (10.14) and (10.15) are identical to those reported in Vandeparre et al. (2007), Bowden et al. (1998), and Cerda and Mahadevan (2003).

10.2.2 Numerical Simulation

Besides the semi-analytical method, the governing equations of wrinkling behavior can be solved through numerical simulation. Analysis of the evolution of wrinkles has been based on numerical simulation (Huang and Suo 2002, Im and Huang 2005, Huang and Im 2006).

If the surface of a polydimethylsiloxane (PDMS) substrate is coated with a thin layer of Au at high temperature, upon cooling, equi-biaxial compressive stress is generated. When a critical condition is achieved, the thin film will wrinkle. Chen and Hutchinson (2004) observed a herringbone-type wrinkling pattern in this system and indicated that the herringbone wrinkles constituted a minimum energy conformation in their numerical simulation.

Im and Huang (Im and Huang 2005, Huang and Im 2006) studied the wrinkling mechanics of an elastic thin film on different substrates such as elastic and/or viscous materials. A mathematic model was developed to explain the wrinkling process. Im and Huang (2008) also numerically studied the wrinkle patterns of isotropic and anisotropic elastic thin films under different stress conditions. According to their simulation for isotropic films, labyrinth wrinkles appeared under equi-biaxial stress conditions, and stripe patterns appeared under uniaxial stress conditions. For anisotropic films, orthogonal wrinkles formed under equi-biaxial stress, and stripe patterns formed under uniaxial stress.

Numerical simulation provides a powerful approach to trace the evolution of wrinkle patterns (Yoo and Lee 2003, Im and Huang 2006). Im and Huang (2008) studied the evolution sequence of wrinkle patterns in a cubic crystal film ($Si_{0.7}Ge_{0.3}$) under equi-biaxial compression. The simulation results revealed two kinds of coarsening processes occurring after the initiation of wrinkling. One is wavelength coarsening and the other is domain size coarsening. Both the wrinkle wavelength and domain size seem to saturate after a long evolution.

10.3 WRINKLING ATOP SMPs

Although most previous experiments were based on conventional polymer substrates such as PDMS and poly(methyl methacrylate) (PMMA) (Bowden et al. 1998, Li et al. 2009), SMP is seemingly the best choice as a substrate for studying wrinkling. This is because we can not only easily control the amount of pre-strain, but also achieve a gradient pre-strain field, utilizing the effect of thermal expansion mismatch (as a conventional technique) and/or the shape memory effect (SME). In our experiments, we deposited a very thin layer of Au atop an SMP sample with/without pre-straining. Subsequently, wrinkles were produced by means of the SME and/or thermal expansion mismatch.

10.3.1 THERMOMECHANICAL PROPERTIES OF SMP SAMPLES

Mechanical properties of soft substrates, for example, the Young's modulus E and Poisson's ratio v, play an important role in the wrinkling phenomenon of an elastic thin film atop a soft substrate. In polymers, these properties appear to be temperature dependent. Therefore, thermomechanical properties of polymers are required for wrinkling study. The glass transition temperature (T_g), melting temperature (T_m) for thermo-plastic polymers, decomposition temperature (T_d), and storage modulus (E') are essential factors in this study. In addition to the thermo-plastic polyurethane SMP (MM3520) from SMP Technologies, we also used a thermo-set polystyrene SMP from Cornerstone Research Group, USA, that is also thermo-responsive because the required stimulus for shape recovery is heat.

Figures 9.16 through 9.18 show the results of differential scanning calorimetry (DSC), thermogravimetry analysis (TGA), dynamic mechanical analysis (DMA), and mechanical tests (at high and low temperatures) of the polystyrene SMP. The results of the same experiments on the polyurethane SMP can be found in Chapter 2.

Note that this polystyrene SMP is hard and brittle at room temperature (glass state); above T_g, it is soft and flexible. As a shape memory material, it can be deformed severely above its T_g. The temporary (deformed) shape can be largely preserved after cooling back (with constraints). Subsequently, upon heating above its T_g, it can fully recover its original shape, i.e., for determining its SME. The recoverable strain may reach a few hundred percent—at least one order higher than its SMA metallic counterparts (Huang 2002).

10.3.2 WRINKLING OF GOLD THIN FILM ATOP FLAT SMP SUBSTRATE

Wrinkling of samples without pre-deformation — The as-received polystyrene SMP was in sheet form with a thickness of about 3 mm. Subsequently, small samples of predetermined size were cut out and well polished to achieve a very smooth surface, first using an 0.3 μm diameter Al_2O_3 suspension (Buehler) and then an 0.05 μm one. After polishing, the average roughness (R_a) was about 10 nm measured by a confocal imaging profiler.

The top surface of the well-polished sample was coated with a gold thin film using an SC7640 gold sputter. The sputtering time was 120 seconds at a current of 20 mA and voltage of 2 kV. The thickness of the Au film was about 30 nm as measured by a confocal imaging profiler.

The profiler was also used to study the surface morphologies of samples after heating to different temperatures (above T_g) for 20 minutes in an oven and then cooling to room temperature (about 22°C) in the air. No wrinkles could be found at heating temperature below 120°C, as shown in Figure 10.6. However, labyrinth wrinkles appeared after heating to 120°C, as shown in Figure 10.7.

The critical compressive stress–strain value for wrinkling is well known. In these experiments, the compressive stress–strain induced by heating and cooling results in thermal mismatch between the elastic thin film (gold) and soft substrate (polystyrene SMP). Because polymers have much higher thermal expansion coefficients than metals, it is clear that the soft substrate expands more than the elastic gold thin film upon heating. As seen in Figure 10.8, with the increase of thermal mismatch, interfacial stress at the thin film–substrate interface builds and grows continuously

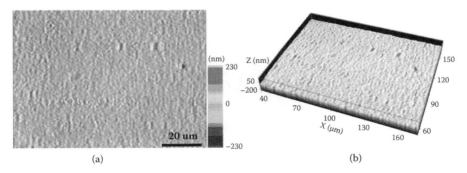

FIGURE 10.6 Surface morphology after heating to 110°C. (a) Two-dimensional image. (b) Three-dimensional image.

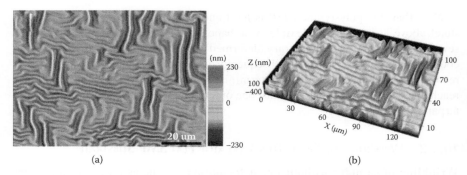

(a)   (b)

FIGURE 10.7 Labyrinth wrinkles after heating to 120°C. (a) Two-dimensional image. (b) Three-dimensional image. (Reprinted from Zhao Y, Cai M, Huang WM et al. EPD Congress, Washington, DC, pp. 167–174, 2010. With permission.)

until a critical level is reached. After that, dislocation between the thin film and substrate occurs and partially reduces the interfacial stress. However, because we never observed debonding, we should assume that the thin film and substrate remain bonded.

Subsequently, to maintain the newly generated boundary condition between the thin film and substrate during cooling, the elastic thin film must contract with the substrate—the thin film actually must contract more than it can. Because the substrate is soft and the thin film is stiffer, the thin film elastically buckles to match the overall deformation of the substrate when the in-plane compressive stress exceeds a

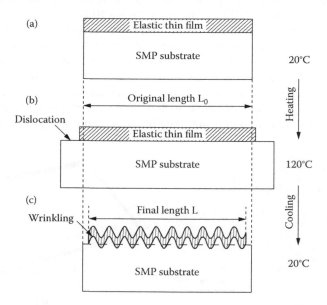

FIGURE 10.8 Wrinkling mechanism. (a) Initial system (stress-free). (b) Dislocation appears at critical temperature upon heating. (c) Substrate returns to original length while thin film is shortened by buckling and wrinkling.

critical value and forms wrinkles atop the surface of a polymer. The SMP substrate is hard to deform at a low-temperature glassy state ($<T_g$). On the other hand, at a high-temperature rubbery state ($>T_g$), it is soft and flexible. Wrinkles should appear before the substrate becomes hard during cooling. Because the in-plane compressive stress is two dimensional and isotropic, the resulting wrinkles are labyrinthine.

Wrinkling of pre-stretched samples — The polystyrene SMP samples were pre-stretched by 2% at 80°C at a strain rate of 0.005/s and then cooled to room temperature. Next, the pre-stretched samples were well polished and coated with a thin gold layer as described above. Because the applied pre-strain was only 2%, a quasi-uniaxial compressive stress along the pre-straining direction was generated in the coated elastic thin film when the substrate was heated above its T_g for shape recovery. Because the substrate was very soft above T_g, the compressive stress at a critical point caused the elastic thin film to buckle in the uniaxial compression direction. Stripe-shaped wrinkles resulted, as shown in Figure 10.9.

Unlike the results from samples that were not pre-stretched, wrinkles occurred at 80°C (slightly above the T_g of the substrate) instead of at 120°C. Assume that at temperatures above the glass transition range, the stiffness of SMP is about a constant. Thus, the fundamental difference between samples with and without pre-stretching is the difference in approach to induce strain mismatch for wrinkling. After pre-stretching, strain mismatch is directly introduced into the SMP substrate. Consequently, shape recovery in the substrate upon heating above the T_g (the SME) is the driving force to induce a compressive stress in the thin film. Without pre-stretching, thermal mismatch is required to reach a critical point for the dislocation at the interface to merge. The sample must be heated to a higher temperature to initialize thermal-mismatch-induced dislocation at the interface; this means that the strain mismatch is introduced by dislocation.

For simplicity, we ignore the influence of Poisson's ratio in this discussion because the pre-strain involved is small. As the compressive stress in the thin film is uniaxial in the pre-stretched samples and biaxial in the samples without pre-stretching, it is expected that wrinkles will be stripe shaped (anisotropic) in the pre-stretched samples

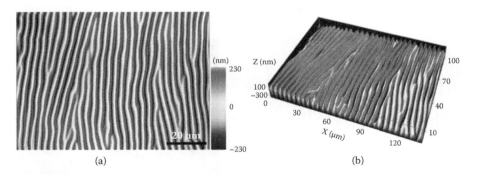

(a)   (b)

FIGURE 10.9 Stripe-shaped wrinkles after heating to 80°C. (a) Two-dimensional image. (b) Three-dimensional image. Pre-stretching was 2% in the horizontal direction. (Reprinted from Zhao Y, Cai M, Huang WM et al. EPD Congress, Washington, DC, pp. 167–174, 2010. With permission.)

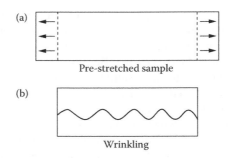

FIGURE 10.10   Formation of wrinkles in uniaxial pre-stretched sample. (a) Pre-stretched sample. (b) After shape recovery.

(Figure 10.10) and labyrinth shaped (isotropic) in the samples without pre-stretching. However, if a pre-stretched sample with stripe-shaped wrinkles is heated to 120°C, the pattern of wrinkles changes to S shapes, as shown in Figure 10.11. This change is the evidence of additional dislocation-induced mismatch strain upon heating to 120°C.

Critical conditions for wrinkling atop flat SMP substrate — The critical conditions for wrinkling and the wavelengths of resulting wrinkles were quantitatively investigated and compared with the experimental results. Buckling is an effective way to release the compressive stress of a thin film coated atop a soft substrate when a critical stress value is reached. Figure 10.12 shows typical sinusoidal-like wrinkles in samples with and without pre-stretching. The gold thin film is about 30 nm thick and the SMP substrate is about 3 mm thick; the wavelengths of wrinkles are at micro scale. Therefore, the deep wrinkling mode ($h \ll \lambda \ll H$) was used. The critical conditions for wrinkling atop a flat substrate in deep wrinkle mode are given by Equation (10.15). Table 10.1 lists the thickness, Young's modulus, and Poisson's ratio of gold film and polystyrene SMP substrate. Hereinafter, s and f represent substrate and film, respectively.

For samples without pre-stretching, wrinkles were induced by membrane stress (biaxial compressive) caused by the thermal expansion mismatch between gold thin film and SMP substrate. According to the theoretical analysis in Section 10.2, the critical wavelength and critical stress for wrinkles atop a flat SMP substrate under

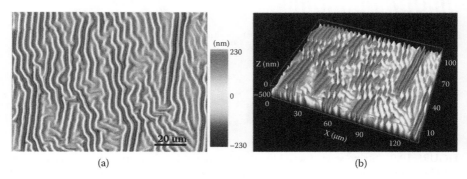

FIGURE 10.11   S-shaped wrinkles in pre-stretched sample after heating to 120°C. (a) Two-dimensional image. (b) Three-dimensional image.

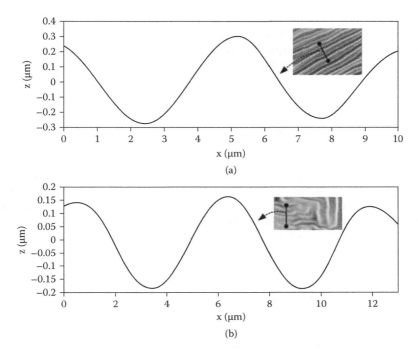

FIGURE 10.12 Wrinkle profiles. (a) Anisotropic. (b) Isotropic.

biaxial compressive stress are given in Equation (10.15), and the critical (compressive) strain in the thin film can be calculated:

$$\varepsilon_c = \frac{\sigma_c}{E_f} = \left(\frac{3\overline{E}_s}{4\overline{E}_f}\right)^{2/3} \tag{10.16}$$

Compressive stress–strain induced by thermal expansion mismatch is related to the thermal expansion coefficient α and temperature variation ΔT. In this system, the top elastic layer is a gold film whose thermal expansion coefficient is 1.41×10^{-5} K^{-1}

TABLE 10.1

Material Properties and Geometry of Gold Layer atop Polystyrene SMP

	E_s (GPa)	v_s	α_s (/°C)	E_f (GPa)	v_f	α_f (/°C)	h (nm)
20°C	1.24	0.37	5×10^{-5}	78	0.42	0.14×10^{-6}	30
>100°C	0.012	0.5					

Source: Buch A (1999). *Pure Metals Properties: A Scientific–Technical Handbook.* ASM International, Cleveland, OH; Mark JE (1999). *Polymer Data Handbook.* Oxford University Press, Oxford.

(Buch 1999). The thermal expansion coefficient of conventional soft polystyrene is ~5 to 8×10^{-5} K^{-1} (Mark 1999)—larger than that of gold. Hence, the compressive strain induced by thermal expansion mismatch is $\varepsilon_c = (\alpha_s - \alpha_f) \ T_c$. Substituting into Equation (10.16), the critical film strain is related to the critical temperature variation, ΔT_c, by

$$T_c = \frac{1}{\alpha_s - \alpha_f} \left(\frac{3\bar{E}_s}{4\bar{E}_f} \right)^{2/3} \tag{10.17}$$

It is well understood that the mechanical properties of polymers are temperature dependent. At low temperatures polymers are in glassy states, and at high temperatures they are in rubbery states. Because $\bar{E}_s$ is much lower in the rubbery state, wrinkling occurs more easily. The theoretical prediction of ΔT_c and λ_c can be calculated based on the DMA result for $\bar{E}_s$ [Equations (10.15) through (10.17) and Table 10.1]. The results were 46°C and 3.8 μm, respectively. The results for ΔT_c and λ_c were 30°C and 5 μm, respectively.

Equation (10.15) indicates that the critical wavelengths of wrinkles depend on the thickness of the thin film. To study the relationship between h and λ_c, samples with different h values were prepared by varying the gold sputtering time. The exact thickness of coating can be estimated as $h = KIVt$, where $K = 0.07$ is an experimentally determined constant; I is the plasma current in mA; V is the applied voltage, in kV; and t is the sputtering time in seconds. Figure 10.13 shows the relationships between h and λ_c based on prediction and experiment.

10.3.3 Wrinkling of Gold Thin Film atop Curved SMP

In Section 10.3.2, we studied wrinkling behavior atop flat SMP substrates and found that self-organized wrinkle patterns may be controlled by varying the stress–strain state. For example, uniaxial stress induces anisotropic stripe-shaped wrinkles and

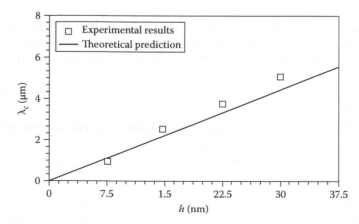

FIGURE 10.13   Relationship between h and λ_c.

biaxial stress induces labyrinthine wrinkles. In many applications, a curved substrate is required for enhanced performance. For instance, wrinkles atop protrusion arrays can significantly improve the hydrophobicity of a surface (Neinhuis and
Barthlott 1997, Jiang et al. 2004). In this section, we investigate wrinkling atop
curved SMP substrates.

Wrinkling atop pre-indented samples — As before, the first step is to well
polish the SMP samples as in Section 10.3.2 after which different-sized spherical
dimples were produced using a microhardness tester with a spherical conical diamond indenter (radius of 20 μm and 90-degree angle). Different-sized indents were
produced by varying the maximum compression forces. Figure 10.14 presents a typical force–depth curve of indentation and a three-dimensional image of the resulting
indent. The applied loading speed was 0.2 μm/second during loading and unloading
with a holding time of 10 seconds when the prescribed maximum load was reached.
All tests were done at room temperature (about 20°).

Subsequently, a layer of gold thin film was deposited atop the SMP substrate using
a gold sputter (Section 10.3.2). The sputtering time was 120 seconds at a current of
20 mA and voltage of 2 kV. The thickness of the gold thin film was about 30 nm.
The prepared samples with different-sized spherical indents were heated to 80°C.
Figure 10.15 shows typical surface profiles of an indent before and after heating to
80°C. It is clear that the SMP almost fully recovers its original flat surface shape
after heating. In addition, wrinkles formed only within the indented area. The reason
is that upon shape recovery, significant compressive stress is induced only within the
indented part of the thin film.

To quantitatively study the critical stress–strain for wrinkling due to the SME,
a series of tests with different-sized indents were conducted. Figure 10.16 presents
the resulting surface morphologies. Few wrinkles may be seen if the diameter of
the produced indent is smaller than ~24 μm. Thus, this size may relate to the critical stress–strain value for wrinkling. We now study the wrinkling behavior of these
indents during shape recovery. As shown in Figure 10.17, the compressive strain in
the coated film results mainly from the longitudinal deformation while the latitudinal

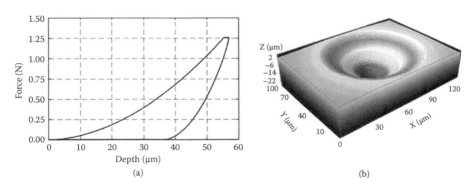

(a)   (b)

FIGURE 10.14   (a) Typical force-versus-depth curve during indentation. (b) Corresponding
three-dimensional image of resulting indent. (Reprinted from from Zhao Y, Cai M, Huang
WM et al. EPD Congress, Washington, DC, pp. 167–174, 2010. With permission.)

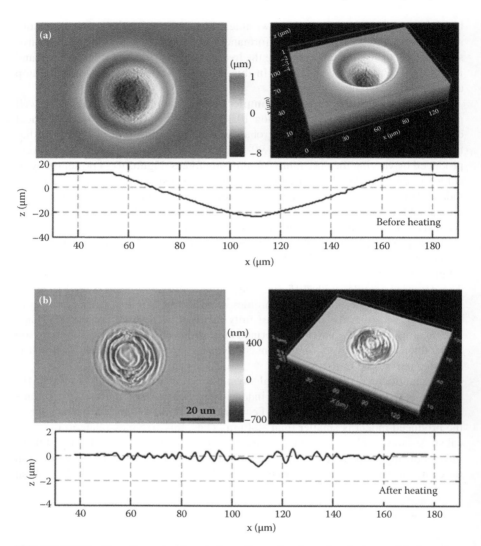

FIGURE 10.15 Two-dimensional and three-dimensional sectional views of indentation. (a) Before heating. (b) After heating.

deformation is small (hence, we should expect ring-shaped wrinkles perpendicular to the longitudinal direction), and can be calculated:

$$\varepsilon = \frac{L}{L_0} - 1 \qquad (10.18)$$

Based on Equations (10.15) and (10.18) and the mechanical properties of the thin film and SMP cited in Table 10.1, the critical size of indent is ~24.4 μm, in good agreement with the experimental result.

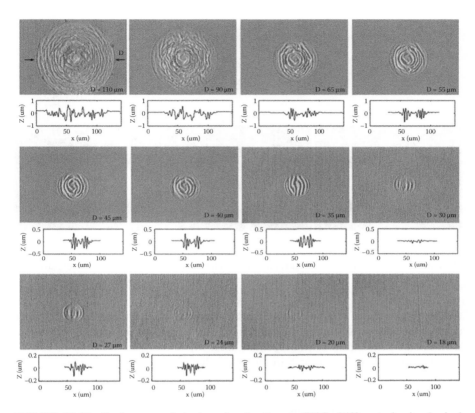

FIGURE 10.16 Surface morphologies after heating to 80°C. Different sized spherical indents were produced before coating with a thin layer of gold.

The next step is investigating the evolution of wrinkle patterns. Figure 10.18 shows the pattern results after heating to different temperatures. After heating to 80°C, shape recovery within the indented area induces compressive stress in the elastic gold film. Ring-shaped wrinkles appear within the indented area. No apparent surface features are noted elsewhere.

Upon heating to 100°C, the SMP substrate is fully in the rubbery state, corresponding to a very low stiffness. Therefore, a lower compressive stress in the thin film is enough for wrinkling. During indentation, the polymer right underneath

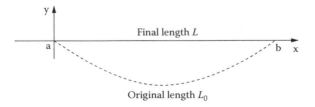

FIGURE 10.17 Shape change during shape recovery.

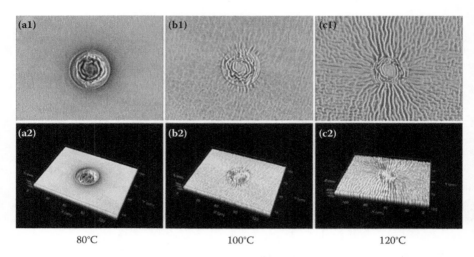

FIGURE 10.18 Evolution of wrinkle pattern after heating to different temperatures: (a) 80°C, (b) 100°C, (c) 120°C. Top: two-dimensional image. Bottom: three-dimensional image.

the indented area is stretched along the longitudinal direction, while the polymer around the indent is slightly stretched, i.e., the tensile strain around the indent is lower than the strain in the indented area. Therefore, after heating to 100°C, radial-shaped wrinkles formed around the indent. In the area farther away from the indent, we can observe some isotropic features that may suggest the start of wrinkling.

Upon heating to 120°C, significant dislocation occurs between the elastic thin film and substrate. During cooling, the compressive stress induced by thermal expansion mismatch between the elastic gold film and SMP substrate is strong enough for wrinkles to emerge even in the area without pre-deformation. The corresponding wrinkle pattern should be isotropic in theory. However, the existing radial-shaped wrinkles act as an initial boundary condition so that the expansion of existing radial-shaped wrinkles is favored within the adjacent area.

Wrinkling atop spherical SMP — Figure 10.19 reveals the preparation procedures for two different groups of samples. In the first step, both groups of samples were well polished according to the process explained in Section 10.3.2. After that, spherical indents were made by a microhardness tester with a spherical conical diamond indenter (see Section 10.3.3 for procedure). Indenters with different sized radii were used to make indents with different radii. The depth of each indent was about 0.3 times the radius of the indenter used.

Subsequently, all samples were carefully polished using 0.3 μm and then 0.05 μm alpha Al_2O_3 suspensions with a DP nap polishing cloth just until the indents disappeared. The surface roughness R_a of the polished SMP samples was about 10 nm measured by a confocal imaging profiler over an area of ~285 × 210 μm.

One group (Group A) of samples was coated with a layer of gold thin film. The other (Group B) was heated to 100°C to form different sized spherical protrusions (curvatures of spherical protrusions were measured individually) and then coated

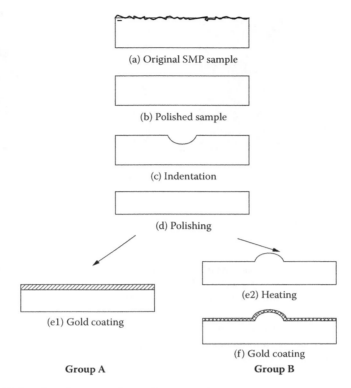

FIGURE 10.19   Sample preparation.

with a thin layer of gold. For both groups, sputtering time was 120 seconds at a current of 20 mA and voltage of 2 kV. The thickness of the gold thin film was about 30 nm.

Wrinkling behavior of Group A — Samples were heated to 80°C and then to 100°C. As shown in Figure 10.20a, a crown-shaped microprotrusion was produced, but we saw no significant wrinkles except some tiny, possibly radial-shaped features that may be small wrinkles around the protrusion. During the formation of a protrusion, the thin film within the indented area actually undergoes a large biaxial expansion, transforming from a flat to a dome shape. After heating to 100°C, we can see radial-shaped wrinkles surrounding the protrusion, but the top surface of the protrusion is still wrinkle-free (Figure 10.20b). Because the thin film within the protrusion was stretched significantly after heating to 80°C, it was difficult to induce further compressive stress even after re-heating to 100°C. The thin film within the surrounding area is different because it was not stretched after heating to 80°C. Heating to 100°C was helpful for fully softening the substrate. Thus radial shaped wrinkles were produced because this area expanded slightly due to previous indentation (as in Figure 10.18b).

For a clearer view, Figure 10.21 illustrates protrusion formation when a well-polished sample is heated. Note that the elastic thin film bends in the longitudinal

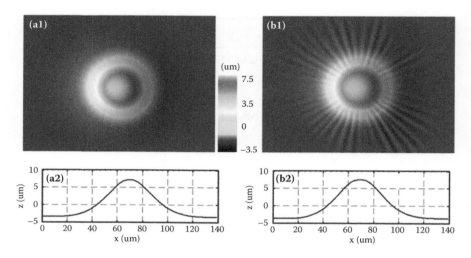

FIGURE 10.20 Typical wrinkles in Group A. (a) Heated to 80°C. (b) Heated to 100°C. Top row: top view. Bottom row: cross section. (Reprinted from from Zhao Y, Cai M, Huang WM et al. EPD Congress, Washington, DC, pp. 167–174, 2010. With permission.)

direction and some contraction appears in the latitudinal direction. Because of the latitudinal deformation, radial-shaped wrinkles appear. The latitudinal strain during shape recovery can be expressed as

$$\varepsilon = \frac{L - L_0}{L_0} = \frac{R}{R_0} - 1 \tag{10.19}$$

where R can be measured by surface scanning of the protrusion, $R_0 = \frac{1}{2}\int_a^b \sqrt{1 + f'^2}\, dx$. See Figure 10.21 for f.

Wrinkling behavior of Group B — The samples were cyclically heated in an oven to pre-determined temperatures (increasing from 100 to 160°C) and cooled in the air to room temperature as soon as each targeted heating temperature was reached. After each heating and cooling cycle, surface morphologies atop the

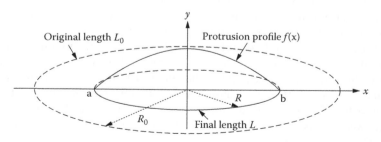

Protrusion formation upon shape recovery

FIGURE 10.21 Protrusion formation.

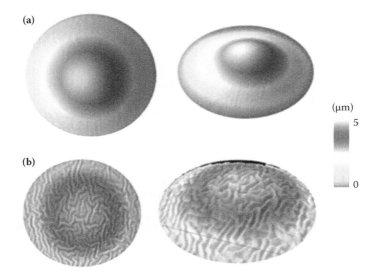

FIGURE 10.22 Surface features atop protrusion. (a) Before heating. (b) After heating.

protrusions were examined using a confocal imaging profiler to check the appearance of wrinkles. Figure 10.22 shows typical surface features before and after heating to 120°C. Figure 10.23 reveals typical surface morphologies atop different sized spherical protrusions after heating to 130°C. We can clearly see that isotropic wrinkles appear atop spherical SMP protrusions after heating only if the protrusion size is large enough.

The relationship between radius of protrusion, R, and critical heating temperature for wrinkling was investigated (Figure 10.24, symbols). The results clearly show the influence of protrusion size on the critical heating temperature for wrinkling. Higher temperature is required to produce wrinkles in smaller-sized protrusions. Figure 10.25 further reveals the relationship of critical wavelength λ_c and radius of protrusion (experimentally measured). Clearly, at $R > 100$ μm, λ_c is about a constant. At $R < 100$ μm, λ_c decreases dramatically with the decrease of R.

Wrinkling on surface of cylindrical SMP — For this test, we used a thermoplastic polyurethane SMP (MM3520) from SMP Technologies. The T_g of this SMP is about 35°C. At room temperature the polymer is relatively stiff, and above the T_g it becomes flexible; it melts fully above 180°C. Figure 10.26 shows the preparation

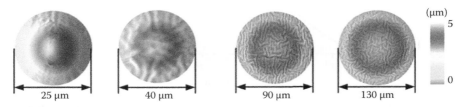

FIGURE 10.23 Surface morphologies atop protrusions after heating to 130°C.

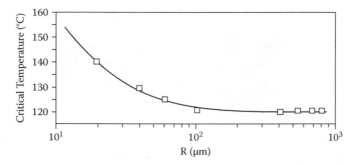

FIGURE 10.24 Effect of radius of protrusion on critical heating temperature for wrinkling. Symbols indicate experimental results; curve indicates data fitting.

of two types of SMP fibers. The as-received material in pellet form was fully melted in an aluminum container. Forceps were used to gently draw a fine thread out of the melting SMP until it broke (same procedure as in Section 9.2). Depending on the pulling speed, SMP fibers or spiral springs of different diameters were obtained at the break point. In general, a straight fiber is produced at a low drawing speed and a spring resulted from a high drawing speed. A thin gold layer tens of nanometers thick was sputter deposited atop the samples, then the samples were heated to the required temperature and cooled back to room temperature. Their surface morphologies were studied using a scanning electron microscope (SEM).

Figure 10.27a reveals wrinkles on the surface of a spring of wire diameter ~40 μm. After coating with a thin gold layer, the spring was heated to 60°C. Regular stripe-shaped wrinkles with wavelengths close to sub-micron scale were observed. Since 60°C is the temperature for shape recovery of this polyurethane SMP, we may logically conclude that further contraction of fiber (due to stretching in fabrication) caused the wrinkles. It should be possible to identify the details of the precise mechanism by simulation in the future.

Nano-scale wrinkles atop SMP sheets were recently been reported and many potential applications were discussed in Fu et al. (2009). Figure 10.27b reveals typical wrinkle patterns at two different scales of a straight SMP fiber. The wavelengths

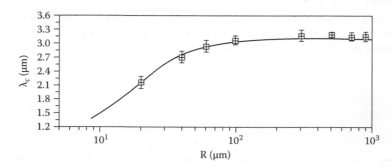

FIGURE 10.25 Relationship between R and λ_c. Symbols indicate experimental results; curve indicates data fitting.

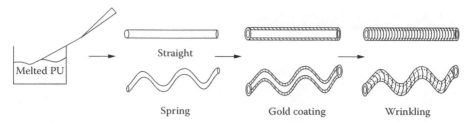

FIGURE 10.26   Preparation of SMP micro fibers and springs.

of these stripe-shaped wrinkles are actually at sub-micron scale. Because the fiber in the experiment was heated only to 60°C, shape recovery may be the mechanism that induces the wrinkles. As in the study of the polystyrene SMP samples mentioned above, the evolution of morphologies on the surfaces of both straight and spiral poly-urethane SMP fibers was investigated under different heating conditions. Both types fibers virtually followed the same trend. Hence, we report only on one.

Figure 10.28a reveals the surface morphologies of both straight and spiral-shaped polyurethane SMP fibers after heating to 25°C, below their T_g. As expected, no vis-ible wrinkles appeared. Figure 10.28b presents the surface morphologies of the fibers

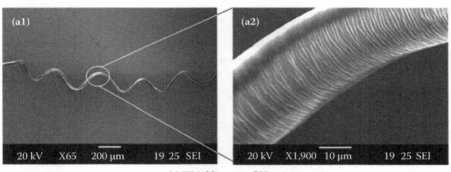

(a) Wrinkling atop PU spring

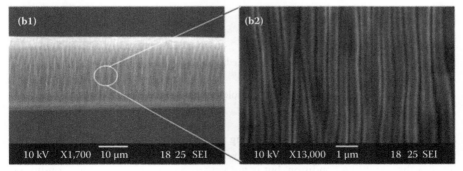

(b) Wrinkling atop PU straight fiber

FIGURE 10.27   Wrinkling on surface of polyurethane SMP fiber.

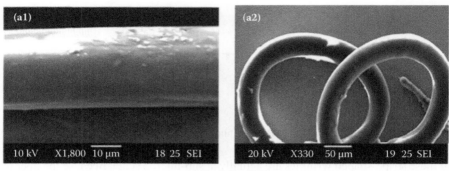

(a) Heated to 25°C

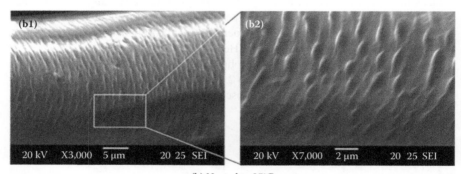

(b) Heated to 35°C

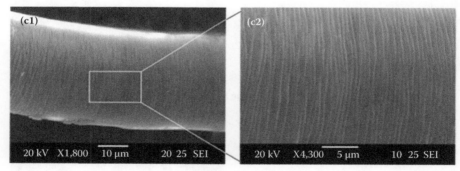

(c) Heated to 60°C

FIGURE 10.28 Evolution of surface morphologies of polyurethane SMP fibers.

after heating to their T_g of 35°C. Wrinkling appears only within some local areas and thus the wrinkles are discrete. Apparently, at this point, the compressive stress just reaches the wrinkling condition, but is not yet strong enough to produce continuous wrinkles over the whole surface. In addition, at T_g, the substrate is not fully softened so that full-scale wrinkling is prevented. After further heating to 60°C, well above

T_g, continuous wrinkles with wavelengths ~700 nm over the whole surface were observed, as shown in Figure 10.28c.

In addition to heating, wrinkles can also be produced in pre-stretched thin-film polyurethane SMPs upon immersing into water (moisture-responsive; Huang et al. 2010).

10.3.4   WRINKLING ATOP PATTERNED SAMPLES

From an engineering application view, it is more useful to have a patterned surface over a large area. For that reason, we extended the technique developed above for large-scale patterning in a cost-effective manner. We used the polystyrene SMP from CRG for this study.

Sample preparation — First, we produced an array of indents atop a piece of flat polystyrene SMP sample (well polished, about 3 mm thick) by compressing an array of 1 mm diameter steel balls that were compactly pre-packed atop the sample. To avoid possible crack formation in the SMP, compression was done at 80°C (above the T_g) so that the SMP was ductile. The loading speed was 0.05 mm/second and the maximum loading depth was 1 mm. The maximum depth was held until the sample cooled back to room temperature (about 20°C). Figure 9.29 shows the experimental setup and result of indentation. The steel balls were removed gently. Figure 9.30a shows the typical morphology of a sample surface. Next, the indented SMP sample was polished to a required depth using alumina polishing suspension with a particle size of 300 nm at first (to quickly achieve the required polishing depth) and then 50 nm to produce a very smooth surface. The well-polished SMP sample was then coated with a thin layer of gold. The sputtering time was 120 seconds at a current of 20 mA and voltage of 2 kV. The thickness of the gold thin film was about 30 nm. Finally, the sample was heated to 100°C for 20 minutes.

Wrinkling — Figure 10.29 shows the typical morphologies atop SMP substrates (with spherical protrusion arrays). Two wrinkle patterns may be seen. One is isotropic wrinkling in the centers of protrusions, and the other is anisotropic wrinkles among the protrusions. Three effects related to wrinkling appear during protrusion formation. The first is the latitudinal compressive stress resulting from protrusion formation. The second is the bending of thin film during protrusion formation, and the third is the compressive stress from thermal expansion mismatch. The first and third effects help induce wrinkles, but the second restrains the thin film from wrinkling. Figure 10.30 illustrates the protrusion and distribution of strain in the elastic thin film corresponding to the individual effect in a typical numerical analysis.

Because the protrusions are sized at macro scale, their curvatures are not large. The strain from bending is somewhat smaller, and less energy can be stored through bending of the thin film. In the center of a protrusion, the latitudinal compressive stress resulting from protrusion formation is also small. Therefore, thermal expansion mismatch plays a major role in wrinkle formation. Within the central area of a protrusion, isotropic thermal strain is larger than the anisotropic latitudinal strain. The isotropic compressive stress induced by the thermal expansion mismatch leads to the formation of isotropic wrinkles. However, the anisotropic latitudinal strain is

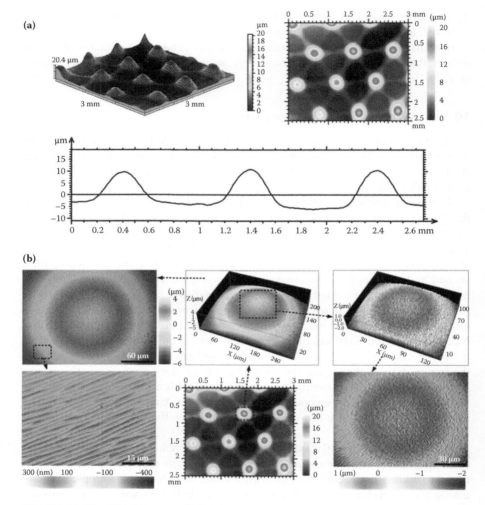

FIGURE 10.29 Wrinkling atop SMP with spherical protrusion arrays (a) and various wrinkle patterns at different locations (b).

larger than the isotropic thermal strain in the area farther away from the center of the protrusion. Consequently, anisotropic radial shaped wrinkles are generated in this area.

This discussion focused on an SMP sample with a crown-shaped protrusion array. This indicates that the sample was deeply polished after indentation (see Section 9.5.2). Figure 10.31 shows the wrinkles atop a piece of SMP with an array of flat-topped protrusions; the sample was less polished after indentation. Note that no prominent wrinkles are visible within the central flat top of the protrusion and that anisotropic radial shaped wrinkles among protrusions are very weak. A closer look reveals many cracks in the thin film, particularly near the flat-top part of the

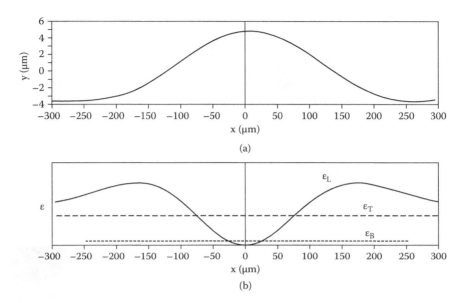

FIGURE 10.30 (a) Surface profile of protrusion. (b) Strains resulting from latitudinal strain during protrusion formation ε_L, strain due to thermal expansion mismatch ε_T, and strain due to bending of thin film ε_B.

protrusion that may have resulted from the huge tensile strain during shape recovery. These cracks effectively release the compressive stress in the film and thus wrinkles can hardly occur.

10.4 SUMMARY

After a brief summary of the basic wrinkling theory of an elastic thin film atop a soft substrate, a systematic investigation of wrinkling atop SMP substrates at different conditions was presented. Different self-organized wrinkle patterns, namely labyrinth, stripe-shaped, S, ring, and radial shaped wrinkles, were produced. Sub-micro-scaled wrinkles were observed. The evolution of wrinkle patterns, wrinkling atop patterned SMP substrates, and size effect were also examined.

The mechanisms of these different types of wrinkles were studied. Isotropic labyrinth-shaped wrinkles resulted from the thermal expansion mismatch between elastic thin film and soft SMP substrate. The S-shaped pattern arises from a combination of thermal mismatch and SME. The other three types resulted from the SME in the SMP substrate. The ease of applying this technique on a large scale to produce different patterns in a cost-effective manner was demonstrated.

ACKNOWLEDGMENT

We thank Y. Zhao for his contributions in drafting and compiling this chapter.

(a)

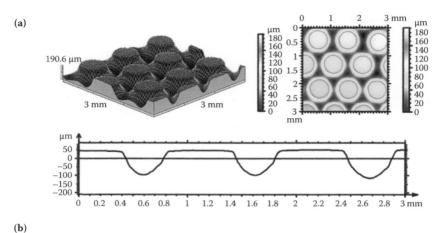

(b)

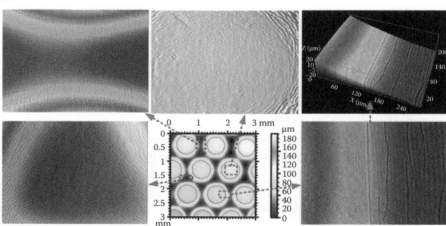

FIGURE 10.31 (a) Wrinkling atop SMP with flat-topped protrusion arrays and wrinkle patterns at different locations. (b) Visible cracks.

REFERENCES

Allen HG (1969). *Analysis and Design of Structural Sandwich Panels*. Pergamon, New York.

Bowden N, Brittain S, Evans AG et al. (1998). Spontaneous formation of ordered structures in thin films of metals supported on an elastomeric polymer. *Nature*, 393, 146–149.

Buch A (1999). *Pure Metals Properties: A Scientific–Technical Handbook*. ASM International, Cleveland, OH.

Cerda E and Mahadevan L (2003). Geometry and physics of wrinkling. *Physical Review Letters*, 90, 074302.

Chan EP and Crosby AJ (2006). Fabricating microlens arrays by surface wrinkling. *Advanced Materials*, 18, 3238–3242.

Chen X and Hutchinson JW (2004). A family of herringbone patterns in thin films. *Scripta Materialia,* 50, 797–801.

Chung JY, Youngblood JP, and Stafford CM (2007). Anisotropic wetting on tunable micro-wrinkled surfaces. *Soft Matter*, 3, 1163–1169.

Chung JY, Nolte AJ, and Stafford CM (2009). Diffusion-controlled, self-organized growth of symmetric wrinkling patterns. *Advanced Materials*, 21, 1358–1362.

Fu CC, Grimes A, Long M et al. (2009). Tunable nanowrinkles on shape memory polymer sheets. *Advanced Materials*, 21, 4472–4476.

Goodier JN and Neou IM (1951). The evaluation of theoretical critical compression in sandwich plates. *Journal of the Aeronautical Sciences*, 18, 649–657.

Gough GS, Elam CF, and Bruyne NDD (1940). The stabilization of a thin sheet by a continuous supporting medium. *Journal of the Royal Aeronautical Society*, 44, 12–43.

Harrison C, Stafford CM, Zhang WH et al. (2004). Sinusoidal phase grating created by a tunably buckled surface. *Applied Physics Letters*, 85, 4016–4018.

Huang W (2002). On the selection of shape memory alloys for actuators. *Materials and Design*, 23, 11–19.

Huang R and Im SH (2006). Dynamics of wrinkle growth and coarsening in stressed thin films. *Physical Review E*, 74, 026214.

Huang R and Suo Z (2002). Wrinkling of a compressed elastic film on a viscous layer. *Journal of Applied Physics*, 91, 1135–1142.

Huang WM, Liu N, Lan X et al. (2009). Formation of protrusive micro/nano patterns atop shape memory polymers. *Materials Science Forum*, 614, 243–248.

Huang WM, Yang B, Zhao Y et al. (2010). Thermo-moisture responsive polyurethane shape-memory polymer and composites: a review. *Journal of Materials Chemistry*, 20, 3367–3381.

Hutchinson JW and Neale KW, Eds. (1985). Wrinkling of curved thin sheet metal. In *Plastic Instability*, Presses Ponts et Chausees, Paris.

Im SH and Huang R (2005). Evolution of wrinkles in elastic-viscoelastic bilayer thin films. *Journal of Applied Mechanics,* 72, 955–961.

Im SH and Huang R (2006). Morphological instability and kinetics of an elastic film on a viscoelastic layer. In *Proceedings of IUTAM Symposium on Size Effects on Material and Structural Behavior at Micron and Nano Scales,* Sun QP and Tong P, Eds., Vol. 142, pp. 41–50.

Im SH and Huang R (2008). Wrinkle patterns of anisotropic crystal films on viscoelastic substrates. *Journal of the Mechanics and Physics of Solids*, 56, 3315–3330.

Jiang L, Zhao Y, and Zhai J (2004). A lotus-leaf-like superhydrophobic surface: a porous microsphere–nanofiber composite film prepared by electrohydrodynamics. *Angewandte Chemie International Edition*, 43, 4338–4341.

Koo WH, Jeong SM, Araoka F et al. (2010). Light extraction from organic light-emitting diodes enhanced by spontaneously formed buckles. *Nature Photon*, 4, 222–226.

Landau L and Lifshitz EM, Eds. (1986). *Theory of Elasticity*, Pergamon, New York.

Li Z, Yang DY, Liu X et al. (2009). Substrate-induced controllable wrinkling for facile nanofabrication. *Macromolecular Rapid Communications*, 30, 1549–1553.

Li B, Huang SQ, and Feng XQ (2010). Buckling and postbuckling of a compressed thin film bonded on a soft elastic layer: three-dimensional analysis. *Archive of Applied Mechanics*, 80, 175–188.

Mark JE (1999). *Polymer Data Handbook*, Oxford University Press, Oxford.

Neinhuis C and Barthlott W (1997). Characterization and distribution of water-repellent, self-cleaning plant surfaces. *Annals of Botany*, 79, 667–677.

Vandeparre H, Leopoldes J, Poulard C et al. (2007). Slippery or sticky boundary conditions: control of wrinkling in metal-capped thin polymer films by selective adhesion to substrates. *Physical Review Letters*, 99, 188302.

Volynskii AL, Bazhenov S, Lebedeva OV et al. (2000). Mechanical buckling instability of thin coatings deposited on soft polymer substrates. *Journal of Materials Science*, 35, 547–554.

Wan CC (1947). Face buckling and core strength requirements in sandwich construction. *Journal of the Aeronautical Sciences*, 14, 531–539.

Wu MJ, Huang WM, Fu YQ et al. (2009). Reversible surface morphology in shape-memory alloy thin films. *Journal of Applied Physics*, 105 (3): 033517.

Yoo PJ and Lee HH (2003). Evolution of a stress-driven pattern in thin bilayer films: spinodal wrinkling. *Physical Review Letters*, 91 (15): 154502.

Zhao Y, Cai M, Huang WM et al. (2010). Patterning atop shape memory polymers and their characterization. EPD Congress, Washington, DC, pp. 167–174.

11 Medical Applications of Polyurethane Shape Memory Polymers

11.1 INTRODUCTION

Minimally invasive surgery (MIS), also known as minimal access surgery, keyhole surgery, and endoscopic surgery, developed in the late 1980s and is regarded as one of the most important achievements in modern medicine. MIS offers tremendous advantages over conventional open surgery. The major benefits include reduced operative trauma, fewer wound complications, shorter hospital stays, and accelerated recoveries. However, MIS is more technically demanding than conventional surgery because the surgical intervention is executed remotely via two-dimensional imaging of the operative field. As a result, a surgeon faces loss of tactile feedback, restricted maneuverability, and less efficient control of major bleeding (Frank et al. 1997, Cuschieri 1999). Shape memory materials provide just-right solutions to many problems in the MIS arena (Yahia 2000, Yoneyama and Miyazaki 2009, Huang et al. 2010a).

In recent years, we have seen a number of medical applications using various types of shape memory polymers (SMPs), particularly thermo-responsive SMPs in cardiovascular implants (Sokolowski et al. 2007, Jung et al. 2010), SMP micro actuators for treating ischemic stroke (Metzger et al. 2002), SMP microfluidic reservoirs (Gall et al. 2004), self-deployable SMP neuronal electrodes (Sharp et al. 2006), SMPs combined with shape memory alloys (SMAs) in endovascular thrombectomy devices (Wilson et al. 2007), SMP dialysis needle adapters for reducing hemodynamic stress in arteriovenous grafts (Ortega et al. 2007), and SMPs for ocular implants (Song et al. 2010). Probably the hottest research topic for medical applications of SMPs is the SMP stent. Braided stents (Kim et al. 2010), solid tube stents (Wache et al. 2003, Chen et al. 2007, Yakacki et al. 2007) and even foam stents (Sokolowski et al. 2007) have been proposed. Other interesting developments include drug eluting stents and biodegradable stents (Chen et al. 2009, Wischke et al. 2010, Wischke and Lendlein 2010).

The triggering mechanisms in these applications—(1) Joule heating of embedded SMA wires (Wilson et al. 2007) and (2) laser heating (Maitland et al. 2002, Metzger et al. 2005, Wilson et al. 2005, Baer et al. 2007)—have been developed as alternative heating methods. The polyurethane SMP invented by Dr. Shunichi Hayashi of SMP Technologies in Japan (Hayashi 1990) has proven to be both biocompatible (Sokolowski et al. 2007) and thermo- and moisture-responsive (Yang

FIGURE 11.1   Typical types of wrinkles atop polyurethane SMP.

et al. 2006). This polyurethane SMP is an excellent shape memory material for bio-medical applications, particularly for MIS (Huang 2010, Huang et al. 2010b, Sun and Huang 2010). Dr. Hayashi and his collaborators have developed a range of such poly-urethane SMP materials with different transition temperatures and forms that are now commercially available. Autochokes for engines, intravenous cannulae, spoon and fork handles for those unable to grasp objects, and sportswear are only a few examples of their applications (Tobushi et al. 1996).

Despite a relatively narrow glass transition temperature (T_g) range (25°C or less for the whole transition range), as reported in Chapter 12, multi-shape memory effects (multi-SMEs) were demonstrated in both solid and foam polyurethane SMPs. These SMPs have also been used as substrates for various types of wrinkles (Huang et al. 2010b). Figure 11.1 shows typical types of wrinkles. Such patterned surfaces alter surface tension and surface roughness (Zhang et al. 2010b), thus alter-ing the strength of adhesion to cells. Pre-patterned SMP substrates (Figure 11.2)

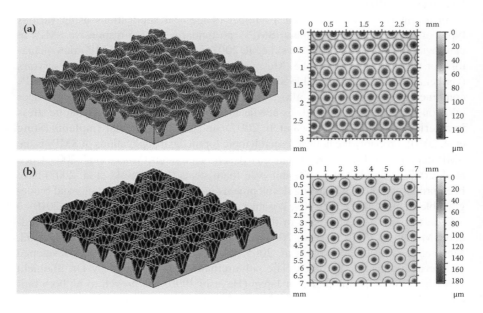

FIGURE 11.2   Typical pre-patterned SMP substrates.

can further enhance the effects. For further details of wrinkling atop SMPs, see Chapter 10.

This chapter presents some recent biomedical applications of this polyurethane SMP, particularly those utilizing its thermo- and moisture-responsive characteristics. Section 11.2 describes the applications based on thermo-responsive features; Section 11.3 reports applications based on the thermo- and moisture-responsive activities; Section 11.4 discusses the possibilities of utilizing the thermal and moisture responses for performing surgery inside living cells. Section 11.5 summarizes the key points of this chapter.

We used the MM5520 and MM3520 polyurethane SMPs (SMP Technologies, Japan) that have T_g values around 55 and 35°C, respectively.

11.2   THERMO-RESPONSIVE FEATURE-BASED DEVICES

In many medical operations, screws are required to fix implants to organs, normally bones. Because the sizes and dimensions of traditional screws are standardized, precise hole sizes are required to ensure firm attachments. As shown in Figure 11.3, a polyurethane SMP rod is pre-stretched and then placed inside an inner-threaded aluminum tube. Upon heating, the SMP rod expands remarkably, and hence the rod engages the inner threads of the tube firmly. If required, the SMP rod (now threaded as a screw) can then be easily removed with a screwdriver. The SMP screw provides a convenient solution in the sense that a single-sized rod fits efficiently into a range of different sized holes simply by heating, with no need for a screwdriver.

SMP tubes may be used as couplers or fixtures. Figure 11.4 demonstrates a micro tag application. The SMP tube was pre-expanded at high temperatures. After mounting on one leg of an ant, it was heated to 40°C for shape recovery. After cooling back to room temperature, the tag hardened and held firmly in place. Although the recovery strain in this SMP was ~100%, the maximum recovery stress induced by heating was limited so it was unnecessary to worry that the ant's leg might be damaged due to over-compression by the recovery force from the SMP tag upon heating. From an application view, it is important to know exactly how firmly a tag may remain in place. This can be determined by a pull-out test. Because an ant's leg is very short, we used a human hair that has about the same diameter as an ant's leg for this experiment.

FIGURE 11.3   SMP rod/screw. (a) Pre-stretched. (b) Fitting well into tube after heating.

FIGURE 11.4 SMP (MM3520) tag. Inset: zoom-in view of part of ant leg.

Figure 11.5a reveals the uniaxial tensile testing results through fracture for four strands of human hair. Despite some minor discrepancies, all the curves virtually share the same trend, i.e., after a short range of elastic deformation (yield stress ~15 MPa), a long plateau followed and a slight hardening was possible before fracture. Figure 11.5b shows the results of three pull-out tests. A custom-designed holder was used to hold the 0.7 mm long SMP tag. Slip occurs at ~0.06 N, which corresponds to a tensile stress of ~2.5 MPa. Good grip between SMP tag and hair was observed.

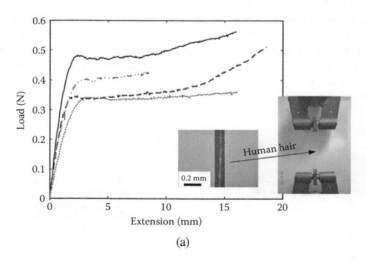

(a)

FIGURE 11.5 (a) Uniaxial tension testing of human hair. (b) Pull-out testing of human hair–SMP tag. Insets: zoom-in views of experimental setup.

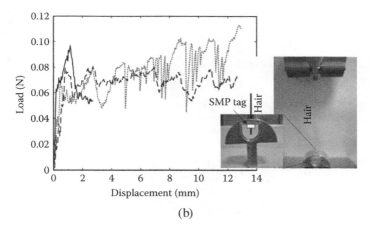

(b)

FIGURE 11.5 (Continued)

Sutures are commonly used during medical operations. However, MIS procedures generally allow only limited space for tying knots to secure sutures in place, and surgeons must tie knots remotely through very small holes. SMPs provide good solutions. As shown in Figure 11.6, a piece of pre-stretched SMP wire is wrapped around a piece of sponge. Upon heating, the SMP wire shrinks and tightens without physical interference. Because its maximum recovery stress is safe for human body tissues, damage caused by over-tightening is no longer a serious issue. If the SMP is biodegradable (Zhang et al. 2010a), sutures will self-degrade and disappear after time, even in a controllable manner (Kelch et al. 2007). Instead of tightening, Figure 11.7 reveals a knot that unravels in 30 seconds upon heating. SMPs may be strengthened by blending with fillers such as attapulgite, a biocompatible clay (Pan et al. 2008, Xu et al. 2009, 2010). As revealed in Figure 11.8, a piece of MM3520 polyurethane SMP loaded with 5 vol.% of original attapulgite clay fully recovered its original shape upon heating to 40°C.

11.3 THERMO- AND MOISTURE-RESPONSIVE FEATURE-BASED DEVICES

The unique thermo- and moisture-responsive abilities of polyurethane SMPs allow a wide range of applications in biomedical engineering. Chapter 3 presents a detailed study of this aspect of polyurethane SMP behavior. Instead of heating for tightening, Figure 11.9 reveals that upon immersion in room-temperature water (about 22°C), a polyurethane SMP (MM3520) knot tightens gradually. Similarly, without heating, a knot self-unravels (Figure 11.10) and a suture automatically tightens (Figure 11.11) upon immersion in room-temperature water.

As discussed in Chapter 3, after immersion, some of the absorbed water (the bound water) remarkably reduces the T_g of this polyurethane SMP. Consequently, the required temperature for shape recovery also lowers. Because the T_g can be reduced up to 30°C, we can utilize this feature to achieve moisture-responsive action

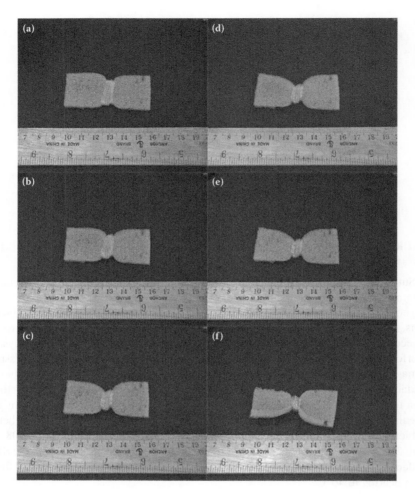

FIGURE 11.6 SMP wire wrapped around sponge tightens upon heating. (Reprinted from Huang WM, Yang B, Liu N et al. *Proceedings of SPIE*, 6423, 64231S, 2007. With permission.)

as demonstrated in Figures 11.9 through 11.11 or realize a gradient T_g by immersing different parts of the polyurethane SMP in water for different periods of time (Huang et al. 2005).

Figure 11.12 shows two pieces of SMP wires (MM3520) with the same diameter (0.4 mm). The left one was pre-immersed in water and the right one was kept dry. After pre-stretching and tying loose knots, both wires were immersed in room-temperature water. Note that the knot in the left piece of wire tightens at a much higher speed than the knot in the right wire.

An alternative approach to utilize the influence of moisture on the T_g of this SMP is demonstrated in Figure 11.13. A piece of 100 μm diameter SMP wire (MM5520) is originally in an S shape (Figure 11.13a). Note that MM5520 has a T_g of 55°C, well above human body temperature. After heating to 70°C, the

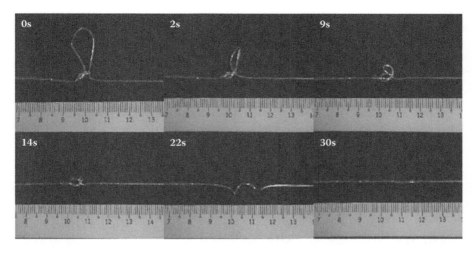

FIGURE 11.7   Self-unraveling knot upon heating.

middle portion of the curved thin wire is straightened (Figure 11.13b). After that, at room temperature (22°C), the wire virtually maintains its temporary shape, even after immersion in room-temperature water for half an hour (Figure 11.13c and d). However, upon heating to 37°C for 2 minutes, the wire recovers its original shape. During immersion in room-temperature water, the absorbed moisture reduces the T_g of the polymer. Upon later heating to 37°C instead of 55°C, it can induce full shape recovery.

Stents represent potentially promising applications for biocompatible polyurethane SMPs (see Figure 1.14 in Chapter 1). Due to the lack of removability, the use of traditional metal stents to treat patients suffering from benign disease was permitted only as a last resort when all other treatment options failed. Biodegradable polymer stents

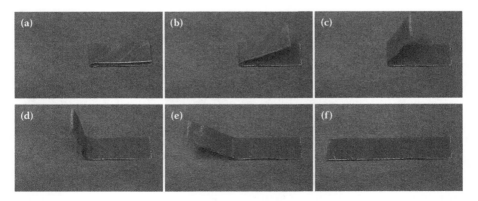

FIGURE 11.8   Shape recovery of piece of polyurethane SMP (MM3520) with 5 vol.% of original attapulgite clay upon heating.

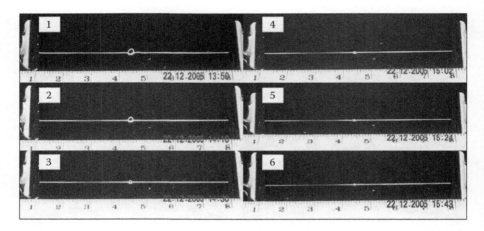

FIGURE 11.9   Self-tightening of SMP knot upon immersion in room-temperature water. (Reprinted from Huang WM, Yang B, Liu N et al. *Proceedings of SPIE*, 6423, 64231S, 2007. With permission.)

based on poly(vinyl alcohol) (PVA), polylactide (PLA), and poly(ethylene glycol) (PEG) are recent inventions (O'Brien and Carroll 2009), but even degradable stents are not without problems. Instant removal is preferred for stents inserted in blood vessels in the brain because degradation may lead to the release of small polymer pieces inside blood vessels—a life-threatening situation. Also, degradation is a relatively long-term process that poses difficulties if instant removal is required for any unforeseen reason.

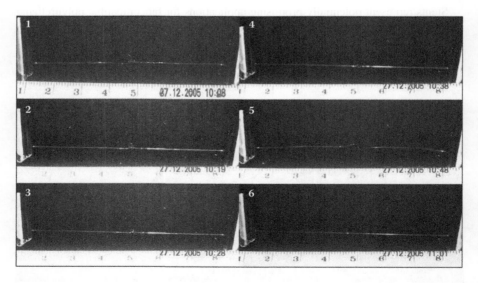

FIGURE 11.10   Self-unraveling of SMP knot upon immersion in room-temperature water. (Reprinted from Huang WM, Yang B, Liu N et al. *Proceedings of SPIE*, 6423, 64231S, 2007. With permission.)

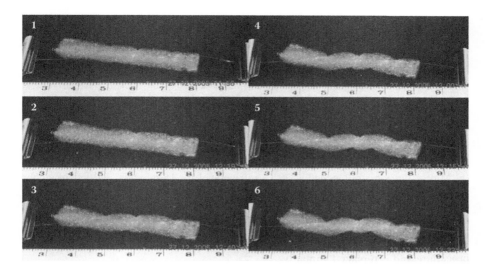

FIGURE 11.11 Self-tightening of SMP wire wrapped around sponge upon immersion in room-temperature water. (Reprinted from Huang WM, Yang B, Liu N et al. *Proceedings of SPIE*, 6423, 64231S, 2007. With permission.)

In bladder and prostate cancer treatments, retractable stents are preferred. As reported in the literature, stenosis may develop at anastomosis sites in up to 18% of bladder and prostate cancer patients, necessitating repeated dilatations to allow reasonable voiding. The current conventional procedure for treating stenosis is mechanical dilations, endoscopic incision, or laser vaporization. Their success rates are low and patients must undergo repeated treatments. A removable stent is a better option

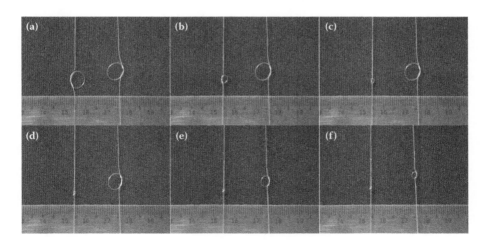

FIGURE 11.12 Self-tightening of two knots of SMP wires (MM3520) by immersion in room-temperature water. The left piece was pre-immersed in water to reduce its T_g.

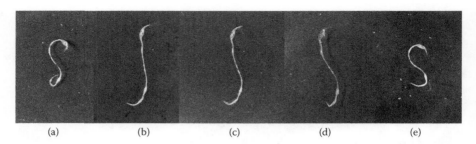

FIGURE 11.13 Shape recovery of 100 μm diameter wire (MM5520). (a) Original shape. (b) Middle portion straightened at 70°C, refrigerated for 15 minutes, and left in air for 10 minutes. (c) Immersion in room-temperature water (22°C). (d) After a half-hour in room-temperature water. (e) Two minutes after immersion in 37°C water. (Reprinted from Sun L and Huang WM. *Materials and Design*, 31, 2684–2689, 2010. With permission.)

because it does not require a highly specialized reconstructive bladder neck repair procedure and is replaced annually.

We propose an approach to utilize the thermo- and moisture-responsive qualities of polyurethane SMPs for retractable stents (Huang et al. 2007). As demonstrated in Figure 11.14, an SMP (MM3520) thin-wall tube was pre-expanded at

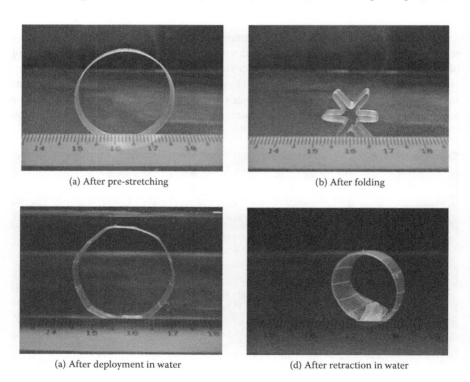

FIGURE 11.14 Retractable SMP stent utilizing moisture-responsive feature. (Reprinted from Huang WM, Yang B, Liu N et al. *Proceedings of SPIE*, 6423, 64231S, 2007. With permission.)

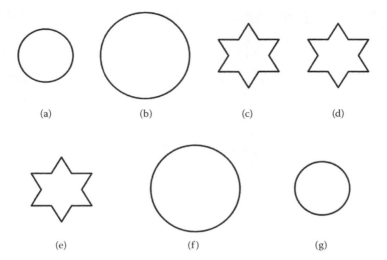

FIGURE 11.15 Retractable stent based on multi-SME concept. (a) Original circular shape. (b) Expanded well above T_g. (c) Further deformed into star shape at T_g. (d) Cooled back to room temperature. (e) Before deployment. (f) After heating to T_g. (g) Upon heating well above T_g (thermo-responsive for quick retraction) or leaving in room-temperature water for moisture-induced gradual recovery.

high temperatures and then cooled back to room temperature. Subsequently, it was mechanically deformed into a star shape. After immersion in room-temperature water, it mechanically expanded back into its circular shape. Note that in biomedical practice, one traditional way to deploy a compacted stent is to use a balloon. This principle is applicable here for mechanical expansion (deployment) of the stent. Because this SMP is also moisture-responsive, it shrinks after a time. As its diameter reduces, the stent can be easily removed.

Combined with multi-SME features, we can achieve retractable stents by means of thermal response for quick retraction or moisture response for slow retraction as illustrated in Figure 11.15. The other major advantages of this concept are

- During deployment, the polymer is not heated far above T_g (where it is in its full rubbery state and very soft); it is heated near T_g to maintain a partial rubbery state. The stiffness of the stent remains high during deployment; in the subsequent cooling process, the stent is hard enough to retain its shape.
- Both deployment and retraction can be achieved by heating.
- A choice of instant or slow retraction is available.

To enhance the performance of SMPs in real applications, various types of fillers can be used to strengthen the material and adjust the shape recovery speed. Attapulgite is a biocompatible clay that has a great ability to absorb water. As shown in Figure 11.16, treated attapulgite has little influence on recovery speed in

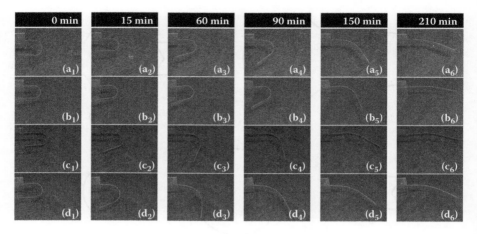

FIGURE 11.16 Shape recovery in polyurethane SMP–attapulgite composites upon immersion in room-temperature water. (a) SMP (MM3520) without clay. (b) SMP with 5 vol.% of treated clay. (c) SMP with 5 vol.% of original clay. (d) SMP with 10 vol.% of original clay.

moisture-induced shape recovery; the original clay does exert influence. This is because after treatment by heating close to 1000°C, the clay largely loses its ability to absorb moisture and/or water. As reported in Pan et al. (2008), treated clay does not alter the T_g of the composites; the original clay can remarkably lower the T_g of the polymer. In addition, with more clay loaded into the SMP, water should more easily penetrate the SMP through the embedded clay serving as micro channels, which also speeds shape recovery.

11.4 TOWARD MICRO MACHINES

Targeted drug release to specified cells at the molecular level is currently an attractive topic in medical research (Rivkin et al. 2010). To achieve a miniaturized mechanical surgeon (Fernandes and Gracias 2009), we need many tiny tools and machines. The concept that "the material is the machine" (Bhattacharya and James 2005) may be the correct approach for these applications. Shape memory materials, particularly SMPs, may be the materials that will realize these concepts (Huang et al. 2010c).

Microdevices can be fabricated by various modern techniques (Ikuta et al. 2003, Miyazaki et al. 2009). With a self-activated device such as a bacterial ratchet motor (Leonardo et al. 2010), chemically propelled nanodimer motor (Tao and Kapral 2008), osmotic motor (Cordova-Figueroa and Brady 2008), or transition metal-based molecular motor (Huc 2006), or even direct actuation via a laser beam outside a cell (Maruo et al. 2003), microdevices may be used for real-time characterization of DNA (Johnson et al. 2007, Smith et al. 2007, Hartl and Hayer-Hartl 2009, Dietz et al. 2009, Han et al. 2010), inside living cells, for example.

In *The Fantastic Voyage* (a 1966 film), a miniaturized submarine, the *Proteus*, was injected into a scientist's body to remove a blood clot in his brain to save his life.

Although it is now possible to remove a blood clot via MIS using various types of shape memory materials (Metzger et al. 2002, Yoneyama and Miyazaki 2009), we are still far away from being able to shrink a real ship to a few micrometers in length. However, technically speaking, we have most of the techniques required to deliver a micro vehicle made of SMPs into a living cell.

As seen in Figure 11.17, a tiny vehicle is deformed into a wire shape and inserted into a cell. If this vehicle is made of a thermo- and moisture-responsive SMP, it will recover its original shape after absorbing enough water inside the cell. Subsequently, it operates within a living cell using, for instance, a laser beam outside the cell to control its motion.

As discussed in Chapter 9, excellent shape memory effect persists in polyurethane SMPs down to tens of nanometers. Because the recoverable strain of a polyurethane SMP is ~100%, the size of the vehicle can be significantly reduced. By using a micro gripper or catheter (Figure 11.18), we should be able to quickly deliver it into a living cell without much damage.

The concept of delivering, for example, a piece of S-shaped SMP wire into a microbe is illustrated in Figure 11.19. Three requirements surround the concept: (1) the SMP wire must be capable of being straightened; (2) the microbe must contain plenty of water; and (3) the microbe must be relatively soft and flexible so that the wire can *freely* recover its original shape. To date, this concept has been experimentally demonstrated at two levels. As shown in Figure 11.20, a piece of 0.4 mm diameter SMP wire (MM3520) in a coiled shape was straightened and delivered into a cavity containing room-temperature water inside a hydrogen gel. After some time, the SMP wire recovered its original coiled shape due to its moisture-responsive feature. In a further step, a spiral-shaped SMP coil (MM5520, wire diameter of 0.3 mm) was straightened at 70°C and cooled to room temperature. It was then injected by syringe into a living jellyfish (Figure 11.21). Unlike the material in Figure 11.20, the SMP did not automatically recover its original shape inside the jellyfish for many days until

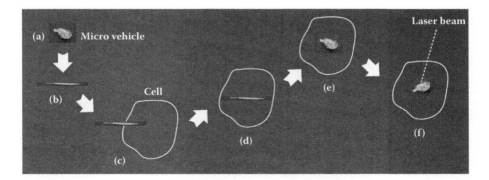

FIGURE 11.17 Delivery of micro vehicle into living cell for inside-cell operation controlled by laser beam. (a) Original shape of micro vehicle. (b) Vehicle after reshaping. (c) Inserting deformed vehicle into cell. (d) Vehicle fully inside cell. (e) Shape recovery of vehicle. (f) Operation of vehicle powered by remote laser beam. (Reprinted from Sun L and Huang WM. *Materials and Design*, 31, 2684–2689, 2010. With permission.)

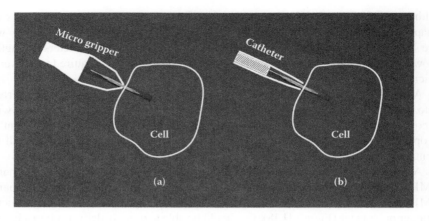

FIGURE 11.18 Delivery of micro device into cell by micro gripper (a) and catheter (b). (Reprinted from Sun L and Huang WM. *Materials and Design*, 31, 2684–2689, 2010. With permission.)

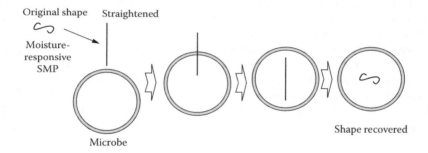

FIGURE 11.19 Delivery of piece of S-shaped wire into microbe.

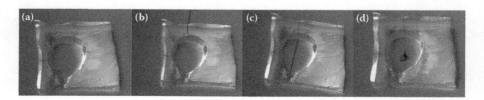

FIGURE 11.20 Delivering coiled SMP wire (0.4 mm diameter) into cavity inside hydrogen gel. (a) Piece of gel with cavity containing room-temperature water. (b) and (c) Delivery of piece of straightened SMP wire into cavity inside gel. (d) Recovery of SMP wire inside cavity. (Reprinted from Sun L and Huang WM. *Materials and Design*, 31, 2684–2689, 2010. With permission.)

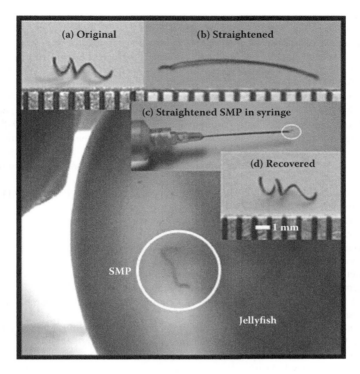

FIGURE 11.21 Delivery of SMP coil into jellyfish by injection. (a) Original coiled shape. (b) Shape after straightening at high temperature. (c) Ready for injection. (d) Recovered shape. (Reprinted from Huang WM, Ding Z, Wang CC et al. *Materials Today*, 13, 54–61, 2010. With permission.)

the water was heated to 37°C. We are currently using micro-sized polyurethane SMP springs (Chapter 9) to test the shape recovery of SMPs inside microbes.

11.5 SUMMARY

We have already observed many medical applications of SMPs. The unique thermo- and moisture-responsive feature of polyurethane SMPs makes them suitable for a wide range of potential applications. The moisture-responsive retractable stent is a typical example among many other possibilities. Based on the multi-SME and shape recovery characteristics of SMPs at sub-micron scale, we expect to see that "the material is the machine" through biomedical applications of these SMPs in the near future.

ACKNOWLEDGMENTS

We would like to thank C.S. Lee, S. Goh, K.C. Chan, Y.S Chan, Y.W. Loh, A.P. Lim, C.T. Low, Y. Zhao, Z. Ding, and N. Liu for their contributions to some of the experiments reported in this chapter. In addition, we appreciate Dr. C.C. Wang's help in compiling this chapter.

REFERENCES

Baer GM, Wilson TS, Benett WJ et al. (2007). Fabrication and in vitro deployment of a laser-activated shape memory polymer vascular stent. *BioMedical Engineering Online*, 6, 43.

Bhattacharya K and James RD (2005). The material is the machine. *Science*, 307, 53–54.

Chen M, Tsai H, Chang Y et al. (2007). Rapidly self-expandable polymeric stents with a shape-memory property. *Biomacromolecules*, 8, 2274–2780.

Chen M, Chang Y, Liu C et al. (2009). The characteristics and *in vivo* suppression of neointimal formation with sirolimus-eluting polymeric stents. *Biomaterials*, 30, 79–88.

Cordova-Figueroa UM and Brady JF (2008). Osmotic propulsion: the osmotic motor. *Physical Review Letters*, 100, 158303.

Cuschieri A. (1999). Technology for minimal access surgery. *British Medical Journal*, 319, 1–6.

Dietz H, Douglas SM, and Shih WM (2009). Folding DNA into twisted and curved nanoscale shapes. *Science*, 325, 725–730.

Fernandes R and Gracias DH (2009). Toward a miniaturized mechanical surgeon. *Materials Today*, 12, 14–20.

Frank T, Hanna G, and Cuschieri A (1997). Technological aspects of minimal access surgery. *Proceedings of the Institution of Mechanical Engineers H*, 211, 129–144.

Gall K, Kreiner P, Turner D et al. (2004). Shape-memory polymer for microelectromechanical systems. *Journal of Microelectromechanical Systems*, 13, 472–483.

Han D, Pal S, Liu Y et al. (2010). Folding and cutting DNA into reconfigurable topological nanostructures. *Nature Nanotechnology*, 5, 712–717.

Hartl FU and Hayer-Hartl M (2009). Converging concepts of protein folding *in vitro* and *in vivo*. *Nature Structural and Molecular Biology*, 16, 574–581.

Hayashi S (1990). Technical report on preliminary investigation of shape memory polymers. Mitsubishi Heavy Industries, Research and Development Center, Nagoya, Japan.

Huang WM (2010). Thermo-moisture responsive polyurethane shape memory polymer for biomedical devices. *Open Medical Devices Journal*, 2, 11–19.

Huang WM, Yang B, An L et al. (2005). Water-driven programmable polyurethane shape memory polymer: demonstration and mechanism. *Applied Physics Letters*, 86, 114105.

Huang WM, Yang B, Liu N et al. (2007). Water-responsive programmable shape memory polymer devices. *Proceedings of SPIE*, 6423, 64231S.

Huang WM, Song CL, and Fu YQ (2010a). A review of recent progress in shape memory materials for minimally invasive surgery. Submitted.

Huang WM, Yang B, Zhao Y et al. (2010b). Thermo-moisture responsive polyurethane shape-memory polymer and composites: a review. *Journal of Materials Chemistry*, 20, 3367–3381.

Huang WM, Ding Z, Wang CC et al. (2010c). Shape memory materials. *Materials Today*, 13, 54–61.

Huc V (2006). Transition metal catalysts for molecular motors: toward molecular lithography. *Journal of Physics: Condensed Matter*, 18, S1909–S1926.

Ikuta K, Maruo S, Hasegawa T et al. (2003). Biochemical IC chips fabricated by hybrid microstereolithography. *Materials Research Society Symposium Proceedings*, 758, 193–204.

Johnson CP, Tang H, Carag C et al. (2007). Forced unfolding of protein within cells. *Science*, 317, 663–666.

Jung F, Wischke C, and Lendlein A (2010). Degradable, multifunctional cardiovascular implants: challenges and hurdles. *MRS Bulletin*, 35, 607–613.

Kelch S, Steuer S, Schmidt AM et al. (2007). Shape-memory polymer networks from oligo[(epsilon-hydroxycaproate)-co-glycolate]dimethacrylates and butyl acrylate with adjustable hydrolytic degradation rate. *Biomacromolecules*, 8, 1018–1027.

Kim JH, Kang TJ, and Yu W (2010). Simulation of mechanical behavior of temperature-responsive braided stents made of shape memory polyurethanes. *Journal of Biomechanics*, 43, 632–643.

Leonardo RD, Angelani L, Dell'Arciprete D et al. (2010). Bacterial ratchet motors. *Proceedings of the National Academy of Sciences of the United States of America*, 107, 9541–9545.

Maitland DJ, Metzger MF, Schumann D et al. (2002). Photothermal properties of shape memory polymer micro-actuators for treating stroke. *Lasers in Surgery and Medicine*, 30, 1–11.

Maruo S, Ikuta K, and Korogi H (2003). Optically driven micromanipulation tools fabricated by two-photon microstereolithography. *Materials Research Society Symposium Proceedings*, 739, 269–274.

Metzger MF, Wilson TS, and Maitland DJ (2005). Laser-activated shape memory polymer microactuator for thrombus removal following ischemic stroke: preliminary *in vitro* analysis. *IEEE Journal of Selected Topics in Quantum Electronics*, 11, 892–900.

Metzger MF, Wilson TS, Schumann D et al. (2002). Mechanical properties of mechanical actuator for treating ischemic stroke. *Biomedical Microdevices*, 4, 89–96.

Miyazaki S, Fu YQ, and Huang WM (2009). *Thin Film Shape Memory Alloys: Fundamentals and Device Applications*. Cambridge University Press, Cambridge.

O'Brien B and Carroll W (2009). The evolution of cardiovascular stent materials and surfaces in response to clinical drivers: a review. *Acta Biomaterialia*, 5, 945–958.

Ortega JM, Wilson TS, Benett WJ et al. (2007). A shape memory polymer dialysis needle adapter for the reduction of hemodynamic stress within arteriovenous grafts. *IEEE Transactions in Biomedical Engineering*, 54, 1722–1724.

Pan GH, Huang WM, Ng ZC et al. (2008), Glass transition temperature of polyurethane shape memory polymer reinforced with treated/non-treated attapulgite (palygorskite) clay in dry and wet conditions. *Smart Materials and Structures*, 17, 045007.

Rivkin I, Cohen K, Koffler J et al. (2010). Paclitaxel clusters coated with hyaluronan as selective tumor-targeted nanovectors. *Biomaterials*, 31, 7016–7114.

Sharp AA, Panchawagh HV, Ortega A et al. (2006). Toward a self-deploying shape memory polymer neuronal electrode. *Journal of Neural Engineering*, 3, L23–L30.

Smith ML, Gourdon D, Little WC et al. (2007). Force-induced unfolding of fibronectin in the extracellular matrix of living cells. *PLoS Biology*, 5, 2243–2254.

Sokolowski W, Metcalfe A, Hayashi S et al. (2007). Medical applications of shape memory polymers. *Biomedical Materials*, 2, S23–S27.

Song L, Hu W, Zhang H et al. (2010). *In vitro* evaluation of chemically cross-linked shape-memory acrylate–methacrylate copolymer networks as ocular implants. *Journal of Physical Chemistry B*, 114, 7172–7178.

Sun L and Huang WM (2010). Thermo/moisture responsive shape-memory polymer for possible surgery/operation inside living cells in future. *Materials and Design*, 31, 2684–2689.

Tao Y and Kapral R (2008). Design of chemically propelled nanodimer motors. *Journal of Chemical Physics*, 128, 164518.

Tobushi H, Hara H, Yamada E et al. (1996). Thermomechanical properties in a thin film of shape memory polymer of polyurethane series. *Smart Materials and Structures*, 5, 483–491.

Wache HM, Tartakowska DJ, Hentrich A et al. (2003). Development of polymer stent with shape memory effect as a drug delivery system. *Journal of Materials Science: Materials in Medicine*, 14, 109–112.

Wilson TS, Benett WJ, Loge JM et al. (2005). Laser-activated shape memory polymer intravascular thrombectomy device. *Optics Express*, 13, 8204–8213.

Wilson TS, Buckley PR, Benett WJ et al. (2007). Prototype fabrication and preliminary *in vitro* testing of a shape memory endovascular thrombectomy device. *IEEE Transactions on Biomedical Engineering*, 54, 1657–1666.

Wischke C and Lendlein A (2010). Shape-memory polymers as drug carriers: a multifunctional system. *Pharmaceutical Research*, 27, 527–529.

Wischke C, Beffe AT, and Lendlein A (2010). Controlled drug release from biodegradable shape-memory polymers. *Advances in Polymer Science*, 226, 177–205.

Xu B, Huang WM, Pei YT et al. (2009). Mechanical properties of attapulgite clay-reinforced polyurethane shape memory nanocomposites. *European Polymer Journal*, 45, 1904–1911.

Xu B, Fu YQ, Huang WM et al. (2010). Thermal–mechanical properties of polyurethane–clay shape memory polymer nanocomposites. *Polymers*, 2, 31–39.

Yahia L (2000). *Shape Memory Implants*. Springer, Heidelberg.

Yakacki CM, Shandas R, Lanning C et al. (2007). Unconstrained recovery characterization of shape-memory polymer networks for cardiovascular applications. *Biomaterials,* 28, 2255–2263.

Yang B, Huang WM, Li C et al. (2006). Effect of moisture on the thermomechanical properties of a polyurethane shape memory polymer. *Polymer*, 47, 1348–1356.

Yoneyama T, Miyazaki S (2009). *Shape Memory Alloys for Biomedical Applications*. Woodhead Publishing and CRC Press, Boca Raton, FL.

Zhang S, Feng Y, Zhang L et al. (2010a). Biodegradable polyester urethane networks for controlled release of aspirin. *Journal of Applied Polymer Science*, 116, 861–867.

Zhang Y, Fan H, Huang WM et al. (2010b). Droplet atop a wrinkled substrate. *Journal of Mechanical Engineering Science C*, 224, 2459–2467.

12 Mechanisms of Multi-Shape and Temperature Memory Effects

12.1 MULTI-SHAPE MEMORY EFFECT AND TEMPERATURE MEMORY EFFECT

After severe and quasi-plastic distortion, shape memory materials (SMMs) can recover their original shapes in the presence of correct stimuli. This ability of returning from temporary to original shape only when a particular stimulus is applied (otherwise the distorted material will maintain the temporary shape forever) is known as the shape memory effect (SME; Huang et al. 2010).

Traditionally, shape recovery means returning from the temporary shape directly to the original shape. Under certain conditions and after pre-programming, shape recovery can be achieved in a step-by-step manner through one or a few intermediate shapes. This is known as the multi-SME (Xie 2010). In essence, the multi-SME allows a piece of SMM to work virtually as a machine, but unconventionally because the material is the machine (Bhattacharya and James 2005). We can produce tiny machines at micron and even sub-micron scale in a simple way and thus significantly widen the applications of SMMs.

In thermo-responsive shape memory alloys (SMAs), the triple-SME (only one intermediate shape) has been achieved by utilizing certain mechanical mechanisms. For instance, based on the buckling phenomenon, a small silicone beam coated with NiTi SMA thin film can bend upward and then downward upon Joule heating (Huang et al. 2004).

The multi-SME can also be realized in SMAs by introducing a gradient transition temperature by means of local thermomechanical treatment or through pre-straining (Sun and Huang 2009). After bending a piece of NiTi SMA strip at different locations into different curvatures, a gradient transition temperature is introduced into the strip. The transition (required shape recovery) temperature is higher if the pre-strain is larger (Huang and Wong 1999). This is because a higher driving force (temperature) is required to initialize the phase transformation in a pre-deformed SMA even without external stress (Huang 1998, 1999). Consequently, as shown in Figure 12.1, the SMA strip returns to its original straight shape gradually upon immersing in gradually heated water. The bottom part recovers first as it is less bent, while the top part straightens later at higher temperatures.

Certainly, a permanent gradient transition temperature can also be attained by local heating and/or severe mechanical treatment. However, in all of these cases, the

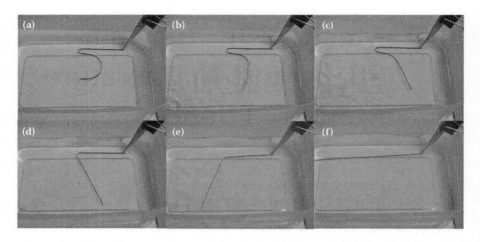

FIGURE 12.1   Shape recovery sequence of pre-deformed piece of NiTi SMA strip upon immersion in gradually heated water.

deformation–transition temperature varies from one cross-section to another along the length direction of a piece of SMA, so that the multi-SME appears in a segment-by-segment manner.

A more recent finding is that under certain conditions and following a specific programming process, SMA strips bend uniformly upward and then downward upon heating (Figure 12.2). In this case, the strain and transition temperature along the length direction of the strip is the same. The subsequent cooling process produces virtually no detectable deformation at all. As a result, a piece of SMA strip treated this way does not exhibit a two-way SME (Huang and Toh 2000), i.e., it cannot switch between high-temperature (austenite) shape and low-temperature (martensite) shape alternatively during thermal cycling.

The mechanisms behind the SME and other shape-memory-related features in thermo-responsive SMAs are the reversible martensitic transformation (between high-temperature stiff austenite and low-temperature soft martensite) and martensite re-orientation (among martensite variants). In shape memory polymers (SMPs) and shape memory hybrids (SMHs), the SME results from the dual-segment/domain (soft–hard) system. In addition, unlike the thermo-responsive SMAs that harden in the presence of heat, the materials in thermo-responsive SMPs and SMHs become soft upon heating. Due to the difference in the mechanism, different SMMs may exhibit unique features or require another processing and programming procedure to achieve seemingly identical features.

The multi-SME has been re-produced in SMHs (Huang et al. 2010). Figure 12.3 presents an example of tri-shape recovery—the triple SME of a piece of SMH made of silicone and two waxes with different melting temperatures. The SMH sample was originally straight. During programming, it was heated to 70°C and then twisted. The twisted shape is held during cooling to 30°C. Subsequently, the sample is bent

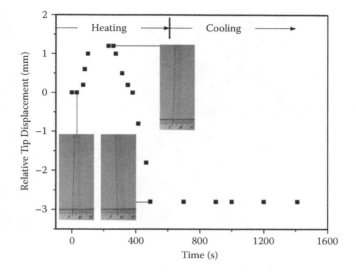

FIGURE 12.2   Up-down motion shape recovery of pre-programmed piece of NiTi SMA strip upon heating in water. (Reprinted from Huang WM, Ding Z, Wang CC et al. *Materials Today*, 13, 54–61, 2010. With permission.)

at 30°C. The resultant shape, which is a combination of twist and bend, is maintained during cooling to 5°C (Figure 12.3a$_1$). In the subsequent thermally induced shape recovery, heating is conducted in two steps. First, the programmed sample (Figure 12.3a$_1$) is immersed in 40°C water. As revealed in the figure (a$_1$ through a$_4$), the sample becomes straight but still twisted. Further heating in 65°C water untwists the sample and it recovers its original shape (Figure 12.3b$_1$ through b$_4$).

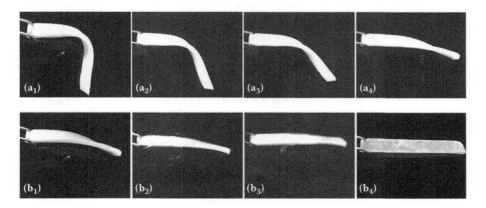

FIGURE 12.3   Shape recovery sequence of piece of pre-programmed SMH. (a) Straightening upon immersing in 40°C water. (b) Untwisting upon immersing in 65°C water.

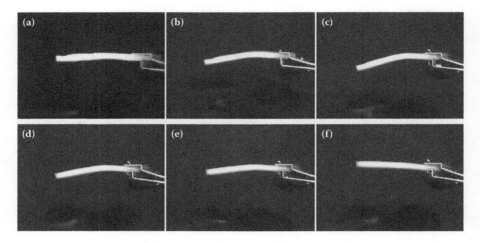

FIGURE 12.4 Up-down motion type shape recovery of a piece of SMH beam upon immersing in hot water. (Reprinted from Huang WM, Ding Z, Wang CC et al. *Materials Today*, 13, 54–61, 2010. With permission.)

We also managed to achieve the triple-SME for a silicone-based SMH filled with only one wax. It showed two distinct peaks in the heating curve, as revealed by differential scanning calorimetry (DSC). As shown in Figure 12.4, after programming, the straight SMH beam bends downward, upward, and then straightens, upon immersing in hot water. The multi-SME (mainly triple shape) in SMPs has been investigated with a number of approaches (Pretsch 2010). Prof. A. Lendlein and his collaborators published extensively on this subject (Bellin et al. 2006, 2007, Behl and Lendlein 2010, Kumar et al. 2010), and recently we have seen significant breakthroughs by Dr. T. Xie (2010).

Like SMAs, by immersing different parts of a piece of polyurethane SMP wire in water for different periods of time, a gradient transition temperature can be achieved because of the significant influence of moisture on the glass transition temperature in this polyurethane SMP (Yang et al. 2006). Thus, the multi-SME can appear even in SMP composites (Huang et al. 2005; Chapters 3 and 5 of this volume).

Some SMPs reveal more than one transition temperature upon heating. For instance, Liu et al. (2005) synthesized a PMMA–PEG semi-interpenetrating (semi-IPN) network that has two transition temperatures that correspond to the melting of PEG crystals (first soft segment) and glass transition of semi-IPN (second soft segment). This is similar to the SMH (Figure 12.3) that contains two waxes with different melting temperatures. If two types of soft segments are sensitive to different types of stimuli, the triple-SME can be achieved by using different stimuli independently. This is a yet-to-explore direction and may reveal significant advantages in many applications.

Theoretically, it is possible to use only one soft-segment for the triple-SME by utilizing the glass transition and melting of these two commonly observed transitions in thermoplastic polymers. This concept is similar to the triple-SME shown in

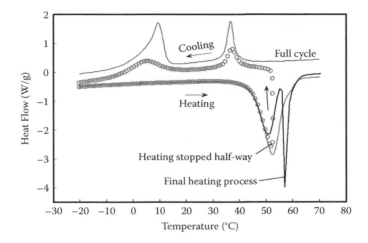

FIGURE 12.5 Typical DSC testing results of a polycrystal NiTi SMA revealing the TME. Grey lines: a complete thermal cycle; o symbols: heating stopped half-way during the martensite to austenite phase transformation; black line: final heating to full austenite. (Reprinted from Sun L, Huang WM, and Cheah JY. *Smart Materials and Structures*, 19, 055005, 2010. With permission.)

Figure 12.4, in which only one wax that has two distinct peaks as observed with DSC is used for the SMH. The alternative is a double-layered SMP composite in which each layer has its own transition temperature (Xie et al. 2009).

Unlike most metals and their alloys, the transition of some polymers occurs over a wide temperature range (Miaudet et al. 2007, Xie 2010) that serve as a series of sub-transitions to achieve the multi-SME (Xie 2010), similar to the process shown in Figure 12.4 for the hybrid. This concept is brilliant because it can be utilized in all SMPs to achieve multi-SME (at least triple-SME) and even in many conventional materials (polymers, metals, alloys, and their composites).

In addition to the well-known shape memory phenomenon—the primary characteristic of SMAs—another interesting phenomenon observed (largely by thermal cycling in DSC [Liu and Huang 2006a, 2006b] and occasionally in mechanical cycling [Airoldi et al. 1998, Wang et al. 2005]) is the temperature memory effect (TME) by which the previous temperature of interruption is memorized and revealed in a later heating process (Figure 12.5). The first report of the TME was probably in NiTi-based SMAs (Airoldi et al. 1993a, 1993b) and then reported for other thermo-responsive SMAs (CuZnAl and CuAlNi; Wang et al. 2006). The mechanism was assumed to result from the generation of new martensite within austenite, instead of the growth of old martensite, during cooling. Based on this speculation, a phenomenological model is able to reproduce the main features of the TME in SMAs (Sun et al. 2010).

The TME was also observed in SMPs, but the appearance was different. Miaudet et al. (2007) reported that in the constrained recovery test of a shape memory nanocomposite with a broad glass transition temperature range, the maximum recovery stress occurred at the temperature at which the material was pre-deformed to the temporary shape. This phenomenon was confirmed by Xie (2010) in another SMP that also has a wide transition temperature range.

This chapter systematically investigates the multi-SME and TME in SMPs. Section 12.2 demonstrates these effects in polyurethane SMPs, and Section 12.3 explains the mechanisms behind both effects in SMPs. Finally, a summary is provided in Section 12.4.

12.2 DEMONSTRATION OF MULTI-SME AND TME IN POLYURETHANE SMP

Polyurethane SMPs from SMP Technologies, Japan, were used in the experiments. The programming procedures and experimental results are presented in detail.

12.2.1 MULTI-SME

A piece of polyurethane SMP foam (MF5520) 48 mm long, 29 mm thick, and 32 mm wide was prepared for the experiment. According to the DSC result (Figure 8.40), its glass transition temperature range is ~25 to 55°C. The steps for programming the foam to demonstrate the triple-SME are

1. Heat foam to 62°C for 30 minutes inside a hot chamber.
2. Compress foam along its thickness (vertical) direction; see top left inset in Figure 12.6. Fix height at this level.
3. Cool chamber to 46°C and hold at this temperature for 30 minutes.

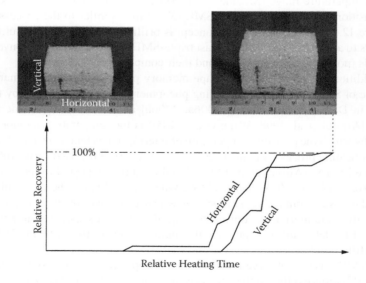

FIGURE 12.6 Shape recovery upon heating of two-dimensionally pre-compressed foam.

4. Compress foam along its length (horizontal) direction and fix length at this distance.
5. Cool chamber to room temperature (about 22°C).

The foam is now ready for the experiment to demonstrate the triple-SME upon heating. It is difficult to ensure a uniform temperature distribution within polymers, particularly in foams, upon heating unless the material is very thin; otherwise, the heating speed must be extremely slow. In our experiment, the heating speed of the hot chamber could not be controlled and the heating was automatic after the chamber temperature was set. The shape recovery process was recorded.

Figure 12.6 shows the curves of the relative recovery in the vertical and horizontal directions against the relative time. As we can see, shape recovery first occurs along the horizontal direction, and then along the vertical. Despite the non-uniform temperature distribution within the foam based on the difficulty of controlling its temperature, Figure 12.6 clearly reveals the triple-SME in the SMP foam.

In the second demonstration, we used a piece of solid polyurethane SMP (MM5520, the same material used in experiments in Chapters 4 and 5; see Figure 5.4 showing DSC results). The sample is in a dog-bone shape (Figure 1.15). The programming procedure is

1. Immerse sample in 50°C water and twist it.
2. Cool water to 46°C with the twisted shape held; bend and hold the sample.
3. Cool the sample in 22°C water.

Figure 12.7a$_1$ reveals the freestanding (temporary) shape of the sample after programming. The free recovery test upon heating was conducted in two steps. In the first step, the sample was immersed in 46°C water. Figure 12.7a$_1$ through a$_4$ shows that the sample straightens but is still twisted. In the second step, the twisted sample was immersed in 50°C water. Figure 12.7b$_1$ through b$_4$ reveals that the sample further untwists—it fully recovers its original shape. These two experiments clearly demonstrated the multi-SME in the polyurethane SMPs upon heating in free recovery tests,

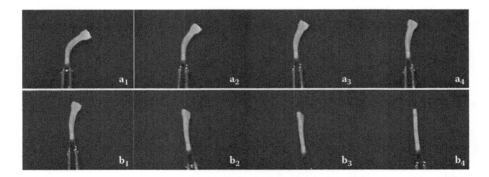

FIGURE 12.7   Triple-SME. (a) Immersion in 46°C water. (b) Further immersion in 50°C water.

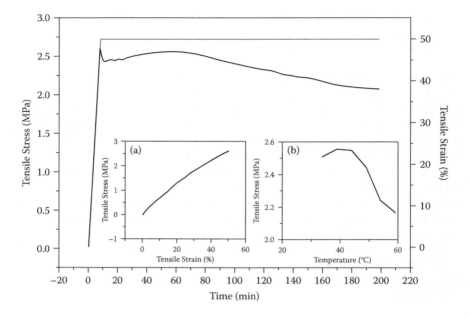

FIGURE 12.8 Experimental results for TME (without unloading).

although multi-SMEs are limited to triple-shape in both cases due to the relatively narrow transition temperature ranges of the materials.

12.2.2 TME

The TME was demonstrated in a piece of solid polyurethane SMP (same piece used to demonstrate the multi-SME in Section 12.2.1, Figure 12.7). The procedure is

1. Stretch sample by 50% at a strain rate of 1×10^{-3}/s inside the hot chamber (46°C) of an Instron 5569. Figure 12.8a shows stress-versus-strain curve during stretching.
2. Cool sample to room temperature (about 22°C) while maintaining the 50% strain.
3. Heat sample from room temperature to 60°C.

This procedure actually is a constrained recovery test without unloading after cooling. Figure 12.8 shows tensile stress and tensile strain against testing time. We can see that the stress (1) increases continuously as the strain increases upon stretching but in a nonlinear fashion; (2) decreases upon cooling while the 50% strain is held; and (3) increases and decreases subsequently so that a virtual temperature peak appears. We re-plotted the results in the format of tensile stress against heating temperature as shown in Figure 12.8b. The maximum stress actually occurred at a temperature very close to that when the sample was 50% pre-stretched. In other words, the previous temperature during programming was revealed.

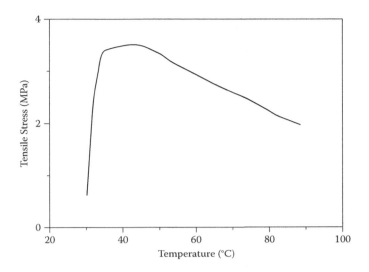

FIGURE 12.9 TME in constrained recovery test (with unloading).

In another test, the 50% pre-stretched sample was unloaded after cooling and then heated for recovery, i.e., following the exact procedure of the conventional constrained recovery test. The tensile stress was plotted against temperature, as shown in Figure 12.9. A peak appeared in the stress-versus-temperature curve at 45.5°C, only 0.5°C lower than the previous programming temperature of 46°C. We can conclude that regardless of unloading conditions (with or without unloading) after programming, a stress peak appears during heating in the constrained recovery test. The maximum stress occurs near the programming temperature. This is the TME that appears when the previous programming temperature is remembered and revealed.

12.3 MECHANISMS

A simple framework has been proposed for the mechanisms generating multi-SME and TME in SMPs (Sun and Huang 2010). We now present a modified version of this framework to more precisely explain both effects. In Sections 12.3.1 and 12.3.2, we will ignore hysteresis to keep explanations simple. The properties and status of a polymer can be uniquely determined by temperature without considering the exact thermal history. The influence of hysteresis is discussed in Section 12.3.3.

12.3.1 Multi-SME

First, we consider a SMP foam that has two transitions: one between T_1 and T_2, and the other (at a lower temperature range) between T_2 and T_3 (see Figure 12.10, vertical axis). Suppose that the foam is always elastic and flexible, just like a two-dimensional elastic spring, within the temperature range discussed here. Two types of fillers are used. Each pore is filled with only one type of filler (Figure 12.10a). At

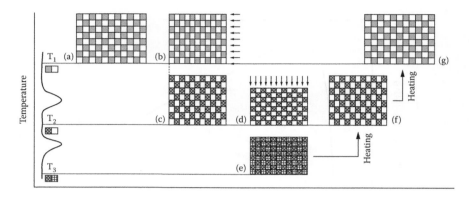

FIGURE 12.10 Two-dimensional programming and free recovery upon heating.

T_3 and below, both types of fillers are stiff. At T_2 and above, one type (Filler I, white) is so soft that its stiffness is ignorable. The other type (Filler II, gray) is still stiff. At T_1 and above, both fillers are very soft.

We now follow the procedure of the triple-SME test on SMP foam (Section 12.2.1, Figure 12.6) except that we first compress the sample in the horizontal direction at T_1 (Figure 12.10b). At T_1, both fillers are soft while the elastic foam remains flexible and may be deformed easily. During deformation of the foam, the elastic energy (potential) is built up in a manner similar to compressing an elastic spring. The sample is cooled down to T_2 at which Filler II becomes stiff; Filler I remains soft. If Filler II is rigid at T_2, the deformed shape is maintained even without application of constraints (Figure 12.10c). This means that the potential energy in the foam is frozen because the stiff Filler II prevents its return to its original shape.

We now compress the sample in the vertical direction (Figure 12.10d). Because Filler II is stiff at this temperature, the deformation is mainly accommodated by Filler I, which is still very soft. More elastic energy is built up in the foam. After further cooling to T_3, both fillers are stiff so that the temporary shape, pre-compressed in both vertical and horizontal directions, can be held without any constraint (Figure 12.10e). Because the foam is compressed in both directions and frozen in this temporary shape, it stores elastic energy.

After the programming process, the sample is heated for free shape recovery. Upon heating to T_2, Filler I becomes soft so that the sample expands vertically, back to its original thickness (Figure 12.10f). The recovery is driven by the elastic energy stored in the elastic foam during programming. Note that only the elastic energy accumulated from stage d to stage e (Figure 12.10) in the elastic foam is released. Any recovery in the horizontal direction is prevented by the still-stiff Filler II. Upon further heating to T_1, Filler II becomes soft as well so that the elastic foam can freely recover its original length (Figure 12.10g). Because the recovery upon heating is in two steps, the result is called triple-SME. This discussion is applicable to most conventional SMPs that exhibit two transitions. The synthesis of such polymers requires strong chemistry and polymer knowledge and many rounds of trial and error.

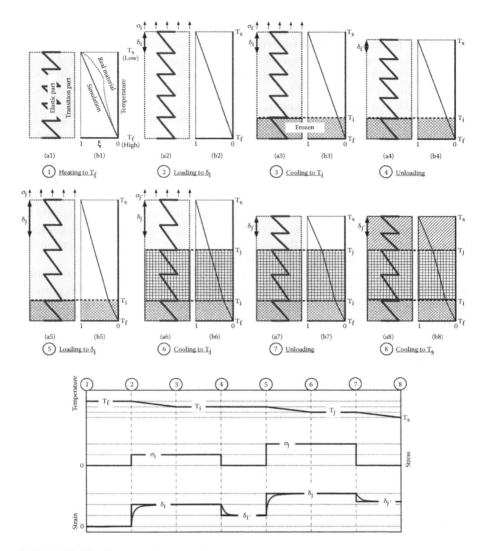

FIGURE 12.11 Programming procedure.

Xie (2010) demonstrated that the multi-SME can be achieved with only one transition if the SMP's transition temperature range is wide enough.

Let us consider a piece of SMP that has only one transition (Figure 12.11a1). Assume its length is one unit and the cross-sectional area is one unit as well. Thermal expansion in both elastic and transition parts is ignored. For simplicity in presentation and discussion, we further assume a linear relationship between the fraction of transition in the transition part and the transition temperature range from T_s and T_f (Figure 12.11b1). Thus, at T_s, which is a lower temperature, the transition part is 100% stiff ($\xi = 1$). At the higher T_f, the transition part is 100% soft ($\xi = 0$). For better illustration, we assume that the properties of the elastic part embedded in the transition part do not change, at least between T_s and T_f (both temperatures are inclusive).

Because the experimental result of the triple-SME test shown in Figure 2 in Xie (2010) served as a benchmark for comparison, the procedure of the experiment was followed closely, although we made some necessary modifications to ensure clear interpretation and better presentation.

First, the SMP is heated to T_f (Figure 12.11) and then stretched to δ_i by applying an instant stress of σ_i. Because the length of the SMP is one unit, the corresponding engineering strain is δ_i. Following the presentation format of Figure 2 in Xie (2010), the temperature, strain, and stress against the experimental steps were plotted as shown at the bottom of Figure 12.11. Note that the horizontal axis is equivalent to the testing time in Figure 2 of Xie (2010). Because of the nature of the viscosity of polymers, δ_i should not reach its maximum instantly in the experiment; it should be reached gradually as reported in Xie (2010). The elongation of the sample causes a tensile stress in the elastic part that may be denoted as σ_i, because the contribution from the transition part that is very soft at this point is negligible.

In the next step, the sample is cooled to T_i. The bottom portion of the transition part hardens (Figure 12.11), freezing the mobility of the elastic part in this portion. The elastic energy in the elastic part is stored and the internal stress, which is tensile, in the elastic part is σ_i. Instead of applying a constant tensile stress as in Xie (2010), a more conventional approach is fixing the length of the sample during cooling. Consequently, the measured tensile stress σ_i' varies (Yang et al. 2006). The tensile stress in the unfrozen portion of the elastic part is also σ_i', while in the frozen portion the stress is still σ_i. After unloading, the top unfrozen portion becomes stress-free (in both elastic and transition parts) and fully recovers its shape. However, in the bottom frozen portion, apart from some tiny elastic recovery (because this portion is frozen and hard), the length is virtually unchanged. Consequently, the internal stress in the elastic part is still σ_i. Again, due to the viscosity of polymers, the sample should gradually shrink to δ_i' in the real experiment.

In the second programming cycle (loading–cooling–unloading; 5 to 7 in Figure 12.11), the middle portion of the sample (T_i to T_j) is frozen. The internal stress in the elastic part within this portion is σ_j, the same as the stress applied to stretch the sample to an overall displacement of δ_j. In the last step, the sample is cooled to T_s and the whole sample is frozen. After unloading at the end of the second programming cycle, the top portion is stress-free for both elastic and transition parts and the internal stress in the top portion of the elastic part remains 0 at T_s. This finishes the programming process.

Comparing the evolution of temperature, stress, and strain during programming (bottom, Figure 12.11) with the measured data in Figure 2 of Xie (2010), we can conclude that both results virtually share the same pattern.

We now start heating the freestanding sample programmed for free shape recovery (thus, stress is always 0). Note that 8′ in Figure 12.12 is essentially identical to 8 in Figure 12.11. Upon heating from T_s to T_j, the top portion softens, but experiences no deformation (9 in Figure 12.12), because the top portion is stress-free in both elastic and transition parts. During further heating to T_i, the middle portion softens,

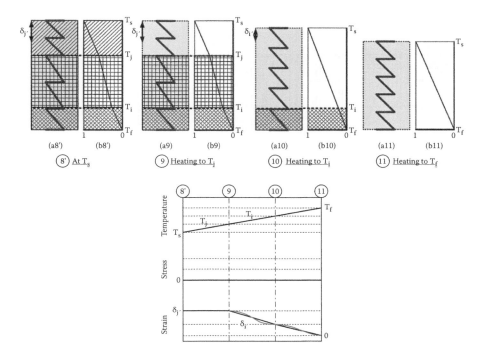

FIGURE 12.12   Free shape recovery upon heating (multi-SME).

releasing the stored elastic energy in the elastic part within this portion. The elastic part returns to its original length, as does the sample within this portion (10 in Figure 12.12). Overall, the displacement of the sample becomes δ_i'. Again, because of viscosity, upon heating, the polymer responds in a slow mode and gradually becomes shorter. From T_i to T_f, the bottom portion becomes stress-free and the sample recovers slowly and eventually returns to its original length. In comparison with the result of Figure 2 in Xie (2010), we see the same pattern of the triple-SME upon heating (Figure 12.12, bottom, strain-versus-experimental step curve).

12.3.2   TME

To further examine the TME, we start with a piece of the freestanding pre-programmed sample (8′ in Figure 12.13) that is identical to 8 in Figure 12.11. In a constrained recovery test, the length of the sample is fixed and the strain is kept constant throughout. Heating to T_j does not generate any stress in a sample because the unfrozen (top) portion is stress-free (9 in Figure 12.13). Upon further heating to T_i, the middle portion is unfrozen, causing the elastic energy in the elastic part within this portion to be re-distributed within the whole soft portion (10 in Figure 12.13). Stress results and increases continuously until this whole soft portion is unfrozen.

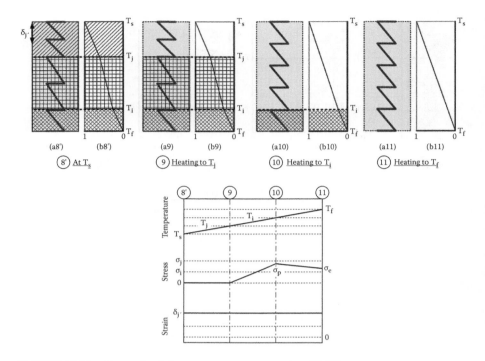

FIGURE 12.13   Constrained shape recovery upon heating (TME).

If the elastic part is linearly elastic, as a first-order estimation, the maximum stress, σ_p, can be expressed as

$$\sigma_p = \sigma_j \times (T_i - T_j)/(T_i - T_s) \tag{12.1}$$

Further heating to T_f causes the re-distribution of the elastic energy, this time within the whole length of the elastic part (11 in Figure 12.13). At T_f, the stress, σ_e, can be estimated by

$$\sigma_e = \sigma_j \times (T_i - T_j)/(T_f - T_s) + \sigma_i \times (T_f - T_i)/(T_f - T_s) \tag{12.2}$$

The condition to observe peak stress at T_i ($\sigma_p > \sigma_e$) for the TME is

$$\sigma_j \times (T_i - T_j) > \sigma_i \times (T_i - T_s) \tag{12.3}$$

For SMPs programmed only once at a temperature within the transition temperature range for the dual-SME (Figure 12.9 in Section 12.2.2; Miaudet et al. 2007, Xie 2010), it is always valid that maximum stress occurs at the temperature at which the SMP is programmed. This is illustrated in detail in Figure 12.14. As we can see, maximum stress, σ_i, occurs at the T_i, programming temperature. Again,

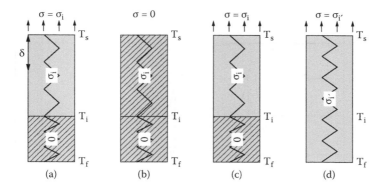

FIGURE 12.14   TME of SMP programmed only once. (a) Stretching at T_i. (b) Cooling to T_s followed by unloading. (c) Heating to T_i for constrained recovery. (d) Heating to T_f.

provided that the elastic part is linearly elastic, the final stress at T_f, σ_i, can be roughly estimated:

$$\sigma_{i'} = \sigma_i \times (T_i - T_s)/(T_f - T_s) \tag{12.4}$$

12.3.3   INFLUENCE OF HYSTERESIS

In the above discussions of the SME and TME, we assume a unique one-to-one relationship between the temperature and properties of SMPs for simplicity. This assumption implies that hysteresis in the polymers, if any, should be small. In practice, some polymers may show significant hysteresis. In theory, hysteresis is an essential part of the energy barrier that must be overcome for any transition or transformation to happen. Energy dissipation occurring during any transition or transformation always plays a role in hysteresis. In some cases, hysteresis may be very small; in other cases, it is substantial and must be considered. For instance, in the phase transformation between water and ice under fixed conditions, both the forward and backward transformations happen at almost the same temperature. On the other hand, hysteresis in SMAs normally cannot be ignored (Ortin and Planes 1988, 1989, 1991). By averaging the energy difference between the forward and reverse martensitic transformations (obtained, for example via DSC), we can estimate the mean energy dissipation during transition (Huang and Xu 2005, An and Huang 2006, Pan and Huang 2006).

If the hysteresis is significant, we can see that the forward and backward transitions occur within different temperature ranges as illustrated in Figure 12.15. T_a and T_b indicate transitions in the cooling process, and T_c and T_d indicate transitions in the heating process. We follow the cooling path (dotted line in Figure 12.15) and program material at each labeled temperature (T_1, T_2, T_3, and T_4) individually. Upon heating, the first shape recovery finishes at $T_{4'}$ (instead of at T_4), because the reverse transition follows a solid curve, as to the other shape recovery steps.

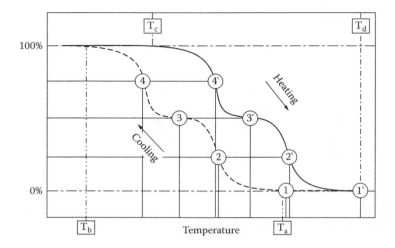

FIGURE 12.15 Programming and heating to obtain multi-SME. Vertical axis is the fraction of transition(s) that may include a single transition or a series of transitions.

If the hysteresis is insignificant, for example, in the SMP foam shown in Figure 12.6 (refer to Figure 8.40 for cyclic DSC result), to achieve a multi-SME (with at least one intermediate shape), the SMP must be programmed in a step-by-step manner during cooling. On the other hand, if hysteresis is significant and sub-transition (transition in incomplete thermal cycling) follows the same mechanism as the TME in some SMAs, it may be possible by carefully arranging the thermomechanical procedure to program in a mixed cooling and heating procedure, although is not ideal and is less effective in practice. The influence of the hysteresis on the TME is the shift in the temperature at which maximum stress occurs. In theory, we still can find the previous programming temperature if the hysteresis can be estimated (for instance, by DSC). In the cases of dual-SMEs, we may program a sample at a temperature within the heating transition range to minimize the influence of hysteresis.

12.4 SUMMARY

The multi-SME and TME in polyurethane SMPs are demonstrated. The mechanisms behind both effects are investigated as well. A framework is applied to reproduce both effects. The resulting patterns are similar to those observed in the experiments. Finally, the influence of hysteresis on both effects is discussed. The framework proposed here can be used as a starting point for developing a full model to simulate the thermomechanical behaviors of SMPs.

ACKNOWLEDGMENTS

We thank Z. Ding, C. Tang, and Dr. C.C. Wang for conducting some of the experiments presented in this chapter, and we thank Z. Ding for help in editing and compiling this chapter.

REFERENCES

Airoldi G, Besseghini S, and Riva G (1993a). Step-wise transformations in shape memory alloys. *Proceedings of the International Conference on Martensitic Transformations*, ICOMAT-92, 1993, pp. 959–964.

Airoldi G, Besseghini S, and Riva G (1993b). Micromemory effects in shape-memory alloys. *Condensed Matter Atomic Molecular and Chemical Physics*, 15, 365–374.

Airoldi G, Corsi A, and Riva G (1998). Step-wise martensite to austenite reversible transformation stimulated by temperature or stress: a comparison in NiTi alloys. *Materials Science and Engineering A*, 241, 233–240.

An L and Huang WM (2006). Transformation characteristics of shape memory alloys in a thermal cycle. *Materials Science and Engineering A*, 420, 220–227.

Behl M and Lendlein A (2010). Triple-shape polymers. *Journal of Materials Chemistry*, 20, 3335–3345.

Bellin I, Kelch S, Langer R et al. (2006). Polymeric triple-shape materials. *Proceedings of the National Academy of Sciences of the United States of America*, 103, 18043–18047.

Bellin I, Kelch S, and Lendlein A (2007). Dual-shape properties of triple-shape polymer networks with crystallizable network segments and grafted side chains. *Journal of Materials Chemistry*, 17, 2885–2891.

Bhattacharya K and James RD (2005). The material is the machine. *Science*, 307, 53–54.

Huang W (1998). Effects of internal stress and martensite variants on phase transformation of NiTi shape memory alloy. *Journal of Materials Science Letters*, 17, 1843–1844.

Huang W (1999). Modified shape memory alloy (SMA) model for SMA wire-based actuator design. *Journal of Intelligent Material Systems and Structures*, 10, 221–231.

Huang W and Toh W (2000). Training two-way shape memory alloy by reheat treatment. *Journal of Materials Science Letters*, 19, 1549–1550.

Huang W and Wong YL (1999). Effects of pre-strain on transformation temperatures of NiTi shape memory alloy. *Journal of Materials Science Letters*, 18, 1797–1798.

Huang WM and Xu W (2005). Hysteresis in shape memory alloys. Is it always a constant? *Journal of Materials Science*, 40, 2985–2986.

Huang WM, Liu QY, He LM et al. (2004). Micro NiTi-Si cantilever with three stable positions. *Sensors and Actuators A*, 114, 118–122.

Huang WM, Yang B, An L et al. (2005). Water-driven programmable polyurethane shape memory polymer: demonstration and mechanism. *Applied Physics Letters*, 86, 114105.

Huang WM, Ding Z, Wang CC et al. (2010). Shape memory materials. *Materials Today*, 13, 54–61.

Kumar UN, Kratz K, Wagermaier W et al. (2010). Non-contact actuation of triple-shape effect in multiphase polymer network nanocomposites in alternating magnetic field. *Journal of Materials Chemistry*, 20, 3404–3415.

Liu N and Huang WM (2006a). Comments on "incomplete transformation induced multiple-step transformation in TiNi shape memory alloys" [*Scripta Materialia*, 2005, 53. 335]. *Scripta Materialia*, 55, 493–495.

Liu N and Huang WM (2006b). DSC study on temperature memory effect of NiTi shape memory alloy. *Transactions of Nonferrous Metals Society of China*, 16, S37–S41.

Liu GQ, Ding XB, Cao YP et al. (2005). Novel shape-memory polymer with two transition temperatures. *Macromolecular Rapid Communications*, 26, 649–652.

Miaudet P, Derre A, Maugey M et al. (2007). Shape and temperature memory of nanocomposites with broadened glass transition. *Science*, 318, 1294–1296.

Ortin J and Planes A (1988). Thermodynamic analysis of thermal measurements in thermoelastic martensitic transformations. *Acta Metallurgica*, 36, 1873–1889.

Ortin J and Planes A (1989). Thermodynamics of thermoelastic martensitic transformations. *Acta Metallurgica*, 37, 1433–1441.

Ortin J and Planes A (1991). Thermodynamics and hysteresis behavior of thermoelastic martensitic transformations. *Journal De Physique IV*, 13–23.

Pan GH and Huang WM (2006). A note on constrained shape memory alloys upon thermal cycling. *Journal of Materials Science*, 41, 7964–7968.

Pretsch T (2010). Triple-shape properties of a thermoresponsive poly(ester urethane). *Smart Materials and Structures*, 19, 015006.

Sun L and Huang WM (2009). Nature of the multistage transformation in shape memory alloys upon heating. *Metal Science and Heat Treatment*, 51, 573–578.

Sun L and Huang WM (2010). Mechanisms of the multi-shape memory effect and temperature memory effect in shape memory polymers. *Soft Matter*, 6, 4403–4406.

Sun L, Huang WM, and Cheah JY (2010). The temperature memory effect and the influence of thermomechanical cycling in shape memory alloys. *Smart Materials and Structures*, 19, 055005.

Wang ZG, Zu XT, Fu YQ et al. (2005). Temperature memory effect in TiNi-based shape memory alloys. *Thermochimica Acta*, 428, 199–205.

Wang ZG, Zu XT, Yu HJ et al. (2006). Temperature memory effect in CuAlNi single crystalline and CuZnAl polycrystalline shape memory alloys. *Thermochimica Acta*, 448, 69–72.

Xie T (2010). Tunable polymer multi-shape memory effect. *Nature*, 464, 267–270.

Xie T, Xiao XC, and Cheng YT (2009). Revealing triple-shape memory effect by polymer bilayers. *Macromolecular Rapid Communications*, 30, 1823–1827.

Yang B, Huang WM, Li C et al. (2006). Effects of moisture on the thermomechanical properties of a polyurethane shape memory polymer. *Polymer*, 47, 1348–1356.

13 Future of Polyurethane Shape Memory Polymers

Although this book focuses on one commercially available polyurethane shape memory polymer (SMP), the main features, mechanisms, and applications discussed here apply largely to other SMPs as well. The great potential represented by the many unique features of SMPs make possible many applications, from outer space to deep within the human body. Based on a unique property called the shape memory effect (SME), SMPs are now re-shaping our approaches to design in many ways because the many limitations of conventional materials and traditional approaches do not apply to SMPs. Because the focus of this book is on a specific SMP, in-depth technical details, potential, and details of various novel applications at different scales are presented and can serve as a solid platform for interested readers to start a fantastic journey to explore this frontier.

As demonstrated in this book, SMPs have the advantages of convenience and low material and fabrication costs (even for complex products with tailorable material properties) using many conventional polymer processing techniques such as injection molding, extrusion, film casting, spin and dip coating, fiber spinning, thermoforming, and foaming. Tailoring the material properties can be realized easily by blending or varying a composition to meet the requirements of a particular application. The unique thermo- and moisture-responsive capabilities of polyurethane SMPs add new dimensions to their applications.

Besides continuously improving the performance of the existing materials, the development of new materials should suit the needs of real applications to achieve sustainability in the long run. In this regard, some major points for the future development of polyurethane SMPs (most of which also apply to other SMPs) are summarized.

13.1 CHARACTERIZATION AND MODELING OF SMPs

Materials scientists are advancing quickly along the frontiers of SMP research by continuing to invent new materials weekly, if not daily. Materials and mechanical engineers lag somewhat behind in thoroughly understanding novel materials. To pave a smooth way for real engineering applications, it is increasingly important for design engineers to fully understand the characteristics of individual SMPs so they can confidently incorporate these materials in their designs.

Based on scientific publications, the polyurethane SMP discussed in detail in this book (from SMP Technologies, Japan) may be the only SMP extensively investigated to determine its thermomechanical and shape memory properties. Prof. Hisaaki Tobushi and his co-workers made significant contributions in this area over many years. However, further study is still needed, in particular of other application-related properties are still lacking substantially from an engineering application view.

To further enable design engineers to achieve optimal designs, we must develop robust material models for numerical simulations. To date, only limited numbers of such simulations have been reported in the literature [e.g., thermomechanical simulation of braided SMP stents; Kim et al. (2010)] and the current material models are not without problems.

The previous constitutive models for SMPs were of two types (Gunes 2009): phenomenological and micro structural. Although phenomenological models (Tobushi et al. 1997, 2001) can usually predict experimental data with reasonable accuracy, they are applicable only to certain simple situations (e.g., uniaxial stress–strain state) and do not provide in-depth insights of fundamental physical and chemical mechanisms. Furthermore, based on the highly complex nature of polymers, micro structural models (Bhattacharyya and Tobushi 2000, Qi and Dunn 2010) derived from the underlying mechanisms at the micro structural level continue to present great difficulties in quantitative predictions for real engineering applications. Other modeling issues, for example, moisture penetration in moisture-responsive SMPs, degradation in degradable SMPs, and stability of performance for long-term applications (discussed in detail in the next section), must be investigated before they are used in real applications. The SMP community must do a lot more work before this fantastic material can progress from being a new material to a conventional one—such as stainless steel—readily available off the shelf for anyone to use.

13.2 STABILITY OF SME

From an application view, it is crucial to ensure that a product or material functions properly over its entire life cycle. Taking active disassembly using SMPs as an example (Figure 13.1) we have to make sure that SMP elements work properly (hold parts firmly together) in normal working environments. At the end of the life of the electrical device, the SMP elements can automatically detach from the top cover upon heating. One may expect electrical devices to function properly for some years; this requires careful examination of the long-term stability of the shape memory effect (SME). Some investigations have been reported (Tey et al. 2001, Pretsch 2010), but they were not conducted systematically and the testing periods were insufficient.

From a design view, different assembly configurations of SMPs and conventional materials (e.g., PMMA, Figure 13.2) result in different strength and stiffness levels under mechanical loading. Figure 13.2 reveals the nominal tensile strength versus nominal tensile strain curves of five SMP–PMMA assembly configurations. Different configurations produce variations of exact loading conditions for the SMP and PMMA at the joining area. For this reason, long-term stability studies should be conducted based on the specific loading needs. In addition, environmental conditions are also very important. For example, for space applications, radiation is an additional concern (Ishizawa et al. 2003). Physical aging is a critical issue for implantable devices (Lorenzo et al. 2009).

The programming procedure is an additional parameter for determining SMP performance (Tobushi et al. 2004, 2008). As revealed in Figure 13.3, the shape fixity

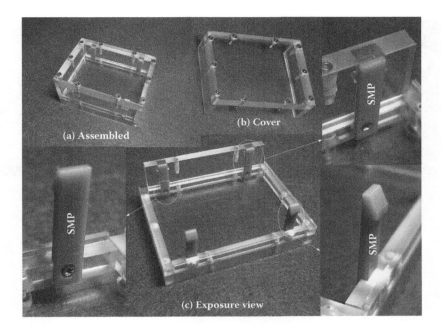

FIGURE 13.1 SMP for active disassembly.

and recovery of SMPs vary based on programming conditions. In this experiment, a 0.1 mm diameter steel ball was used to make indents at 15°C above the glass transition temperature, T_g, with different holding times (1 minute to 1 hour). The inset in Figure 13.3 shows a typical result after indentation and cooling back to room temperature (about 22°C). The measured maximum indentation depth is almost a linear function of holding time. A prolonged indentation holding time produces a deeper indent. Subsequently, the sample was heated to $T_g + 15°C$ for 15 minutes. Clearly, the shorter the holding time, the more shape recovery occurs in terms of the maximum depth of the indent. Further investigation of this topic is required to find a way to achieve the best performance.

13.3 CYCLIC ACTUATION

Cyclic actuation is required by many mechanisms for repeated and continuous motion. As we know, three types of working principles apply to shape memory alloy (SMA)-based actuators (Huang 2002). In the first type (Figure 13.4a), the SMA element is pre-stretched (block P moves right) and then heated for shape recovery (P moves left). If a conventional spring is connected to the pre-stretched SMA (Figure 13.4b), upon heating, P moves left; during cooling, it moves right. If the conventional elastic spring is replaced by another piece of SMA element (Figure 13.4c), P moves to the side on which the SMA is heated.

In these cases, the SMA remembers only the high-temperature shape. This is a one-way SME that is intrinsic to all SMAs. Cyclic actuation is achievable in actuators

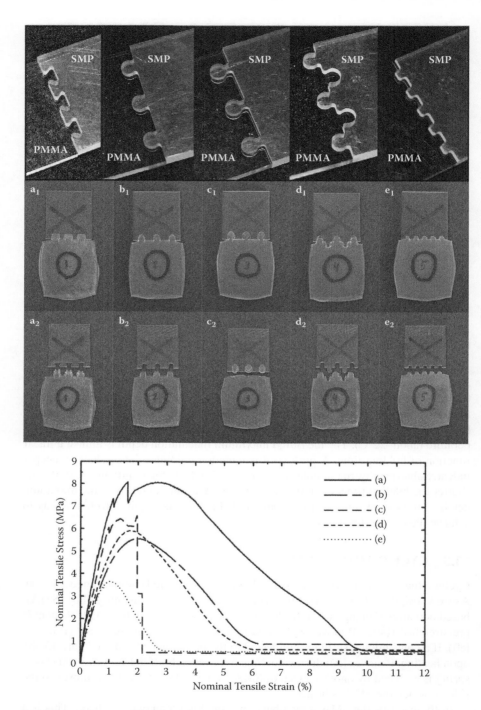

FIGURE 13.2   Tensile strength of PMMA and SMP in different assembly configurations.

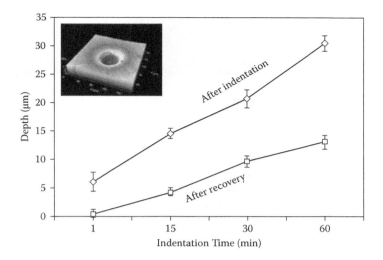

FIGURE 13.3 Shape fixity and recovery of SMP via indentation and heating recovery test.

using the one-way SME in SMAs by three approaches: (1) against a force, (2) against an elastic spring, or (3) against another piece of SMA.

As an alternative for cyclic actuations, SMAs can be trained to have two-way SME: they can remember both the high- and low-temperature shapes (Huang and Toh 2000). As shown in Figure 13.5, a piece of C-shaped SMA wire closes upon heating and opens upon cooling. This process is highly repeatable. We may call this activity material two-way SME; Figure 13.4 shows a mechanical two-way SME. Fundamentally, both SMEs follow the same working mechanism but at different scales. In general, mechanical two-way SME is more powerful and reliable and is widely used in actuators. Material two-way SME has much smaller recoverable strain and is easy to degrade against a substantial force; it is more applicable to integrated compact sensors.

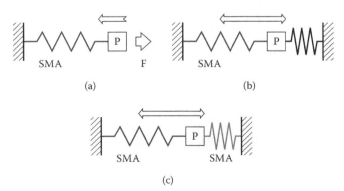

FIGURE 13.4 Working principles of SMA-actuated devices. (Reprinted from Huang W. *Materials and Design*, 23, 11–19, 2002. With permission.)

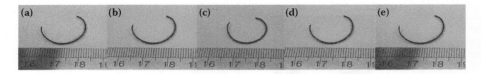

FIGURE 13.5   Two-way SME in SMA upon thermal cycling. (a–c) Heating. (c–e) Cooling.

In polymeric materials, cyclic actuation has been realized via special mechanisms based on a physical phenomenon. As demonstrated in Figure 13.6, a piece of SMP material pushes a thin elastic beam downward upon heating and moves back (up) during cooling. This is based on a careful design of the polymeric material (a composite) and utilizing significant thermal expansion in one of the filler materials.

According to Figure 13.6, we can sketch the evolution of stiffness during thermal cycling of a material that is suitable for cyclic actuation as shown in Figure 13.7. Apparently, because SMAs are stiff at high temperatures (austenite phase) and soft at low temperatures (martensite phase), they are suitable for cyclic actuation. The above-mentioned polymeric material should share the same characteristic during thermal cycling, although the underlying mechanism is totally different. Stiffness (vertical axis of Figure 13.7) is a term normally used by structural and mechanical engineers. If we present Figure 13.7 in a format that is more familiar to materials

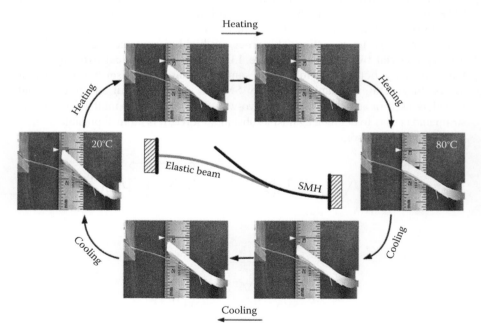

FIGURE 13.6   Cyclic actuation of a piece of shape memory polymeric material against an elastic beam in thermal cycling. (Reprinted from Huang WM, Ding Z, Wang CC et al. *Materials Today*, 13, 54–61, 2010. With permission.)

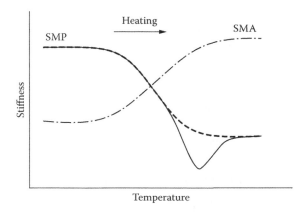

FIGURE 13.7 Changes of stiffness in thermo-responsive SMA or SMP upon heating.

researchers in general and polymer researchers in particular, the vertical axis may be elastic modulus or storage modulus. From a mechanical view, if the dynamic mechanical analysis (DMA) of a material shows the same pattern as that of the SMA in Figure 13.7, the material may have the potential for cyclic actuation.

Unfortunately, most polymers are soft at high temperatures and hard at low temperatures due to glass transition or melting, i.e., the storage modulus decreases monotonously with the increase in temperature. Hence, SMPs are generally not suitable for cyclic actuation except for a few that show V-shaped curves in DMA results upon heating (thin solid line in Figure 13.7) so that the storage modulus increases upon heating within a certain temperature range. The V curve has been reported in Du and Zhang (2010), Qin and Mather (2009), and Chung et al. (2008), but not all materials are workable for cyclic actuation. According to the mechanical mechanism of cyclic actuation, the stiffness of a material must increase upon heating, even within only a small temperature range; a V shape in a DMA curve that basically measures storage modulus by cyclic loading with a tiny force is not fully equivalent to this requirement.

Polymers are more complex than alloys and metals in many ways. As reported in Hornbogen (1978), reversible phase transformation in PTFE ~19°C results in a small shape change due to a volume variation during the phase transformation. According to the mechanical mechanism of cyclic actuation, this polymer can be utilized for cyclic actuation through a proper mechanism design based on the volume variation during phase transformation.

In fact, mechanism design is a generic approach for almost all SMPs to achieve cyclic actuation. The laminate–bilayer (Tamagawa 2010, Chen et al. 2008) and SMA–SMP composite are popular designs (Tobushi et al. 2009). The major disadvantage of the laminate approach is that the reversible strain is very limited and deformation is small unless a very thin structure is used, greatly limiting the actuation force. The SMA–SMP presents some difficulties in fabrication because SMAs must be placed or embedded precisely and firmly.

High actuation speed is another important aspect of many cyclic actuation applications. For thermo-responsive SMPs, the most effective approach to increase actuation speed is to narrow the transition temperature range. A range less than 5°C has been achieved in an SMP through the concept of composite design. In addition, high thermal conductivity is useful for quick heating and cooling. Blending with highly thermal conductive fillers is another way to achieve the high speed. We expect composites to play an important role here. Finally, another concern in cyclic actuation applications is SMP degradation. This issue needs further investigation. SMPs exhibiting cyclic actuation are highly desired in engineering applications from macro to micro scale, but this aspect is relatively less explored and deserves more attention from the SMP community (Rousseau 2008).

13.4 ALTERNATIVE ACTUATION TECHNIQUES

In an editorial for *eXPRESS Polymer Letters* in 2008, as a personal view, Prof. Byung K. Kim listed a few stimuli for actuating SMPs (Kim 2008). Pressure-response is one stimulus. We propose two approaches to achieve this and preliminary results have been obtained. As with electro- and magneto-responsive SMPs, our first approach blended intrinsically thermo-responsive SMPs with self-heating fillers. As illustrated in Figure 13.8, pressing one end of a sample generates self heat in the filler so that the SMP recovers its original shape. The other approach does not rely on the thermo-responsive feature for shape recovery and no additional fillers are required. The approach is truly and naturally pressure-responsive. As illustrated in Figure 13.9, by further compressing a pre-compressed sample, the material is able to recover its original shape. Preliminary results reveal an 86% recovery ratio in a 26% pre-compressed sample.

As illustrated in Figure 13.10, we demonstrated the possibility of triggering shape recovery in SMPs by heating or cooling. Cooling for shape recovery is a new approach and opposite to the conventional heating approach in thermo-responsive SMPs. A material that can be triggered for shape recovery by either heating or cooling should present an additional degree of freedom in engineering applications. Sound was the

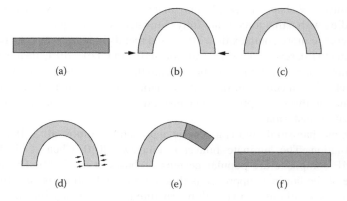

FIGURE 13.8 Shape recovery in thermo-responsive SMP via pressure-induced self-heating. (a) Original shape. (b) Deformed shape at high temperatures. (c) After cooling back to room temperature. (d) Pressing one end of the sample. (e) Self-heating triggers shape recovery. (f) Recovered shape.

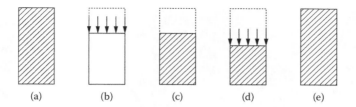

FIGURE 13.9 Pressure-responsive for shape recovery by means of compression. (a) Original shape. (b) Compressed at high temperatures. (c) After cooling back to room temperature. (d) Applying impact load. (e) Recovered shape.

only stimulus mentioned by Prof. Kim in 2008 (Kim 2008), but exploration has yet to be realized; this is another challenge for the SMP community.

13.5 MULTIPLE FUNCTIONS

Multi-function is a trend in current materials research. In SMPs, single or multiple (re-)actions can be triggered in response to one or more stimuli following a prescribed procedure. In fact, the working principle of SMPs ensures that they have the integrated functions of sensing and actuation even down to sub-micron scale. They can be utilized as intelligent machines even down to nanometer size.

The concept that "the material is the machine" was discussed by Profs. K. Bhattacharya and R.D. James (2005). They identified SMAs as potential candidates for the mission, but because of better flexibility and versatility, SMPs appear more promising for realizing this concept. As an example, a range of biomedical applications can be developed from a technology platform based on SMPs (Lendlein et al. 2010, Wischke and Lendlein 2010). Controlled drug release and targeted drug release (injection of drug only into certain cells) can be achieved by SMPs at sub-micron scale. Recent work by Dr. T. Xie (2010) on the multi-SME of SMPs yielded a more powerful technique allowing SMPs of arbitrary sizes to operate as intelligent machines with sensing and actuation functions. As revealed in Figure 13.11, we demonstrated the retraction function of a polyurethane SMP stent based on the multi-SME. This concept is applicable to any SMP stent, even degradable types. In case of an emergency, a degradable stent can be removed instantly if required. In addition to working as intelligent machines, other functions can also be developed in polyurethane SMPs. A few typical examples are discussed in this section.

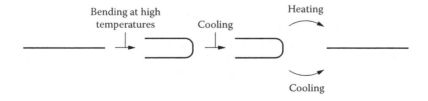

FIGURE 13.10 Shape recovery of SMP triggered by heating or cooling.

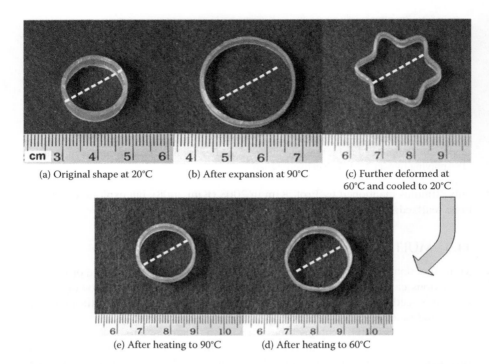

(a) Original shape at 20°C   (b) After expansion at 90°C   (c) Further deformed at
                                                                  60°C and cooled to 20°C

(e) After heating to 90°C   (d) After heating to 60°C

FIGURE 13.11   Retractable polyurethane SMP stent (glass transition temperature range ~40 to 70°C) based on multi-SME.

The original polyurethane SMP is not an ideal biodegradable material, but it is highly biocompatible (Cabanlit et al. 2007). A few biodegradable polyurethane SMPs have been synthesized recently (Pereira and Orefice 2007, Eglin et al. 2010). They are useful for cellular surgery in which a tiny machine made of a biodegradable SMP is deformed into a small size and injected into a living cell. It then recovers its original shape to properly accomplish the desired mission before it is fully dissolved (Huang et al. 2010, Sun and Huang 2010). Soil fungal communities involved in the biodegradation of polyester polyurethanes have been investigated (Cosgrove et al. 2007). This technique is highly useful for recycling. As demonstrated by Zhang et al. (2010), azobenzene-containing polyurethane SMPs exhibit photoisomerization functions that may be utilized for drug release and optical data storage.

In the *Terminator* series of science fiction movies, future robots quickly heal any injury and regain their strength again and again. The material to accomplish this feat does not yet exist, but self-healing materials on a limited scale have attracted great attention due to scores of potential applications (Toohey et al. 2007, Hayes et al. 2007, Amendola and Meneghetti 2009). For self-healing purposes, the two technical requirements are (1) shape recovery and (2) strength recovery. In recent years, studies of self-healing SMPs focused on utilizing the SME for shape recovery; strength recovery is not so well explored (Margraf et al. 2008). Huang et al. (2010) demonstrated repeated instant healing in a rubber-like shape memory polymeric

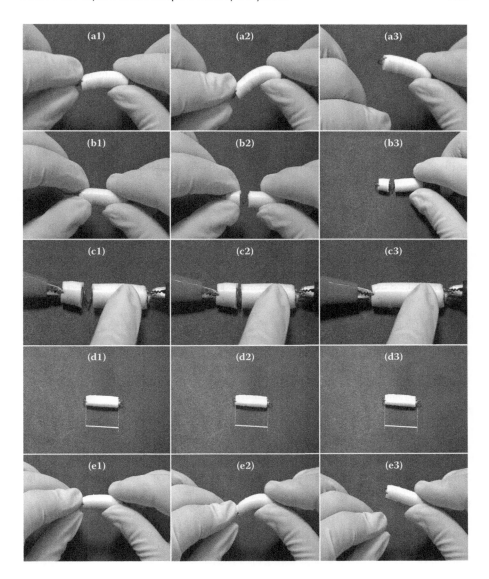

FIGURE 13.12 Instant self-healing of rubber-like shape memory polymeric material (with SMA spring embedded inside). (a) Sample is rubber-like in response to bending at room temperature. (b) Sample breaks into two parts only when applied pulling force exceeds limit; the two separated parts remain linked by SMA spring. (c) Upon Joule heating of SMA spring, the two parts are joined. (d) During cooling back to room temperature, the sample shrinks slightly. (e) Sample becomes virtually one piece and is again rubber-like in response to bending.

material (Figure 13.12). Materials design instead of materials synthesis has been cited as an alternative for self-healing compounds (Kessler 2007). We believe that through proper mechanisms and materials designs, polyurethane SMPs and their composites with fantastic self-healing abilities might become the materials for future Terminators.

ACKNOWLEDGMENTS

We thank J. Wei, H. Purnawali, N.W. Khun, and Dr. C.C. Wang for conducting some of the experiments reported in this chapter. We also thank Dr. Wang for help in compiling and editing this chapter.

REFERENCES

Amendola V and Meneghetti M (2009). Self-healing at the nanoscale. *Nanoscale*, 1, 74–88.
Bhattacharyya A and Tobushi H (2000). Analysis of the isothermal mechanical responsive of a shape memory polymer rheological model. *Polymer Engineering and Science*, 40, 2498–2510.
Bhattacharya K and James RD (2005). The material is the machine. *Science*, 307, 53–54.
Cabanlit M, Maitland D, Wilson T et al. (2007). Polyurethane shape-memory polymers demonstrate functional biocompatibility *in vitro*. *Macromolecular Bioscience*, 7, 48–55.
Chen S, Hu J, Zhuo H et al. (2008). Two-way shape memory effect in polymer laminates. *Material Letters*, 62, 4088–4090.
Chung T, Romo-Uribe A, and Mather PT (2008). Two-way reversible shape memory in a semicrystalline network. *Macromolecules*, 41, 184–192.
Cosgrove L, McGeechan PL, Robson GD et al. (2007). Fungal communities associated with degradation of polyester polyurethane in soil. *Applied and Environmental Microbiology*, 73, 5817–5824.
Du H and Zhang J (2010). Solvent induced shape recovery of shape memory polymer based on chemically cross-linked poly(vinyl alcohol). *Soft Matter*, 6, 3370–3376.
Eglin D, Grad S, Gogolewski S et al. (2010). Farsenol-modified biodegradable polyurethanes for cartilage tissue engineering. *Journal of Biomedical Materials Research A*, 92, 393–408.
Gunes IS (2009). Analysis of shape memory properties of polyurethane nanocomposites. PhD dissertation, University of Akron, OH.
Hayes SA, Jones FR, Marshiya K et al. (2007). A self-healing thermosetting composite material. *Composites A*, 38, 1116–1120.
Hornbogen E (1978). Shape change during the 19°C phase transformation of PTFE. *Progress in Colloid and Polymer Science*, 64, 125–131.
Huang W (2002). On the selection of shape memory alloys for actuators. *Materials and Design*, 23, 11–19.
Huang W and Toh W (2000). Training two-way shape memory alloy by reheat treatment. *Journal of Materials Science Letters*, 19, 1549–1550.
Huang WM, Ding Z, Wang CC et al. (2010). Shape memory materials. *Materials Today*, 13, 54–61.
Ishizawa J, Imagawa K, Minami S et al. (2003). Research on application of shape memory polymers to space inflatable systems. *Proceedings of Seventh International Symposium on Artificial Intelligence, Robotics and Automation in Space*, Nara, Japan, May 19–23.
Kessler MR (2007). Self-healing: a new paradigm in materials design. *Journal of Aerospace Engineering G*, 221, 479–495.

Kim BK (2008). Shape memory polymers and their future developments. *eXPRESS Polymer Letters*, 2, 614.

Kim JH, Kang TJ, and Yu WR (2010). Simulation of mechanical behavior of temperature-responsive braided stents made of shape memory polyurethane. *Journal of Biomechanics*, 43, 632–643.

Lendlein A, Behl M, Hiebl B et al. (2010). Shape-memory polymers as a technology platform for biomedical applications. *Expert Review of Medical Devices*, 7, 357–379.

Lorenzo V, Diaz-Lantada A, Lafont P et al. (2009). Physical aging of PU-based shape memory polymer: influence on their applicability to the development of medical devices. *Materials and Design*, 30, 2431–2434.

Margraf TW Jr, Barnell TJ, Havens E et al. (2008). Reflexive composites: self-healing composite structures. *Proceedings of SPIE*, 6932, 693211.

Qi HJ and Dunn ML (2010). Thermomechanical behavior and modeling approaches. In *Shape-Memory Polymers and Multifunctional Composites*, Leng J and Du S, Eds., Taylor & Francis, New York, pp. 65–90.

Qin H and Mather PT (2009). Combined one-way and two-way shape memory in a glass-forming nematic network. *Macromolecules*, 42, 273–280.

Pereira I and Orefice RL (2007). SAXS analysis on shape memory biodegradable polyurethane. Activity report, Brazilian Synchrotron Light Laboratory.

Pretsch T (2010). Review on the functional determinations and durability of shape memory polymers. *Polymers*, 2, 120–158.

Rousseau IA (2008). Challenges of shape memory polymers: a review of the progress toward overcoming SMP's limitations. *Polymer Engineering and Science,* 48, 2075–2089.

Sun L and Huang WM (2010). Thermo/moisture responsive shape-memory polymer for possible surgery/operation inside living cells in future. *Materials and Design*, 31, 2684–2689.

Tamagawa H (2010). Thermo-responsive two-way shape changeable polymeric laminate. *Materials Letters*, 64, 749–751.

Tey SJ, Huang WM, and Sokolwski WM (2001). On the effects of long term storage in cold hibernated elastic memory (CHEM) polyurethane foam. *Smart Materials and Structures*, 10, 321–325.

Tobushi H, Hashimoto T, Hayashi S et al. (1997). Thermomechanical constitutive modeling in shape memory polymer of polyurethane series. *Journal of Intelligent Material Systems and Structures*, 8, 711–718.

Tobushi H, Okumura K, Hayashi S et al. (2001). Thermomechanical constitutive model of shape memory polymer. *Mechanics of Materials*, 33, 545–554.

Tobushi H, Matsui R, Hayashi S et al. (2004). The influence of shape-holding conditions on shape recovery of polyurethane-shape memory polymer foams. *Smart Materials and Structures*, 13, 881–887.

Tobushi H, Hayashi S, Hoshio K et al. (2008). Shape recovery and irrecoverable strain control in polyurethane shape-memory polymer. *Science and Technology of Advanced Materials*, 9, 015009.

Tobushi H, Hayashi S, Sugimoto Y et al. (2009). Two-way bending properties of shape memory composite with SMA and SMP. *Materials*, 2, 1180–1192.

Toohey KS, Sottos NR, Lewis JA et al. (2007). Self-healing materials with microvascular networks. *Nature Materials*, 6, 581–585.

Wischke C and Lendlein A (2010). Shape-memory polymers as drug carriers: a multifunctional system. *Pharmaceutical Research*, 27, 527–529.

Xie T (2010). Tunable polymer multi-shape memory effect. *Nature*, 464, 267–270.

Zhang Y, Wang C, Pei X et al. (2010). Shape memory polyurethane containing azo exhibiting photoisomerization function. *Journal of Materials Chemistry*, 20, 9976–9981.

Index

3D nanomaterials, 150

A

ABS, 3
Absorbed water, 61
Acrylonitrile butadiene styrene (ABS), 3
Active disassembly, 10–12
Actuation by moisture, 16–17, 69
Actuation speed, 349–350
Additives for property tailoring, 14–15
Aging and property effects, 15
Aircraft component morphing, 12–13
Alloys, *see* Shape memory alloy (SMA)
Alternative magnetic field, 117
    actuator for biomedical devices, 167
Alumina nanofiller, 154
    thermal conductivity modifier, 173
Amine-functionalized silica particles, 155
Amino functionalized MWCNTs, 164
Aneurysm treatment, 20, 186
Anionic polymerization, 152
Anisotropic wrinkles, 275
Arteriovenous grafts, 307
Aspect ratio of filler, 163–164
Atom transfer radical polymerization
    (ATRP), 152
Attapulgite clay additive, 16, 150, 311
    moisture effects, 317–318
    pretreatment, 157
Austenitic transformation, 329
Autochokes, 308
Azobenzene-containing polyurethane SMPs, 352

B

Bacterial ratchet motor, 318
BET, 72
Biaxial compressive stress for wrinkle
    induction, 288–289
Biocompatibility, 4, 186, 307, 311, 313, 352
    of attapulgite clay, 317
    of graphene, 171
Biodegradability, 4
Biodegradable SMPs, 352
Biodegradable stents, 307
Biomarker monitoring, 173
Biomedical engineering applications, 13, 18–19
    micro-machines, 318–321
    multi-function, 351

porous polymer preparation for, 185
    SMP foams, 186–187
Black body principle, 82
Bladder cancer treatments, 315–317
BN, 173
Bonding of CNTs, 152, 164
Bonding of graphene, 171
Boron nitride (BN) additive, 173
Bound water, 60; *see also* Moisture effects
Braided stents, 15, 307
Braille paper, 9–10
Brittle crushing of SMP foams, 224
Brunauer, Emmett and Teller (BET) surface
    area, 72
Bubble formation
    heating effects, 208
    reversibility, 208
    size control, 200, 206–208
    water in polyurethane, 196–200
Buckling
    of sheets, 275–276
    for wrinkle induction, 288
Bucky onions, 154
Bulging test, 7
Butterfly-like feature, 257–261

C

Carbon black (CB) additive, 15, 16; *see also*
    Carbon nanoparticle fillers; Carbon
    powders
    with CNF, 167
    for conductivity, 70–71
    mechanical property effects, 167
    with nickel powder, 139–143
Carbon dioxide expansion for porous polymer
    preparation, 185
Carbon nanofibers (CNFs), 15, 165–167
    with CNTs, 167
    competing property effects, 170
Carbon nanopaper, 167
Carbon nanoparticle fillers, 72, 168–169; *see
    also* Specific materials
Carbon nanotubes (CNTs), 15, 147
    acid treatment, 164
    as additives, 151
    alignment, 163–164
    bonding to polymers, 152
    with carbon black, 167
    with carbon nanofiber, 167

357

For Product Safety Concerns and Information please contact our EU
representative GPSR@taylorandfrancis.com Taylor & Francis Verlag GmbH,
Kaufingerstraße 24, 80331 München, Germany

Printed and bound by CPI Group (UK) Ltd, Croydon, CR0 4YY
01/05/2025
01858480-0001